AF540291

Breeding and Seed Production of Fin Fish & Shell Fish

THE AUTHORS

Dr. P.C. THOMAS

Born in the year 1939 at Kulanada, Kerala, Dr. Thomas had his early education at Kerala. He has obtained the M.Sc. degree in Zoology with specialisation in Fish and Fisheries in the year 1961 and Ph.D. degree in Zoology (fish endocrinology) in the year 1981 from Benaras Hindu University. He is the author of several research papers published in national and international journals. He is a teacher, researcher and academic administrator of good standing. He has played a key role in developing the academic programmes and infrastructural facilities of the college of Fisheries under Orissa University of Agriculture and Technology which is widely acclaimed. He started his professional career as a teacher in Bhadrak College, Orissa in the year 1961 and later rose to the position of Professor Aquaculture and Director, College of Fisheries, OUAT from which capacity he retired in September, 1999. He was appointed as emeritus professor by the Indian Council of Agricultural Research, at CIFA, Kausalyaganga, Bhubaneswar. He has wide teaching and practical experience in the field of fish breeding and seed production technology. He is associated with several professional societies concerned with Fisheries research and development.

Dr. (Mrs.) KANTA DAS MOHAPATRA

Born on 23rd February, 1961, Dr. (Mrs.) Kanta Das Mohapatra, had obtained both M.Sc. and Ph.D. degrees in Zoology from Utkal University, Vanivihar. She joined the Orissa State Fisheries department and worked there for five years. She took diploma in Fishery Science from Central Institute of Fisheries Education, Mumbai, in the year 1997. She joined ARS Service (ICAR) and was posted in the Fish genetics unit of CIFA, Kausalyaganga. Since then, she is working on several research projects in the field of fish genetics including "the Indo Norwegian Collaboration project on genetic improvement of rohu for growth through selective breeding". Her contribution in the production of genetically improved variety of rohu "Jayanthi" under this project is well known. She has published a number of research papers in fish genetics in national / international journals.

Dr. SURESH CHANDRA RATH

Dr. Suresh Chandra Rath, born in 1958 at BiraRam, Chandrapur, Sakhigopal, Puri district, Orissa took his M.Sc. and Ph.D. in Zoology from Utkal University, Bhubaneswar. Dr. Rath started his professional career as hatchery manager, in a World Bank assisted carp hatchery in Orissa. Later on he joined as Technical Officer at Central Institute of Freshwater Aquaculture, Kausalyaganga, Bhubaneswar. He, as a team member , has made significant contribution in developing many breakthrough technologies viz. multiple breeding of Indian major carps, non-stripping hatchery breeding of grass carp, cryo preservation of carp milt and its utilization. Dr. Rath has provided technical guidance to establish several modern eco carp hatcheries. He has more than 30 research papers in national and international journals, besides many other publications in shape of book chapters, articles and booklets in the field of freshwater aquaculture education and extension. Dr. Rath has bagged the prestigious one time team award of Indian Council of Agricultural Research (ICAR) in the year 1998 for his contribution to the research on cryopreservation of carp milt as a tool for improving broodstock. Trained in Israel, Dr. Rath is presently working as Training Associate at Trainers' Training Centre (TTC) of CIFA, Bhubaneswar.

Breeding and Seed Production of FIN FISH & SHELL FISH

Dr. P.C. THOMAS
Former Professor, Aquaculture & Director
College of Fisheries (OUAT) & Emeritus Scientist (ICAR)
CIFA, Bhubaneswar

Dr. SURESH Ch. RATH
Training Associate,
TTC, CIFA

Dr. (Mrs.) KANTA DAS MOHAPATRA
Senior Scientist, (CIFA)
Kausalyaganga

Foreword by
Dr. T.V.R. PILLAY

2020
Daya Publishing House®
A Division of
Astral International Pvt. Ltd.
New Delhi – 110 002

Print on Demand Edition, 2020

ISBN 978-81-7035-308-9 (Hardbound)

***Front Page**: Magnified view of developing eggs of common carps attached to hydrilla twig.*

***Back Page**: Background–Enlarged view of water hardened carp eggs. 3 photographs as inset: (a) Handfull of water hardened carp eggs, (b) Circular incubation pool for major carps eggs, (c) Glass jar incubation units.*

Published by : **Daya Publishing House®**
A Division of
Astral International Pvt. Ltd.
– ISO 9001:2015 Certified Company –
4736/23, Ansari Road, Darya Ganj
New Delhi - 110 002
Phone: 011-4354 9197, 2327 8134
E-mail: info@astralint.com
Website: www.astralint.com

Digitally Printed at : **Replika Press Pvt. Ltd.**

This book is dedicated to the loving memory of
My Parents

Late M.M. Cherian & Rachel Cherian

who have left for their heavenly abode and to
My Dear Wife

Susheela

Dr. P.C. Thomas

Acknowledgements

I am indebted to my co-author Dr. (Mrs.) Kanta Das Mohapatra, for her contribution in the chapter on genetic upgradation of fish seed and to Dr. S.C. Rath for his contribution in the chapter on on the seed production technology of major carps. He has also arranged for several photographs and illustrations for the book.

I have much pleasure in expressing my deep sense of gratitude to Dr. S.D. Gupta, Principal Scientist, CIFA for critically going through the manuscript on the chapter on the seed production technology of major carps and for valuable suggestions from his long years of experience in fish breeding. A number of authors / publishers have kindly permitted to use some of the photographs / illustrations from their publications which are duly acknowledged in the text. Among them are Blackwell publishing, Osney Mead, Oxford; Information division, Food and Agriculture Organization of United Nations, Rome, Italy; Director, Centre for Tropical and Subtropical Aquaculture, the Oceanic Institute, Hawaii; Director TungKang Marine Laboratory, Taiwan Fishery Research Institute, Taiwan and Dr. D.G.M. Jamieson, Dept. of Zoology, University of Queensland, Australia.

I am indebted to Dr. T.V.R. Pillay, Former Programme Director, FAO Aquaculture Development and Coordination Programme, Bangalore for writing the foreward for the book.

I am grateful to the Indian Council of Agriculture Research for financial assistance under the emeritus scientist scheme. This book is completed during the tenure of the scheme. I want to place on record my deep sense of gratidude to Dr. S. Ayyappan, Former Director,CIFA and present DDG, ICAR (Fisheries) and to Dr. C. Saha, and Dr. R.K. Jana former Directors, CIFA and Dr. K. Janaki Ram, Director, CIFA for providing all facilities at CIFA, Kausalyaganga.

Assistance rendered by Sri. Girish Mishra, Gigabyte Computers, IRC Village, Nayapalli in typing the manuscript is hereby gratefully acknowledged. Several other friends and well wishers have cooperated in the preparation of this text book for which I am grateful to them.

Foreword

Aquaculture is known to be a multidisciplinary science, which has proved its relevance in the present scenario of limited expansion of the fisheries sector in meeting the food security of surging human populations. The growth of aquaculture at an average annual rate of almost 10% during the last decade affords ample proof of the importance of the sub-sector in world food supply; compared with 3% for terrestrial livestock and 1.5 % for capture fisheries. More than 80% of total aquaculture yield is produced in low-income food-deficit countries, thus playing a vital role in alleviation of poverty and generation of wealth for people living in developing countries. The great diversity of the sub-sector, which covers very small-scale to very large- scale enterprises, contribute significantly to a wide range of development needs.

Among the major problems facing aquaculture today is the lack of knowledge to close life cycles and mass-produce seedlings of target species. This has come to the fore in attempts to domesticate species and diversify export of aquatic products. Dr. Thomas has made a significant service to aquaculture technology by compiling relevant information on the reproduction of culturable species and hatchery production of several freshwater, brackishwater and marine species of finfish, shellfish, crustaceans and their food organisms. It has been experimentally proven that healthy seedlings are the key to successful aquaculture. An important factor that has contributed to the collapse of large-scale farming of species like shrimps in a number of countries in Asia, is the outbreak of viral diseases that could be traced to the deficiencies of hatchery management. It can be expected that through genetic manipulation, resistance to diseases can be developed in hatcheries, in order to make the necessary improvements in farming.

The basic information contained in this book should be of help in developing practical technologies in the production of disease-free hatchlings and management of aquatic farming operations. The extent of coverage may help students and researchers to acquire information on the present state of knowledge of the subject , and select areas they can concentrate on in the future.

T.V.R. Pillay
Former Programme Director
FAO Aquaculture Development &
Coordination Programme
Bangalore, India

Preface

Quality fish seed is the prime requirement for aquafarming of any species of finfish or shell fish. A few years back, the only source for procurement of fish seed was from its natural spawning ground. By this method pure stock of seed of selected fish in adequate quantity could not be made available for farming. Last three decades have witnessed a phenomenal growth in the technology of seed production of different species of fin fishes and shell fishes as a result of which, at present, seed of many species of cultural value are available on a commercial scale.

As far as the major carps that contribute the largest share in the global aquaculture production are concerned, the technique of induced breeding through the administration of pituitary hormones coupled with the improved provisions of incubation devices paved the way for large scale production of fry-stock for aquaculture. This optimised production of eggs per Kg. body weight of brood fish, enhanced rate of fertilization and hatching rate and also maximized survival of hatchlings. Development and use of synthetic compounds for induced spawning in fish is yet another milestone in this field. Improved techniques of brood husbandry, use of prophylactics in hatchery operations, strict vigil on the maintenance of water quality parameters, scientific methods of nursery pond preparation to ensure optimum production of live feed biomass and standardisation of feeding strategies decreased the mortality rate of hatchlings which is the most vulnerable stage in the life cycle of the fish. It also increased the recovery rate of fry up to even 95% and also decreased the duration of nursery crop cycle. In a nutshell, by this time, package of practices have been evolved to produce and supply huge quantities of major carp fry to meet the market demand.

Perhaps, another area in which considerable progress has been made is "Shrimp seed production". Design, construction and operation of commercial hatcheries for tiger shrimp seed production has aided the development of tiger shrimp farming as an industry along the coastal areas of India and many other countries. This has generated employment , utilized waste land to produce and draw wealth from water and resulted in the establishment of several processing units and enhanced the export of forzen shrimp thereby generating considerable foreign exchange for the nation.

In order to minimize the environmental stress and disease outbreak due to repeated cropping of a single species in a given area; it has become imperative to diversify the aquaculture production introducing different species of fishes and shell fishes and practice crop rotation. This also caters to the varying consumer preferences and consequent market demand. In the fresh water sector introduction for culture of air breathing fishes such as Clarias, Heteropneustes, Anabas, murrels etc.; silurids such as Wallago, Pungasius, Mystus, channel catfish; cyprinids such as Puntius; Cichlids such as Tilapia ; coldwater fishes like trouts and mahseer and fresh water prawns such as giant fresh water prawn and Indian river prawn etc. are noteworthy in this regard. In the brackish water farming sector fishes like pearlspot, mullets, milkfish, thread fins, mudcrab etc are other important candidate species. In the marine sector fishes such as seabass, flat fish, tunas, seer fish, salmon, pomfrets and crustaceans such as *P.indicus* (white prawn), *P.merguiensis* (banana prawn), *Metapenaeus monoceros, M.dobsonii, M.brevicornis;* crabs such as N*eptunus pelagicus, N.sanguinolentis,* lobsters and molluscans such as mussels, clams, edible oysters, pearl oysters, abalone, etc. are important.

During the course of my teaching career that spanned about 38 years I have felt an acute shortage of text books in several areas, particularly in Fishery Science while offering courses at both undergraduate and post graduate levels. Seed production technology is one of the areas that called for my immediate attention. In this book, I have tried to compile all the available literature on the seed production technology of a number of species of fin fishes such as Indian major carps, common carps, grass carp, silver carp; cat fishes such as Clarias, Channel cat fish and Heteropneustes; Cichlids such as Tilapias; salt water fishes such as milk fish, mullets, sea bass and grouper, cold water fishes like trouts and mahseer. Breeding of several genera of ornamental fishes has been described in a separate chapter. Shell fishes dealt in the book include tiger shrimp, fresh water prawn, mud crab, oysters, mussels, clam and sea-cucumber. As any hatchery is designed to take the full advantages of the reproductive potential of a given species knowledge on the reproduction is a prerequisite for any fish breeder. In view of this, a chapter on reproduction in fish has been included and besides details of reproductive biology such as age and size at first sexual maturity, sexual dimorphism, fecundity, courtship and mating, site of reproduction, style or reproduction etc. of each species is described prior to the description of hatchery techniques. A chapter on the culture of live feed organism is included in view of the total dependence of the larva / fry of all species of finfish and shell fish on live feed during the initial stages of larval development. Further, chapters on genetic upgradation of fish seed and cryopreservation of fish gametes have been incorporated to familiarize the students and researchers on the recent innovations to generate fish seed of superior qualities.

I have tried my level best to provide an up-to-date state of the art knowledge on the subject. I hope the book shall be of interest and benefit to students, teachers, researchers , entrepreneurs and policy makers in their pursuit of attaining excellence in respective areas.

Dr. P.C. Thomas

Contents

Chapter 1

Reproduction in Fishes

Introduction

To reproduce and to multiply in a geometrical ratio is the innate capacity of every organism. Among the 21723 living species of fishes existing in the world at present, a wide variety of patterns of reproduction such as unisexuality, bisexuality, hermaphroditism and parthenogenesis are observed. Jhingran and Gopalakrishnan (1974) listed 314 species of teleosts utilized for aquaculture throughout the world. Eversince this, more and more number of species of fishes are being added to the list for aquaculture. As indiscriminate fishing from natural habitats is depleting the natural fish resources, there is need for evolving species specific technologies to culture fish and shellfish of commercial value. As basic requirement for culture of any species of fish is availability of quality seed and as the availability of seed primarily depends on the efficient functioning of the reproductive organs, this chapter is devoted to a review of the various aspects of reproduction in fish. In order to evolve and standardise an efficient hatchery system for producing the young fish of good quality in sufficient numbers of a given species, it is essential to know the details of its natural breeding. This involves the study of (i) sexuality of a given fish (ii) sexual dimorphism (iii) reproductive cycle (iv) time/ season/ place of reproduction (v) fecundity (vi) courtship behaviour and spawning pattern (vii) parental care, if any and (viii) larval life history. Further, several other details of factors affecting reproduction such as (i) environmental factors (light, temperature, waterquality parameters etc.) (ii) intrinsic physiological factors such as hormonal/ neurohormonal control on reproduction and (iii) nutritional requirements of the fish such as natural food and feeding pattern, balanced diet required for optimum gonadal growth etc are also to be studied.

SEXUALITY IN FISHES

Unisexual Fishes

Unisexual fishes are those that produce offsprings of one sex only. It is a rare phenomenon among fishes. Hubbs and Hubbs (1932, 1946) were the first to report all female population of the fish, *Poecilia (Mollienisia) formosa*. The females of *P. formosa* mates with males of other species of Poecilia such as *P. latipinna/P. mexicana.* During mating the sperm is introduced into the body of female formosa, but the sperm only triggers the process of development of the egg, the male genome do not fuse with female genome. Thus the offsprings so produced shall be all gynogenetic offsprings and hence shall be all females. Gynogenesis has been also reported in the natural populations of silver crucian carp (in Central Asia, Western Siberia and Europe) and in *Carassius auratus gibelio* which also produces all female progeny. In East Asia silver crucian carp population has both males and females.

Bisexual Fishes (gonochorism)

In most of the fishes sexes are separate *i.e.* there are male and female fishes. Such fishes are called bisexual (gonochoristic) fishes. Gonochoristic fishes have been classified into 2 categories such as (i) *undifferentiated gonochoristic fishes* and (ii) *differentiated gonochorists* based on their embryonic development. In case of fishes, during development, genital ridge (gonodal blastema) is characterised by the absence of medullary portion. The future ovary/testes is developed from the same cortex unlike other higher animals where testes is developed from medulla and ovary from cortex. In case of undifferentiated gonochorists, during initial stages of development, gonadal tissue remain in the indifferent stage, then developes into ovary and then into either testes or ovary. In the differentiated gonochorists, indifferent gonad directly develops into either ovary or testes. Occurrence of intersex are common among undifferentiated gonochorists where as in differentiated ones intersexes are absent. *Anguilla anguilla, Salmo gairdnerii* are examples of undifferentiated gonochorists where as platy fish *(Xiphophorus maculatus)* medaka *(Oryzias latipes)*, common carp, gold fish etc. are differentiated gonochorists.

Hermaphroditism

A number of fishes have been reported to be hermaphrodites. A hermaphrodite is one in which both male and female sex organs are present in the same individual. Two types of hermaphroditism have been distinguished such as (1) synchronous hermaphroditism and (2) metagonous hermaphroditism (asynchronous/ sequential/ consecutive hermaphroditism). In synchronous hermaphroditism male and female reproductive organs ripen at the same time. In consecutive hermaphrodites, either male or female reproductive organ develop at a given time. If an organism functions as female first and then undergoes sex reversal to become male, it is called protogynous hermaphroditism. If an organism is male first but later changes to a female it is called protandrous hermaphroditism. Atz (1964) has identified 13 families of teleosts belonging to 5 orders exhibiting hermaphroditism. The main groups are four families of the order perciformes such as serranidae, sparidae, centracanthidae and labridae, four families of the order myctophiformes, a few species of the order cyprinodontiformes, stomiatioids of salmoniformes, and a few species of order synbranchiformes and scorpaeniformes.

Synchronous hermaphroditism

The gonad of these fishes is divided into ovarian and testicular areas- the ovary developing oocytes and testis developing spermatozoa. Examples are (1)

Serranus scriba (2) *S. cabrilla* (3) *S. subligerius* and (4) *Hepatus hepatus.* Some authors have suggested the possibility of self fertilization also in these fishes. If an individual fish is kept separately, it can release its own eggs and then eject its own milt and fertilize the eggs. If two or more fish are kept together, they may form spawning pairs– when one releases egg the other releases the sperm and later viceversa.

Consecutive hermaphroditism (Sequential hermaphroditism)

Protandrous hermaphrodites

The giant perch/sea bass *(Lates calcarifer)* of family centropomidae and order Perciformes - a commercially important fish of cultural significance is a prot androus hermaphrodite. Other fishes of this group are *Sparus auratus* (Mediterranian bream), *Sargus sargus* and *Pagellus mormyrus*.

Protogynous hermaphodite

Pagellus erythrinus and *Spondyliosoma cantharus* of the family sparidae and nine species belonging to genus *Epinephelus, Mycteroperca, Alphes, Petrometopon, Cephalopholis* and Atlantic sea bass of family serranide and L*abrus turdus* and *L. merula* of family labridae and synbranchoid eel M*onopterus albus* are protogynous hermaphrodites.

In consecutive hermaphrodites, the juvenile gonad is usually ambisexual - ovarian and testicular rudiments are present in the same fish. The age and size at which sex reversal takes place in a given population within a species may also vary from individual to individual. Sometimes only 50% of the population may undergo sex change.

SEXUAL DIMORPHISM

Most of the fishes exhibit sexual dimorphism or secondary sexual characters by which sexes can be distinguished from each other. In a few fishes secondary sexual characters are discernible throughout the life span where as in some others they are discernible only during the breeding season. Secondary sexual characters serve several functions such as (a) recognition of opposite sex by the members of a given sex (b) helping in the act of copulation such as sexual embrace (c) transfer of spermatozoa from male to female and (d) facilitating parental care. External morphological differences between male and female pertain to the following features.

Size of the Fish

Some species of fishes exhibit size differences between sexes. For example, in Tilapia, males grow faster and bigger than the females, but in major carps the females grow faster than male. In some fishes male fish are too small compared to female and are called dwarf males. Dwarf males have been reported in some fresh water salmonids. Dwarf males of Angler fish - a deep sea fish-live as parasite on female fish. In fishes such as *Silurus glanis, Pseudobagrus fulvidraco* where male is larger than female, large size of male is considered a protective adaptation as the male plays an active role in parental care.

Length/shape/texture of Fins

In several species of fishes, sexual differences are found in the shape/size of the fins. In *Labeo dero* dorsal fin of male is different from that of the female - anterior part of the dorsal fin is elongated in the form of a lobe in male which is not found in female. In fantail darter *(Etheostoma flabellare)* dorsal fin spines end in fleshy knobs (like a club) in case of male but in female the same are absent. In Tibetan loach *(Nemachilus stoliezkai)* the pectoral fin of males has a number of fin rays thickened in the form of spine where as in females it is absent. In case of Tench, *Tinca tinca,* the pelvic fin of a male has strong spines, which are not found in females. Pectoral fins are longer in males and also rough at the innerside during breeding season in case of major carps. In males of live bearers like sword tail last finrays of anal fin of male is modified to form an intromittent organ "gonopodium" having a groove (Fig. 1.1) which is absent in female. In fishes belonging to Jenynsidae and Anablepidae anal fin is modified to form a tube opening at its tip which works as a penis during mating. In round tailed paradise fish *(Macropodus chinensis)* anal and dorsal fins are more pointed in male but in female they are some what rounded.

Colouration

Some species of fishes exhibit sexual dichromatism i.e. different colour pattern in different sexes. During breeding season male develops brighter colouration in fishes such as salmonids and in some cyprinids. Males may also develop tubercles, on the head. In many aquarium fishes such as Siamese fighter fish males are more brightly coloured than female.

Genital Papilla

In a few fishes like Clarias, Heteropneustes etc. males have prominent genital papilla, particularly during breeding season which are elongated and cylindrical in male but small, button shaped in female (Fig. 1.2).

Ovipositor

In a few fishes like bitterlings (*Rhodeus sericeus*) the ready to spawn females develop long, pink tubes (called ovipositor) from genital apperture region which is used for laying eggs into the mantle cavity through the inhalent siphon of freshwater mussel.

Shape of Head

A few species exhibit sexual differences in the shape of the head. In male salmon, the snout develops like a hook whereas in female the same are not noticed (Fig. 1.3).

REPRODUCTIVE CYCLE

A few fishes breed only once in their long life span *eg.* Pacific salmon (Oncorhynchus). The fresh water eel, *Anguilla anguilla* breeds only once in 10-14 years. Most of the fishes breed in one season every year, they are called seasonal breeders, but some breed in all seasons, through out the year, and they are called year round spawners. Seasonal breeders exhibit rhythmic changes in the structure and physiology of ovary and testes in different seasons. These changes are demarcated

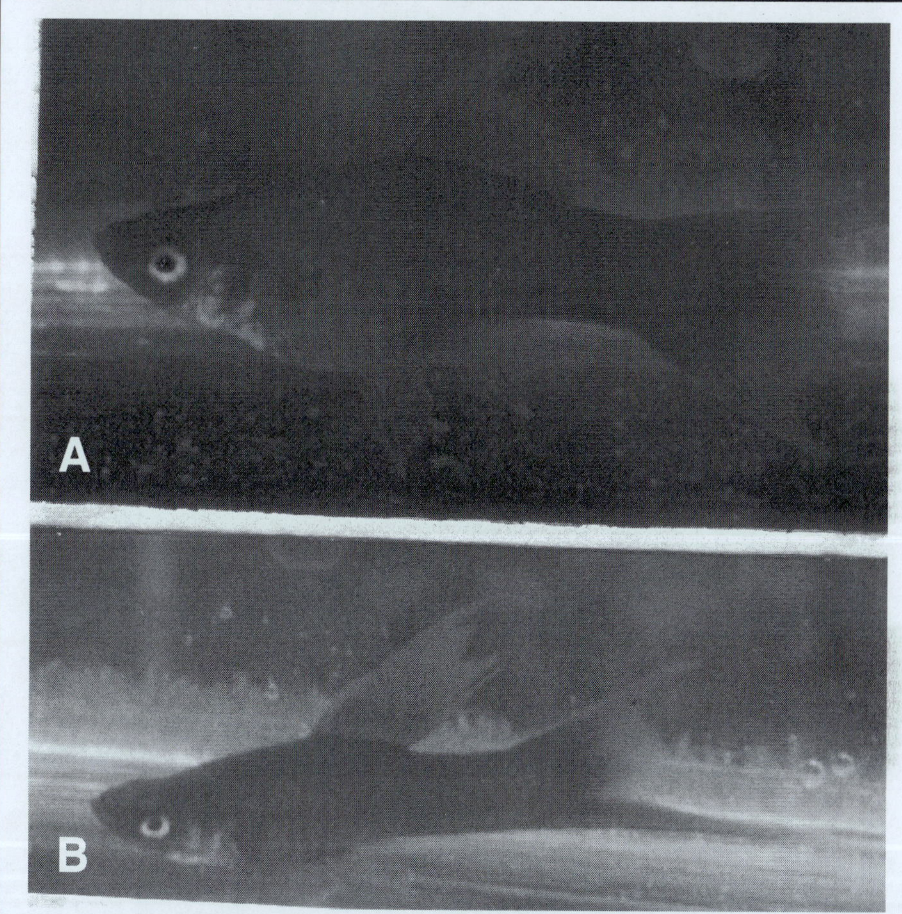

Fig. 1.1: Sexual Dimorphism in a Poecilid Fish, Sword tail (*Xiphophorus helleri*)
(a) Female sword tail, (b) Male sword tail

Note the presence of gonopodium modified from anal fin in male which is absent in female. Shape and structure of dorsal fin and caudal fin are also different in both sexes.

Fig. 1.2: Sexual Dimorphism in Magur (*Clarias batrachus*)
Genital papilla is Button Shaped in Female (Left) but Elongated in Male (Right).

Fig. 1.3: Sexual Dimorphism in Salmon, *Salmo salar.* (a) Male, (b) Female
Note: The Hooked Jaw in Male
(Reproduced with permission from *Trevor Meyer*)

into five phases such as (1) resting phase (2) preparatory phase (3) maturing phase (4) spawning phase (5) post spawning phase (recovery phase).

Ovarian Cycle (Figs. 1.4 & 1.5)

Resting Phase

Usually it extends from November to January in most of the teleosts of tropical climate. The gonadosomatic index will be the lowest during this period and the ovary is in the immature state containing nests of oogonial cells stage 1 oocyte at different phases of growth and a few stage II oocytes. Primary oocytes develop from oogonial cells after a series of mitotic divisions. These oocytes have larger germinal vesicles which are chromophobic, relatively small sized ooplasm which is chromophilic and undifferentiated oocyte envelopes. During this stage the first meiotic division of nucleus is initiated and the same is arrested at pachytene stage. These oocytes are also characterised by the presence of yolk nucleus (Balbiani's vitelline body) in the ooplasm. Initially, these bodies are found adjacent to the nucleus and as it grows, gets separated from the wall of the nucleus and get scattered in the ooplasm after undergoing fragmentation. Later they disappear. Yolk nucleus is rich in protein, lipid, carbohydrate and Golgibodies. In spite of it universal occurrence and several histochemical and electron microscopic studies, the precise role of these bodies is not yet clearly understood. According to some workers, the yolk nucleus provides the basic cytoplasmic machinery for future vitellogenesis that takes place inside the ooplasm.

Stage II Oocytes are larger in size. The oocyte envelopes get differentiated to the outer most thecal layer, middle granulosa layer and the innermost chorion or zona pellucida or oolemma.

Preparatory Phase

This phase is noticed from February to March in most of the fishes. In places where winter is not severe, it may be observed from the latter part of January itself. Oocytes in the preparatory phase (stage III oocytes) are characterised by further growth in the size and development of cortical alveolii at the periphery of ooplasm. This is the beginning of vitellogenesis in the oocyte cytoplasm. Oocyte envelopes get differentiated with the thecal and granulosa layers assuming the steroidogenic function. Cells of the granulosa layers start producing the female hormone, estradiol. Side by side synthesis of vitellogenin takes place in the liver under the influence of estradiol. Vitellogenin is released into the blood from liver and the same gets incorported into the ooplasm through the oocyte envelope. Vetellogenin is the precursor for yolk and the same is deposited in the oocyte cytoplasm. This process of formation of yolk in the ocplasm is called vitellogenesis.

Ultrastructural studies have revealed the existence of a follicle - oocyte connection/communication through numerous microvillii arising from the surface of ooplasm which project through the pores of zona pellucida to granulosa layer. The granulosa cells in turn produce cytoplasmic processes, which are in touch with the microvillii. This facilitates communication between follicle cells and ooplasm (Fig. 1.13).

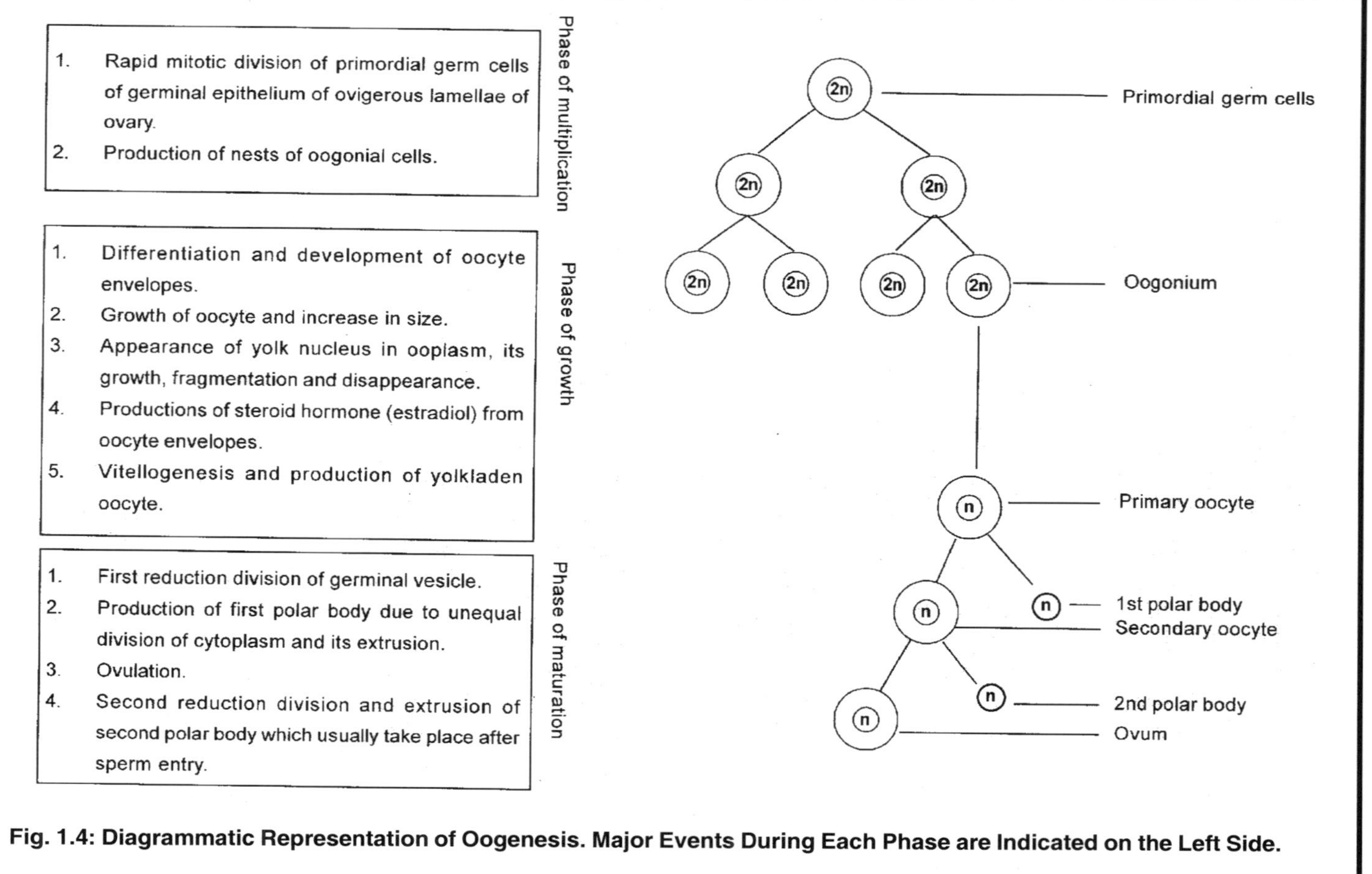

Fig. 1.4: Diagrammatic Representation of Oogenesis. Major Events During Each Phase are Indicated on the Left Side.

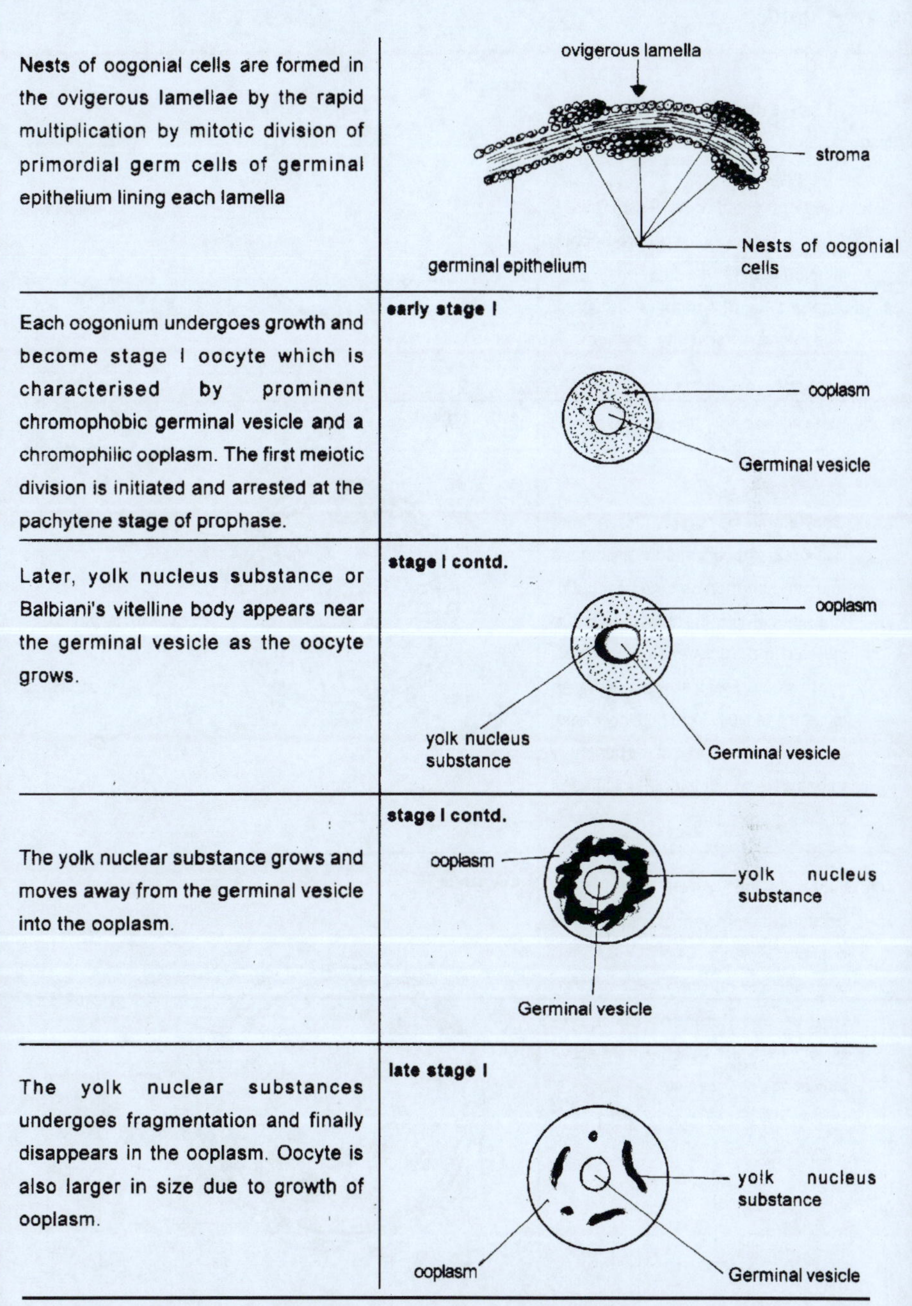

Fig. 1.5: Schematic Drawings of Different Stages* in the Development of Oocytes of a Teleost. The Main Events in Each Stage are Indicated in the Left Hand Column.

Fig. 1.5–Contd...

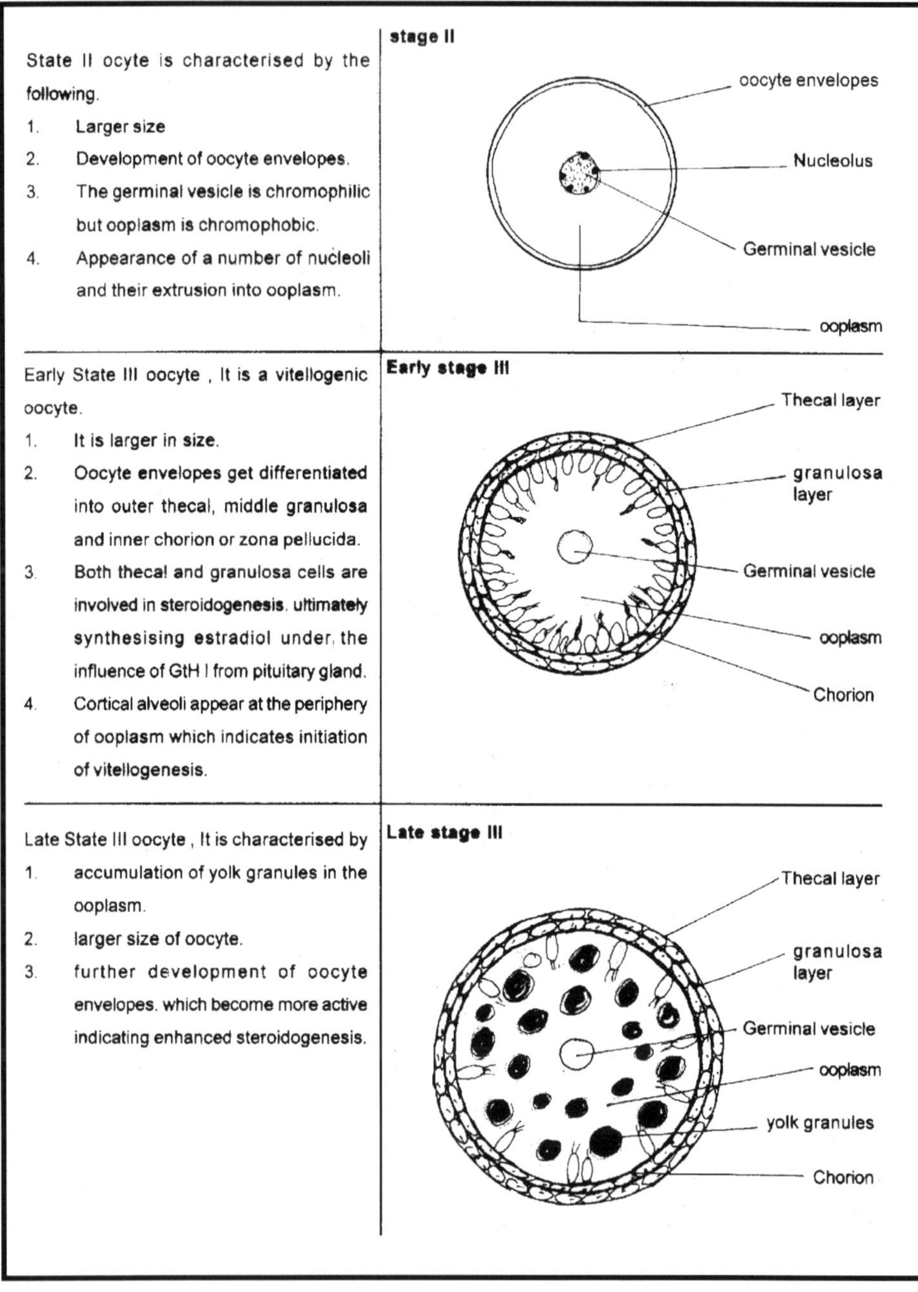

State II ocyte is characterised by the following.

1. Larger size
2. Development of oocyte envelopes.
3. The germinal vesicle is chromophilic but ooplasm is chromophobic.
4. Appearance of a number of nucleoli and their extrusion into ooplasm.

Early State III oocyte , It is a vitellogenic oocyte.

1. It is larger in size.
2. Oocyte envelopes get differentiated into outer thecal, middle granulosa and inner chorion or zona pellucida.
3. Both thecal and granulosa cells are involved in steroidogenesis, ultimately synthesising estradiol under the influence of GtH I from pituitary gland.
4. Cortical alveoli appear at the periphery of ooplasm which indicates initiation of vitellogenesis.

Late State III oocyte , It is characterised by

1. accumulation of yolk granules in the ooplasm.
2. larger size of oocyte.
3. further development of oocyte envelopes. which become more active indicating enhanced steroidogenesis.

Fig. 1.5–Contd...

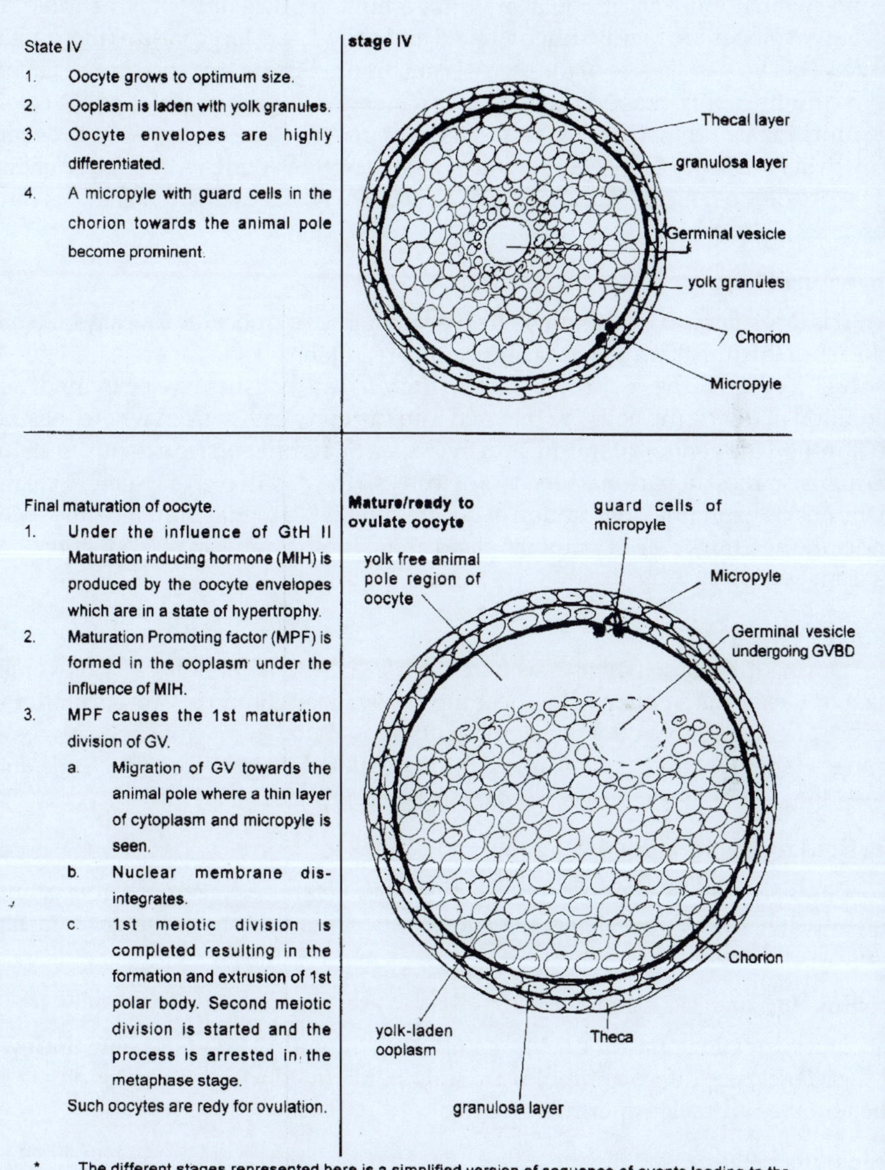

State IV

1. Oocyte grows to optimum size.
2. Ooplasm is laden with yolk granules.
3. Oocyte envelopes are highly differentiated.
4. A micropyle with guard cells in the chorion towards the animal pole become prominent.

Final maturation of oocyte.

1. Under the influence of GtH II Maturation inducing hormone (MIH) is produced by the oocyte envelopes which are in a state of hypertrophy.
2. Maturation Promoting factor (MPF) is formed in the ooplasm under the influence of MIH.
3. MPF causes the 1st maturation division of GV.
   a. Migration of GV towards the animal pole where a thin layer of cytoplasm and micropyle is seen.
   b. Nuclear membrane dis-integrates.
   c. 1st meiotic division is completed resulting in the formation and extrusion of 1st polar body. Second meiotic division is started and the process is arrested in the metaphase stage.

   Such oocytes are redy for ovulation.

* The different stages represented here is a simplified version of sequence of events leading to the development of a mature/ready to ovulate egg. Many workers have identified the same into VII or VIII stages. However main changes in the oocyte envelopes, ooplasm and germinal vesicle remain same

Maturing Phase or Prespawning Phase

This phase extends from March to May and is characterised by the intensive process of vitellogenesis by which ooplasm of an oocyte is loaded with yolk granules leaving a small portion of ooplasm with germinal vesicle at the animal pole. The oocyte envelopes are highly differentiated and oocytes are large in size (these are the stage IV oocytes). The size of the ovary is maximum during this season and GSI will be maximum at this phase. The ovary gets distended, highly vascularised to cater to the nutritional and metabolic requirements of oocytes. The ovary undergoes rapid growth in size due to the accumulation of yolky oöcytes. Along with stage IV oocytes, a few previtellogenic oocytes and oocytes with early stages of vitellogenesis can be observed in the ovary.

Spawning Phase

It is characterised by the gravid ovary containing ripe oocytes. During spawning, follicles of the fully ripe oocytes rupture, as a result the oocytes are released into the ovocoel in case of fishes with cystovarian ovary. In case of fishes having gymnovarian condition of ovary, the eggs are released into the body cavity from where they pass out through the genital aperature into the water. Spawning phase extends from June to August in most of seasonal breeders of tropics. (For details of preovulatory changes in the oocytes and final maturation of oocytes under hormonal influence, the section under the heading "role of pituitary gland and hypothalamus in reproduction" may be seen)

Post spawning Phase

During this phase which is seen from early/late September to early November, the ovary exhibits a collapsed appearance as evacuated follicles are seen after the release of eggs. If any yolky oocytes remain unovulated, it may undergo regressive changes/atresia, finally reabsorbing the unovulated oocyte. Oogonial cells and a few stage I/stage II oocytes also can be seen at this phase.

Testicular Cycle (Fig. 1.6)

Parallel to the ovarian cyclic changes taking place in female, the primary reproductive organ of male *i.e.* testis also undergoes rhythmic changes which are summarised below:

Resting Phase

During resting phase the testis remain in an inmature state, GSI being the lowest at this stage. Seminiferous tubules are solid being filled with spermatogonial cells. Such tubules are called spermatic cords.

Preparatory Phase

During this stage, further steps in spermatogenesis can be noticed. As a result primary spermatocytes and secondary spermatocytes and spermatids are produced. Spermatids possess haploid number of chromosomes.

Mature Phase

During this phase spermatids undergo further development and develop into

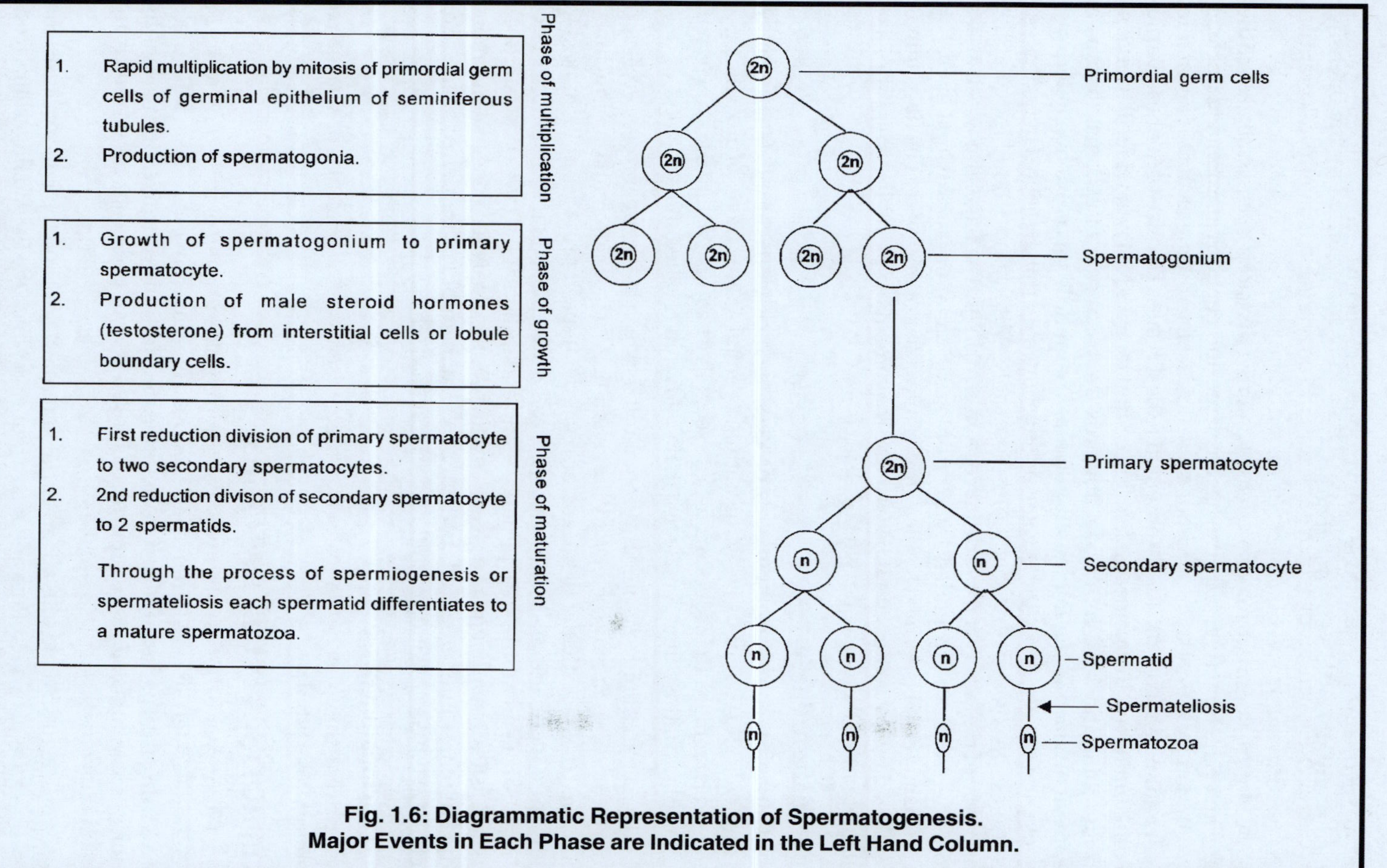

Fig. 1.6: Diagrammatic Representation of Spermatogenesis. Major Events in Each Phase are Indicated in the Left Hand Column.

mature spermatozoa. The process of development of spermatozoa from spermatid is called spermateleosis or spermiogenesis.

Structure of a Mature Spermatozoa of a Teleost

A mature spermatozoa of a teleost has three divisions such as head, middle piece and flagellum (Fig. 1.7). The head is large and contains a prominent nucleus surrounded by a thin layer of cytoplasm. A conical acrosome at the tip noticed in higher animals is conspicuously absent in teleost fishes. The size of the head varies in different fishes. In channel cat fish it is 2.3 mm long and 2.4 mm wide, in trout it is 2.5 mm long and 1.5-2 mm wide. Towards the basal side of nucleus, there is a depression which is the implantation fossa. The middle piece extends anteriorly, enveloping up to half of the nucleus in a collar like arrangement. There are two centrioles- proxinal centriole and distal centriole- Mitochondria are located at the periphery of midpiece, at some distance from centrioles. The flagellum which arises from the fossa is surrounded by a cytoplasmic canal at its proximal end. The cytoplasmic canal is formed by the cytoplasmic collar of midpiece. The flagellum is supported by axonemes and it helps in the movement of sperm. In some fishes like channel catfish there may be two flagellae.

Spermiation Phase

During the courtship and mating process, the male ejects the milt (spermatozoa in seminal fluid) out of its body through the genital aperature to fertilize the eggs released by female. In most of the fishes except live bearers fertilization is external, taking place in the water. A male fish that yields milt is called milter.

The seminal fluid - the fluid medium through which the spermatozoa are released from the body - is mostly secreted by the cells lining the vasa efferentia (sperm ducts) found across the testicular tissue and also vas deferens. It provides nourishment to the mature spermatozoa. Both spermatogenesis and seminal fluid secretion are under the control of gonadotropin of pituitary gland and testicular hormone (testosterone). In some fishes lobule boundary cells of seminiferous tubule are secreting male hormone while in others interstitial cells present between the tubules are steroidogenic (For details on the composition of milt please refer chapter 12.)

Post Spermiation Phase

During this phase, the testis is characterised by the presence of evacuated seminiferous tubules.

REPRODUCTIVE STRATEGIES

Freshwater fishes usually spawn either in confined water, or in flowing water or in inundated terrains. Different species chose different spawning sites within these areas. Major carps usually spawn in inundated terrains during rainy season. Its choice of spawning site is advantageous for its survival and propagation which are noted below.

(a) The fresh flood kills all terrestrial fauna and flora and their decay causes the growth of microflora and micro fauna on which the fry and fingerlings feed. There shall be abundant natural food for the fry in these areas.

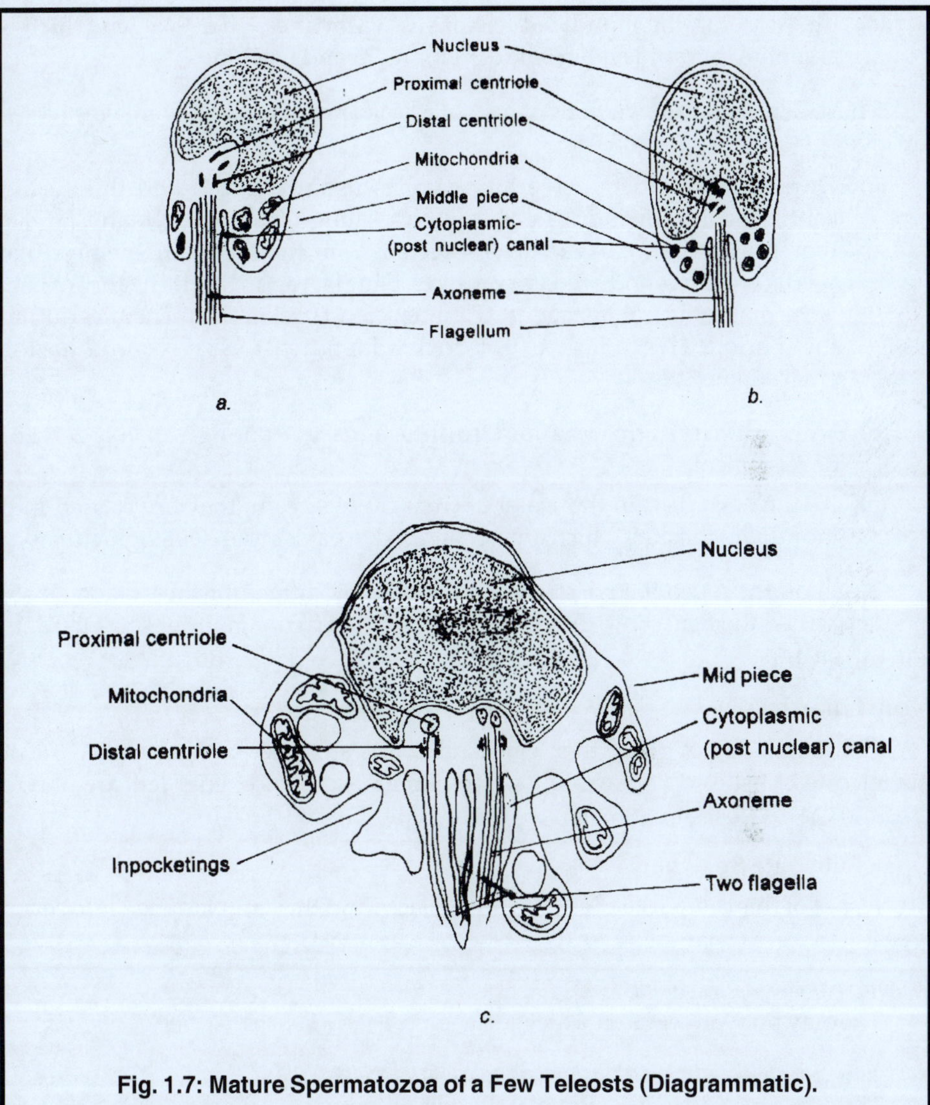

Fig. 1.7: Mature Spermatozoa of a Few Teleosts (Diagrammatic).

(a) Sagittal Section of Spermatozoa of *Carassius auratus*, (b) Longitudinal Section of Spermatozoa of *Clarias senegalensis*, (c) Longitudinal Section of Spermatozoa of Channel cat fish *Ictalurus punctatus* (Reproduced with permission from Jamieson, B.G.M., 1991).

(b) This area will be free of natural enemy of spawn and fry of carps as fresh flood kills all terrestrial fauna.

(c) The water in the inundated terrains is warm and rich in oxygen which is essential for rapid embryonic development and hatching.

Those fishes that spawn in flowing waters also have certain advantages which are given below:

(a) During rainy season, riverine water is usually turbid and this turbid continuously flowing water has a concealing effect on eggs and larvae, thereby, eggs and larvae are protected from their natural enemies like predatory fishes and predatory insects. Egg/larvae, in the life history of fish, is the most vulnerable stage as it cannot defend itself from its enemies during this period. Hence it is a protective adaptation to survive in a hostile environment.

(b) Non adhesive, floating and rolling eggs get enough of oxygen for development.

(c) As the eggs drift in the water current downstream, they are carried into inundated terrains which are rich in food organisms of developing fry.

Based on the reproductive strategies adopted by different fishes, Balon (1975, 1981) classified the fishes into three categories such as (i) nonguarders (ii) guarders and (iii) bearers.

Non Guarders

Fishes that do not guard their eggs and larvae are called nonguarders. 78% of the teleostean fishes do not exhibit active parental care. Non guarders are of two types. a) Open substrate spawners, b) Brood hiders

Open Substrate Spawners

Fishes of this category lay eggs in open places and they are classified into seven types.

Pelagic spawners (Pelagophils)

They lay buoyant eggs on water surface. *eg. Lates niloticus, Ctenopoma muriei*

Rock and gravel spawners with pelagic larvae (Lithopelagophils)

The eggs are laid on rock or gravel. To start with eggs may have adhesive chorion but soon some eggs may become buoyant and hatchlings are also pelagic *eg. Prochilodus.*

Rock and gravel spawners with benthic larva (lithophils)

Early hatched embryos are photophobic and they hide under stones etc. *eg. Labeo.*

Nonobligatory plant spawners (Phytolithophils)

They lay adhesive eggs on submerged items including aquatic plants. Hatchlings have cement glands by which they remain attached to leaves of aquatic plants or other substrates. Larvae are photophobic *eg.* Common carp, *Rutilus rutilus.*

Phytophils (Obligatory plant spawners)

Adhesive egg envelopes stick to submerged live or dead plants. Incubation period may be prolonged. Hatchlings are with cement glands and are not photophobic. *eg. Puntius gonionotus*

Sand spawners (Psammophils)

They spawn adhesive eggs in running water on sand or fine roots over sand. Free hatchlings are without cement glands. *eg. Gobio gobio*

Terrestrial spawners (aerophils)

Small adhesive eggs are scattered outside water in damp soil. *eg. Brycon petrosus*

Brood Hiders

They lay eggs in hidden places. They are of five types:

Beach spawners (aeropsammophils)

They spawn above the high tide line. Zygote develop in damp sand and hatch upon the movement of waves and it becomes pelagic afterwards.

Annual fishes (Xerophils)

The eggs are laid and buried in the bottom mud. When the pond bottom gets dried, the adult fishes die but the eggs remain dormant for many months till the next season when water refils the pond after rain. *eg. Nothobranchias*

Rock and gravel spawners (Lithophils)

They lay eggs in depressions of gravel (redds) or interstices of rocks. They usually have large quanitity of yolk. Hatchling is called alevin. *eg.* Salmonid sp.

Cave spawners (Speleophils)

They lay eggs in cave like areas.

Spawning in live invertebrates (Ostrocophils)

Female deposits the eggs in the body cavity of other invertebrates like mussels, crab, sponges, etc. by the help of ovipositor. The female bitterling, *Rhodeus sericeus* lays the eggs in to the mantle cavity of fresh water mussel by introducing the ovipositor in to the inhalent siphon of the mussel.

Guarders

In these fishes either male or female or both the parents guard the eggs and larvae after laying. 22% of the teleost fishes have been found to exhibit active parental care out of wshich male alone guards the egg in some or female alone in some others while both male and female guard the eggs in a few. Guarders are classified into two types.

(a) Substrate choosers, and (b) Nest spawners.

Substrate Choosers

Fishes of this category only select the spawning site, clean the area and lay the eggs. They are of four categories:

Pelagic spawners

They lay nonadhesive, positively buoyant eggs which are guarded by parents in the water surface. *eg. Channa, Anabas.*

Above water spawners (aerophils)

Eggs are laid on objects found above the water. The male guards the eggs in water just below the object containing eggs and splashes water by which eggs are kept moist periodically. *eg. Copeina arnoldi*

Plant spawners

Adgesvie eggs are attached to aquatic plants. The parent fish stands guarding the eggs. Hatchlings are without cement gland. *eg. Polypterus*

Rock spawners

They spawn strongly adhesive eggs attached to rocks at one pole, in clusters. Larvae are pelagic. *eg. Loricaria parva.*

Nest Spawners

They lay eggs in nests prepared by either of the parent or both. Eight categories of nest spawners have been identified:

Froth nesters (Bubble nest builders or aphrophils)

The parent fish, usually males, produce mucus bubbles from its mouth and when it forms a cluster floating on the water surface eggs are deposited by the female and kept in the bubble nest mass, eg. *Siamess fighter fish. Gouramis*

Hole nesters

Holes are prepared by burrowing in the bottom or horizontal holes are prepared in elevated floors. *eg.* Several cichlids, *Cottus aleoticus*

Miscellaneous material nesters (polyphils)

Nests are prepared out of miscellaneous materials available in water. *eg. Notopterus chitala, Hoplias malabaricus.*

Rock and gravel nesters

Nest is made of rocks and gravel. *eg. Ambloplites rupestris*

Anemone nesters

Some fishes like Amphiprion sp. lay clusters of adhesive eggs at the base of sea anemone. The eggs are coated by the mucus produced by the parents so that nematocystss of sea anemmone will not damage the eggs.

Plant material nesters, glue making

Nest is made up of plant material, which are bound together by a viscid thread spinned from the secretion of kidney of parent fish. *eg. Gasterosteus aculeatus.*

Plant material nesters - nonglue making

Nest prepared out of plant materials are not bound by glue secreted from the kidney of fish. *eg. Micropterus salmoides*

Sand nesters

These fishes make sand nest. Their eggs have adhesive chorion and hence eggs with sand grain attached to it may be washed off gradually. *eg. Cichlasoma nicaraguense.*

Bearers

Fishes that carry the eggs in their body during incubation are called bearers. They are of two types:

(a) External bearers, and (b) Internal bearers.

External Bearers

Those fishes that carry their eggs on the external parts of the body including gill cavity, mouth cavity etc. are called external bearers. They are of seven categories:

Transfer brooders

These fishes carry their eggs for sometime in the pelvic fin or as a cluster hanging from genital pore and then the eggs are deposited. After depositing eggs they are similar to nonguarders. *eg. Oryzias latipes, Callichthys, Corydoras.*

Skin brooders

Eggs are carried in clusters or as a layer on the spongy skin of parent fish. *eg. Bunocephalus, Platystachus cotylephorus.*

Forehead brooders

Eggs carried on the anterior part of head region. *eg. Kurtius gullivers*

Gill chamber brooders

Eggs are incubated in the gill cavity of fish. *eg. Typhlichthys subterraneus*

Pouch brooders

These fishes incubate the eggs in in a pouch. In fishes like *Syngnathus* (pipe fish) and *Hippocampus* (sea horse) pouch is iocated externally. In these fishes males have a pouch in the body in to which the female lays the eggs. The eggs are incubated in this pouch. In fishes like Brazilian cat fish (*Loricaria typus)* males develop an enlarged lower lip which is formed in to a pouch. Eggs are incubated in this labial pouch.

Mouth brooders

Eggs are incubated in the mouth. A number of species of Tilapia incubate their eggs in the mouth cavity of either the male parent or female parent or both. In species like *Oreochromis mossambicus, O. niloticus, O. aureus* and *O. machrochir* the mother keeps the egg in their mouth for incubation. Even after hatching the fry are cared for a few days by the mother. When the fry are disturbed they take shelter in the mouth of the mother. In case of *O. macrocephalus* the male incubates the egg in the mouth. In *Sarotherodon galileus* both male and female incubate the eggs in their mouth. So also marine cat fishes belonging to the family Ariidae males incubate the eggs in their mouth.

Intestinal brooders

Eggs are incubated inside the intestine. *eg. Tachysurus barbus.*

Internal Bearers

They are classified into following:

Facultative internal bearers

These are fishes in which internal fertilization takes place only accidentally in normally oviparous forms when the gonopores of male and female are in close apposition. Zygotes are retained within the reproductive system of female where early stages of embryonic development up to cleavage takes place. *e. g. Rivulus marmoratus.*

Obligate lecithotrophic live bearers

In these fishes fertilization is internal, gestation takes place in the reproductive system of female, but there is no maternal embryonic nutrient transfer as the developing embryo sustains itself on the yolk of the oocytes. *eg. Poeciliopsis monacha, Poecilia reticulata.*

Viviparous trophoderms

Fertilization is internal. Gestation takes place in the reproductive system of female but there is "maternal-embryonic nutrient transfer". Either the secretory material of the mother is ingested by the fetus or placental analogues arising from the embryo absorb the nutrients from the mother. *eg. Anableps dowii.*

FECUNDITY AND SPAWNING

In a few species of fishes, the female releases all the eggs that mature in a particular season within a short time interval. Such species of fishes are called total spawners. Salmon and trouts are total spawners. In case of some others, the female spawns several times during a given breeding season and such fishes are called batch spawners/multiple spawners/ fractional spawners. In total spawners gonado somatic index (GSI) shall be very high (20-30%) before spawning, but in case of batch spawners, GSI shall be relatively low (2-14%). Some like three spine stickle back is an exception to this rule, as, though it is a batch spawner, spawning at intervals of 3-5 days, it has GSI of '20' prior to spawning. In multiple spawners, only a small portion of the eggs became ripe and ovulate in one spawning. Hence the name fractional/ partial spawners. Fractional spawning is characteristic of tropical and subtropical fishes. In temperate climates fractional spawners are less and in Arctic region they are completely absent. Fractional spawning can be considered as an adaptation to abundant food availablity for the larvae as in tropics plankton is available at all time and hence there shall be no larval mortality due to inadequate food. It can be also considered an adaptation for survival as total spawners may face extinction in case its offsprings face adverse condition whereas in fractional spawners, even if one batch perish another batch can survive and multiply.

The term fecundity refers to the number of eggs a female lays in a specified period of time and this depends on the number of eggs per spawning and the number of spawnings. Batch fecundity: It refers to the number of eggs produced by a female per spawning-Breeding season fecundity refers to the number of eggs produced by a female in a particular breeding season. Lifetime fecundity refers to the total number of eggs laid by a female through out its life span. Total/absolute individual fecundity refers to the total number of eggs contained in the ovary of a fish. The term relative fecundity denotes the number of eggs per unit weight or length of fish.

Fecundity varies from species to species. Different individuals within a given species also may exhibit variation in fecundity depending on the size, age, nutritional status and environmental factors. Fecundity of a given female in two consecutive years may not also be same. Generally, fishes that d not exhibit active parental care produce more number of eggs. An ocean sunfish, a pelagic open water spawner, (*Mola mola*) produce 28, 000, 000 eggs in one season where as Gasterosteus which exhibits parental care produce only 30-100 eggs. In live bearers such as guppies it is less than 2 dozen.

In majority of the fishes, fecundity increases with age till it reaches senility, after which fecundity starts decreasing. For example in *Cyprinus carpio*, the fecundity increases from the length group 35 cm onwards, till it reaches 60cm, thereafter it decreases. The fecundity expressed in thousand between length group 35-40, 40-45, 45-50, 50-55, 55-60, 60-65, 65-70 are 181, 229, 375, 428. 5, 550, 525, 525 respectively. This indicates that in *C. carpio* senility sets in when it reaches a length of 60cm.

COURTSHIP AND MATING

Courtship activity involves species/sex recognition, mutual orientation to spawning site and synchronisation of the activities of both sexes so that male and female gametes are released at the appropriate time. Courtship in fishes may be a temporary association of ripe males and females or it may be long term association between pair in monogamous species. The following mating patterns are exhibited by fishes.

Promiscuity

Both sexes will have multiple partners and has only a single breeding season. These fishes exhibit poorly developed courtship *eg. Gasterosteus, Clupea.*

Monogamy

When mating partners remain together always it is called monogamy. In these fishes courtship is elabourate and time consuming *eg. Cichlasoma, Serranus.* According to some, *Channa striatus* is also a monogamous fish.

Polygamy

Either male or female will have multiple partners in a given breeding season.

Polygyny

If the male has multiple partners in a given season it is called polygyny. *eg. Oreochromis, Gasterosteus*

Polyandry

If the female has multiple partners in a given season it is referred to as polyandry. *eg.* Indian major carps, *Amiphiprius.*

Courtship is usually preceded by the choice of a mate. Secondary sexual characters may play a role in mate choice. For example, female three spine stickle back shows preference for males whose red breeding colour is more demarcated. This is because brighter males are more effective in defending their nest from predators. Some species of fishes like sockeye salmon, and Indian major carps shows preference to mate with the opposite sex of same size but in some species they prefer to mate with opposite sex of larger size.

STYLE OF REPRODUCTION

Based on their style of reproduction, fishes can be classified into two categories such as (a)Oviparous those that lay eggs. (b)Live bearers. Live brearers - fishes that give birth to young ones. Live bearers are either ovoviviparous or viviparous. Ovoviviparous fishes are those in which fertilization is internal, but during gestation there is no "maternal-embryonic" exchange of nutrients, developing embryo sustain itself on the yolk reserve of the oocyte. In viviparous fishes fertilization is internal and during gestation, there is maternal-embryonic tranfer of nutrients. Development of young within the female and parturtion occurs only in 2 orders of teleosts such as cyprinodontiformes (family poecilidae, goodeidae, anablepidae, jenynsiidae) and Perciformes.

Fertilization, Gestation and Parturition in Live Bearers

In live bearers, once the male introduces the sperms into the reproductive tract of the female, it may remain embedded in the ovarian epithelium for a prolonged period extending even to many months. A mature ovum is fertilized usually when it is inside the follicle. In some cases, the fertilized ovum may undergo development to hatchling inside the follicle itself. Then it is called follicular gestation. Follicular gestation is found in Poecilidae and Anablepidae. In some other cases, soon after fertilization inside the follicle, the fertilized egg fall into the ovarian cavity and gestation takes place in this cavity. This type of gestation is called ovarian gestation. In some other cases, the fertilization and gestation takes place in the ovarian cavity.

Gestation within the Ovarian Cavity (Ovarian Gestation)

Fertilization and Gestation in the Ovarian Cavity

Soon after ovulation, the mature eggs fall into the ovarian cavity where it is fertilized by the sperm which are remaining embedded in the follicular epithelium. The development of zygote to the hatchling stage take place inside the ovary. The embryo depends on its own yolk reserve for nutrition. *eg. Sebastodes paucispinis.*

Fertilization in the Follicle but Gestation in the Ovarian Cavity

Sperm is stored in the ovary for six months. Egg, when mature, is fertilized by the sperm while the egg is inside follide. The fertilized eggs are released into the ovarian cavity where gestation takes place. Egg is not rich in nutrient reserves. Ovigerous folds are highly glandular and they secrete a fluid into the ovarian cavity. The developing embryo derives its nourishment from this secretion. In addition to the secretion entering into the alimentary canal of embryo through the mouth, it also has vascular extension of vertical fins which assist in nutrition and respiration. These vascular extension arise when the embryo reaches 15 mm long. *eg. Cymatogaster agregata.*

In *Lermichthys multiradiatus*, a Goodeid fish having ovarian gestation, rectal region of embryo develop trophotaenia which is highly vascular and the same is bathed in the secretions of ovarian cavity and trophotaenia absorbs the nutrients.

In Jenynsiids, which have ovarian gestation, the ovarian lamellae develop vascular folds, which remain in close contact with the pharyngeal and mouth cavities of the embryo. As it is in close contact with the gill it is called 'branchial placenta' or trophonemata of maternal origin.

The above fishes can be termed as viviparous fishes.

Follicular Gestation

In this category both fertilization and gestation take place in ovarian follicle. In Guppy, Black molly and Sword tail yolk present in the egg is sufficient for embryonic nutrition. The yolk sac is expanded over the head as a neck strap. The portal system of yolk sac is in intimate association with the vascular wall of ovarian follicle. This is primarily meant to facilitate exchange of gasses and nitrogenous wastes but its role in transfer of nutrients from mother to embryo is not yet established.

Heterandria formosa is another fish with follicular gestation. During gestation faetal-maternal exchanges are done through a pericardial sac which is highly expanded and vascular. This is found around the anterior part of the embryo. Further, expanded coclimic cavity of the embryo called belly sac on the ventral side is in intimate association with fingerlike projection of follicular wall which is called "follicular pseudoplacenta".

In *Anableps anableps*, in addition to villi from follicular wall, numerous vascular bulbs appear in the portal circulation of belly sac. Enlargement of gut helps in the digestion of follicular fluid taken into the alimentary canal.

Super Foetation

In many poeccilids where follicular gestation is the rule, a phenomenon called super fetation (*i. e.* existence of many broods of different ages within the ovary in different follicles) is noticed. As sperms embedded in the ovarian epithelium remain alive up to 10 months after mating and as the eggs ripen successively, each ripened ovum is fertilized by the sperm remaining embedded in the ovarian epithelium successively even when previous embryos develop in the follicles. In these fishes, parturition takes place at 10 days intervals usually.

HORMONAL/NEUROHORMONAL CONTROL OF REPRODUCTION

Neurohormones of hypothalamus and hormones of pituitary, gonads and many other endocrine glands are believed to be involved in the regulation of reproduction in fishes.

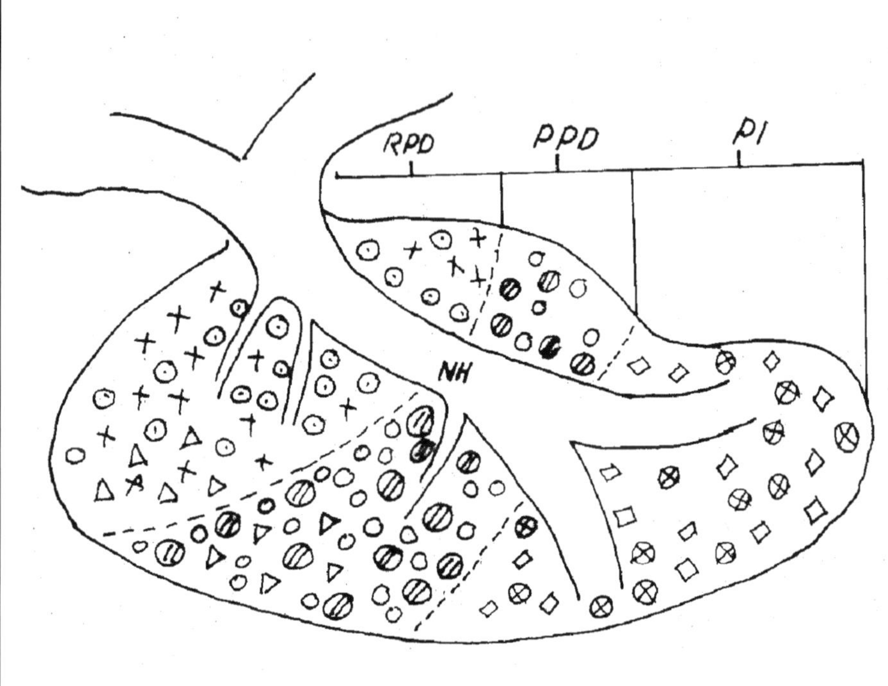

Fig. 1.8: Diagram Showing the Component Parts of the Pituitary Gland and the Distribution Pattern of Different Cell Types of *Labeo rohita* (modified from Jose & Sathyanesan, 1972)

- ⊙ Prolactin cells
- + ACTH (Adreno corticotrophic hormone) cells
- △ Thyrotrophs
- ◍ Somatotrophs
- ○ Gonadotrophs
- ⊕ MSH (Melanocyte stimulating hormone) cells
- □ PAS+ cells (somatolactin)

RPD-Rostral pars distalis

PPD-Proximal pars distalis

P.I.-Pars intermedia

N.H.-Neurohypophysis

Role of Pituitary Gland and Hypothalamus in Reproduction

The pituitary (hypophysis)is an endocrine gland of dual origin found on the ventral side of mid brain attached to it by means of stalk (Fig. 1.8). This has two components such as neurohypophysis and adenohypophysis. During embryonic development neurohypophysis arises as a downward evagination from floor of 3rd ventricle. The adenohypophysis is derived from the embryonic pouch, Rathke's pouch, arising from the roof of the buccal cavity as an outward evagination. The neurohypophysis consists of nerve fibres, arising from the perikaryon of neurons, which are located in the hypothalmus. This ciuster of neurosecretory neurons is called preoptic nucleus or nucleus preopticus (NPO). These neurons secrete nannopeptides such as arginine vasotocin and isotocin (icthyotocin) which function like vasopressin and oxytocin of higher animals. Though isotocin is found in teleosts, sharks have valitocin and aspargtocin. Ray fishes have glumitocin and lung fishes have mesotocin.

The neurohypophysis interdigitates into all areas of adenohypophysis but major interdigitations are found in pars intermedia. The neuroscretory materials from the neurons of preoptic nucleus are transported through the axons and stored in the axonal terminals ending at the junction between neurohypophysis and adenohypophysis which is called neuroadeno interface. This interface is highly vascular and it is the centre for storage and release of neurohormones. Hence it is called as neurohaemal area. Apart from NPO there are several other clusters of neurosecretory neurons in the hypothalamus out of which nucleus lateralis tuberis (NLT) is extensively studied. In case of teleosts these neurons also send their axonal projections to neuro-adeno interface. The secretion of these neurons control the adenoohypophysial cells. Hence they are called adenohypophysiotropic neurons or neurons secreting releasing hormones or inhibiting hormones. The releasing hormones are meant to stimulate the release of adenohypophysial hormones and inhibiting factors (hormones) are meant to inhibit the secretion of adenohypophysial cells. Three adenohypophysiotropic hormones isolated so far from hypothalamus are GnRH (Gonadotropin releasing hormone), TRH (Thyrotropin releasing-hormone) and somtostatin (Growth hormone inhibiting hormone). All these neurohormones are peptides. GnRH is a decapeptide and it controls the synthesis and release of gonadotropin from gonadotrophs. TRH is a tripeptide and somatostatin is a tetradecapeptide. (amino acid sequence of GnRH molecules isolated and purified from fishes is given in chapter 2).

The adenohypophysis of a telost has three regions such as rostral parts distalis (RPD), proximal pars distalis (PPD) and pars intermedia (PI). The RPD has two types of cells such as ACTH cells and prolactin cells. Adrenocorticotropic hormone (ACTH) controls the secretion of steroidogenic cells of interrenal gland. Prolactin cells produce Prolactin hormone, which is concerned with osmoregulation. It is mainly concerned with the freshwater survival of fish preventing the efflux of ions from the gill to the surrounding water. PPD has 3 types of cells *viz.* (i) gonadotrophs (GtH), secreting gonadotropin which controls gonadal function. (ii) Thyrotrophs, secreting thyroid stimulating hormone (TSH) which controls thyroid follicle cells to produce T_4/T_3 (Thyroxin/Triiodothyronin) (iii) GH cells (growth hormone cells) secretes growth hormone which controls the normal growth. The pars intermedia (PI) has two types of cells *viz.* (i) MSH cells secreting melanocyte stimulating hormone and (ii) PAS^{+} somatolactin cells secreting somatolactin. The function of somatolactin is not yet clear.

A knowledge of the reproductive physiology of fish is an essential prerequisite for making the seed production of fish successful and viable on a commercial scale. Major carps are seasonal breeders and it exhibits circannual rhythm in the gonadal development. In India, generally gonads of the fish exhibit resting phase from October/ November to February, preparatory phase from latter part of February to March, sexually active phase from April to August and an inhibitory phase from August/ September to October/ November. The ovarian development consists of multiplication of ooginal cells to produce oocytes, growth of oocytes and vitellogenesis. There is also increase in the hydration and vascularisation of the ovary.

Intricate mechanisms involved in yhe neuroendocrine regulation of reproduction in teleosts is depicted in Fig. 1.9. Environmental clues such as increasing temperature and increasing light duration of day-night cycle are received by CNS which controls the secretion of GnRH. A number of monoamines have been implicated in the synthesis and secretion of GnRH. For example dopaminergic neurons are shown to exert inhibitory effect GnRH neurons. GnRH in turn stimulates the gonadotroph cells of pituitary to produce gonadotropins. There are two types of gonadotropins such as gonadotropin-I (GtH-I) and gonadotropin-II (GtH–II), former is secreted during vitellogenic phase of oocyte development where as the latter is secreted during postvitellogenic phase. GtH-I induces the secretion of estradiol from follicle cells (thecal cells and granulosa cells). Both thecal cells and granulosa cells have receptors for GtH-1. GtH-1 stimulates the thecal cells to produce testeosterone from the cholesterol precursor. By side chain cleavage of cholesteral molecule pregnenolone is produced which in turn is changed to progesterone and then progesterone to 17α OH progesterone which is then converted to androstenedione and then to testosterone (Fig. 1.13 &1. 15). Each step is catalysed by a specific engyme. The synthesis of these engymes are controlled by GtH-1 molecule. The testosterone so produced in the thecal cells enters into the granulosa cells. The granulosa cells also possess GtH-1 receptor. Binding of GtH-1 to the receptor of granulosa cell induces the conversion of testosterone to estradiol. This is also catalysed by an aromatase enzyme which is under the control of GtH-1. Once the estradiol is synthesised, these molecues are liberated into blood circulation. Estradiol in turn acts in the liver cells to produce vitellogenin, which is the precursor of yolk protein. Vitellogenin is carried to oocytes through circulation and incorporated into the cytoplasm of oocytes where vitellogenin is transformed in to yolk protein- lipovitellin and phosvitin and deposited as yolk granules or platelets. (Fig. 1.10). GtH-I also causes the differentiation and development of follicular envelopes.

In a post vitellogenic oocytes, the nucleus or germinal vesicle which is in the pachytene stage of prophase of first meiotic division undergoes further steps in meiosis under the influence of GtH-II which is secreted at this stage from gonadotrophs of pituitary giand. GtH-II stimulates the follicle cells of oocyte to produce a hormone called MIH (Maturation Inducing Hormone) which is 17α, 20β dihydroxy – 4- Pregnene – 3 one (17α - 20β dihydroxy pogesterone) from 17α-OH Progesterone. Thus there is

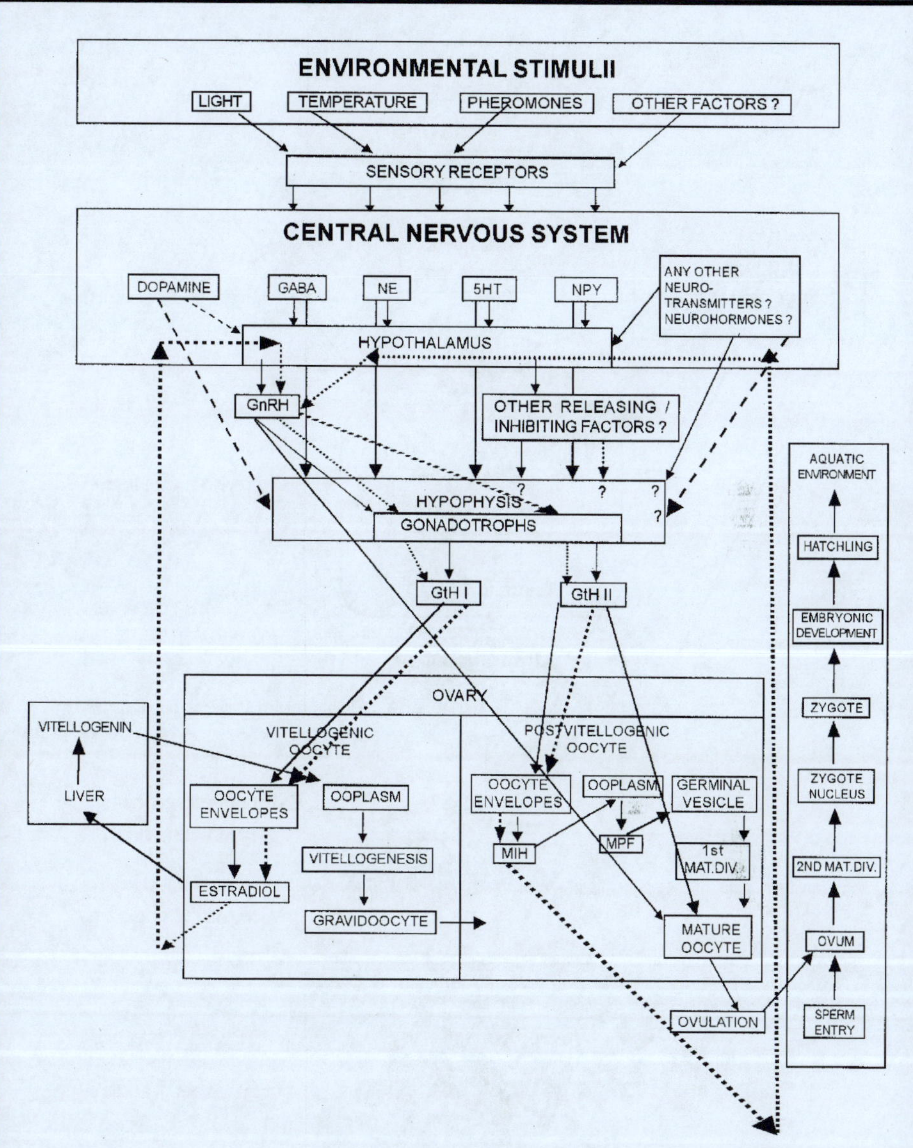

Fig. 1.9: Diagrammatic Representation of the Endocrine & Neuroendocrine Regulation of Reproduction in Teleosts.

Solid Line (—) Indicates Stimulatory Effect. Broken Line (- -) Indicates Inhibitory Effect. Dotted Line (....) Indicates Negative Feed Back Inhibition, GABA - Gamma Amino Butyric Acid. NE - Norepinephrine, 5HT - 5 Hydroxy tryptamine (serotonin), NPY - Neuropeptide Y, GnRH - Gonadotropin Releasing Hormone GtH I - Gonadotropin I GtH II - Gonadotropin II, MIH - Maturation Inducing Hormone, MPF - Maturation Promoting Factor.

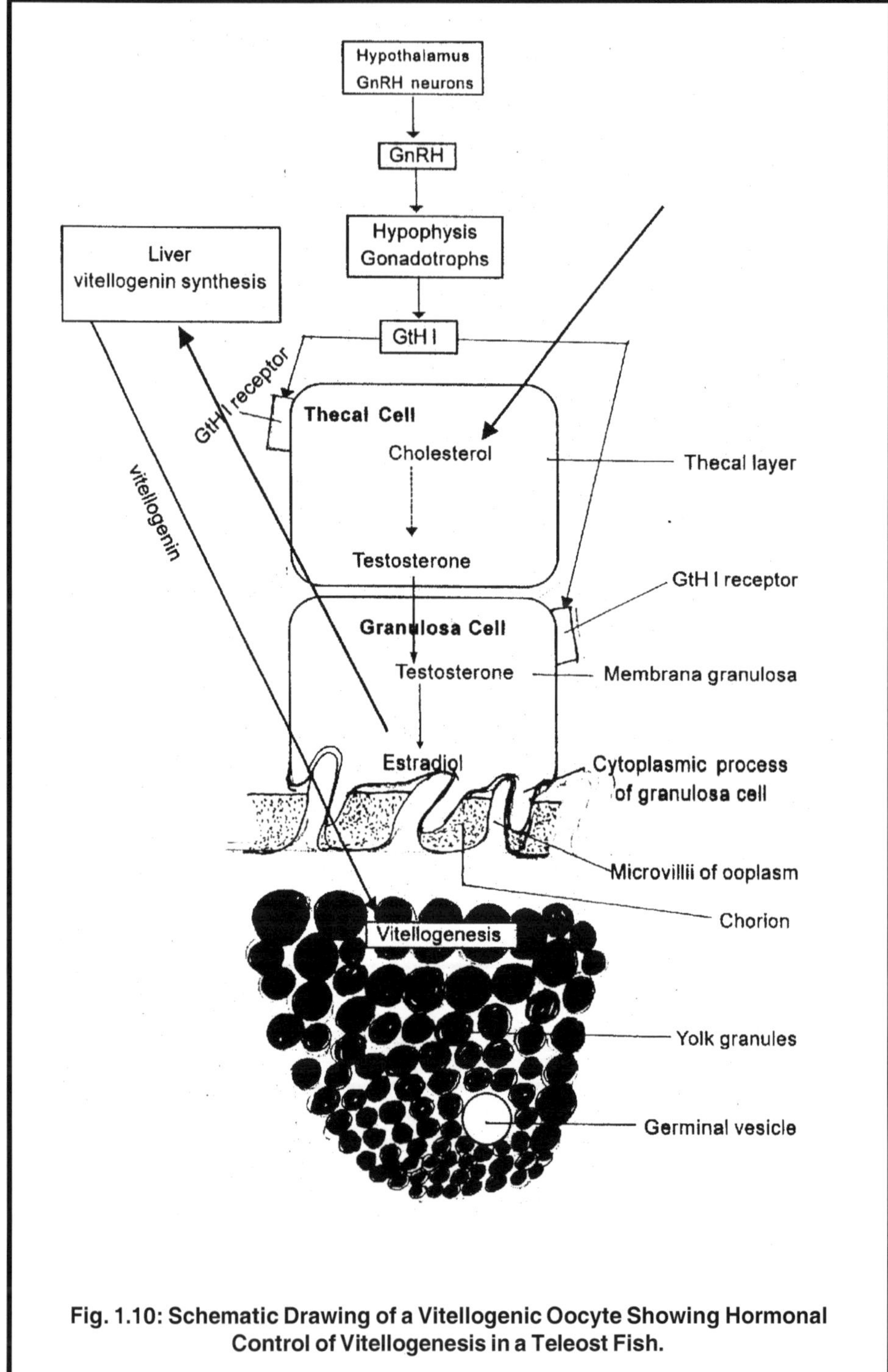

Fig. 1.10: Schematic Drawing of a Vitellogenic Oocyte Showing Hormonal Control of Vitellogenesis in a Teleost Fish.

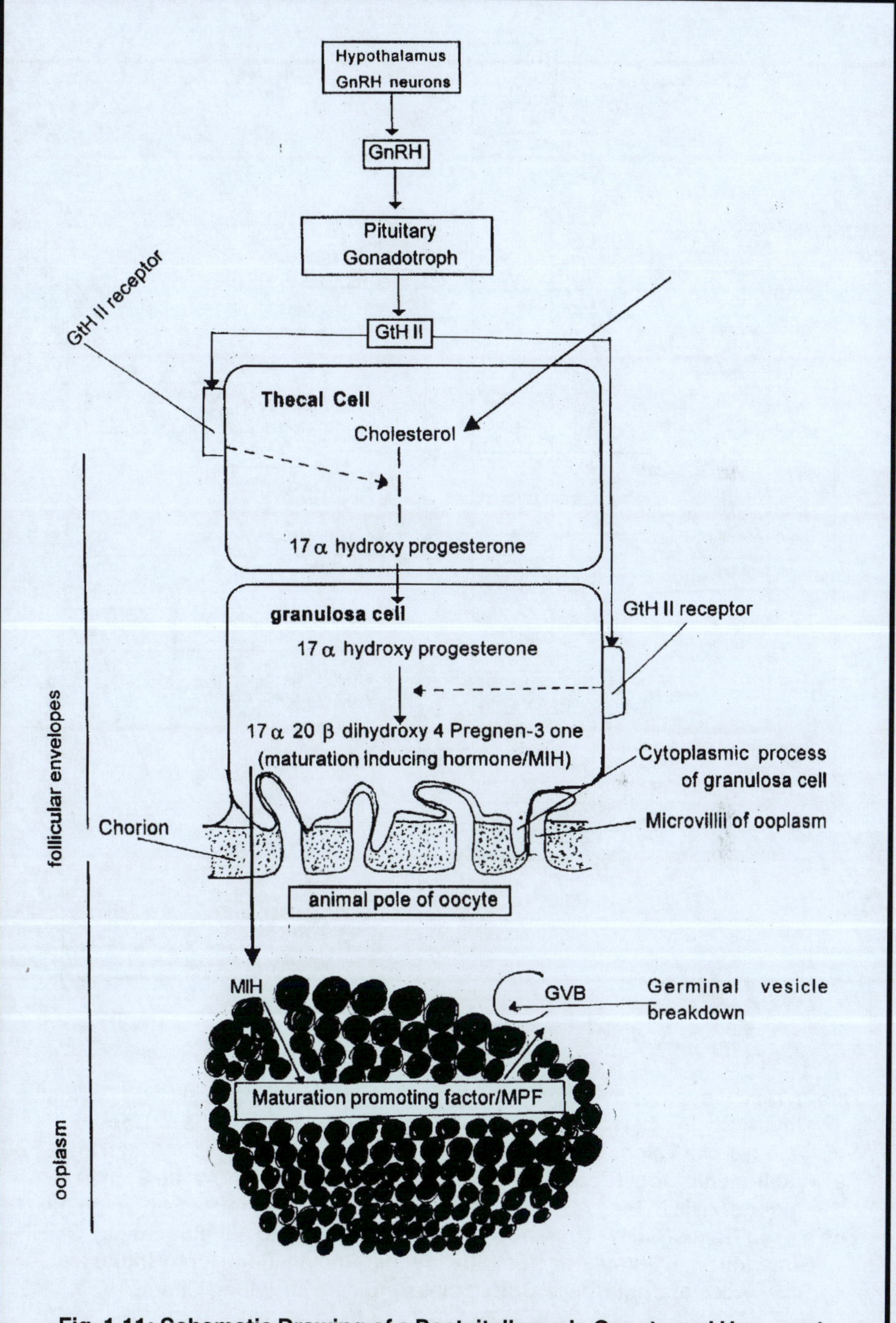

Fig. 1.11: Schematic Drawing of a Postvitellogenic Oocyte and Hormonal Control of Final Maturation in a Teleost Oocyte.

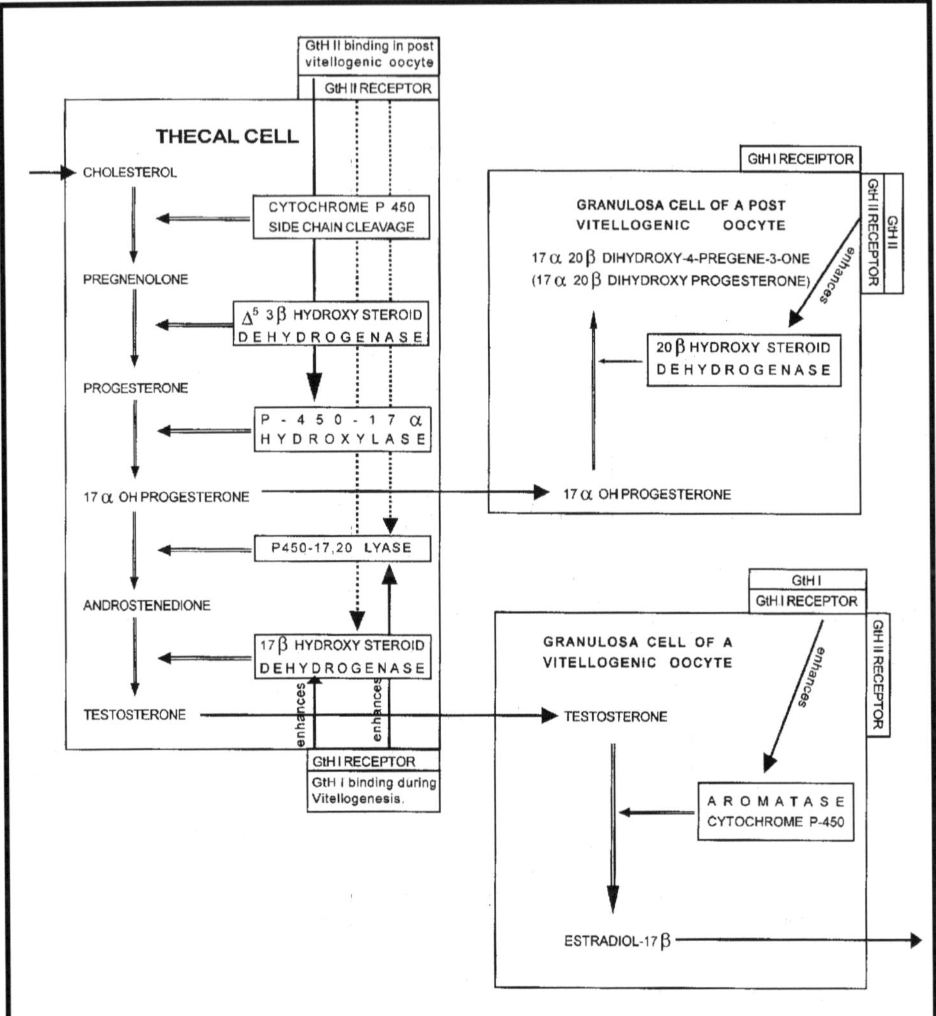

Fig. 1.12: Biosynthetic Pathway of Steroid Hormones (Estradiol and 17α, 20β, Dihydroxy Progesterone) in the Thecal and Granulosa Cells of Oocyte Envelopes of a Teleost. Granulosa Cell of Vitellogenic Oocyte and of a Post Vitellogenic Oocyte are Represented Side by Side to Show the Shift in Steroidogenesis in the Latter. Enzyme Catalysing Each Step is Shown on the Left side in Boxes. GtH I - Gonadotropin I Secreted During Vitellogenesis. GtH II Gonadotrpin II Stimulates Post Vitellogenic Oocyte. Solid Lines Indicate Accelerating Effect, Dotted Lines Indicate Inhibiting Effect.

a shift in steroidogenesis at this stage. (Figs. 1.11 & 1.12) MIH stimulates the production of MPF (Maturation Promoting Factor) in the cytoplasm of oocyte. MPF is formed by the union of a molecule of cyclin B and cdc 2 kinase (Fig. 1.11). MPF brings about the following changes or final maturation of oocytes.

(a) Migration of germinal vesicle to the animal pole of the oocyte towards the micropyle.

(b) GVB (germinal vesicle breakdown) germinal vesicle membrne breaks down and its contents mix in the ooplasm.

(c) Condensation of chromosomes and their allignment in the metaphase spindle.

(d) Division of nucleus and extrusion of 1st polar body)

(e) Realignment of chromosomes in the metaphase spindle of the second meiotic division.

Side by side with the above nuclear changes, cytoplasm also exhibits following

(a) Coalescence of yolk globules and lipid droplets and increase in the transluscency.

(b) Rapid increase in oocyte size.

At this stage, oocyte is considered as mature, and it remains at this stage till fertilization. During spawning, this mature oocyte surrounded by chorion is liberated by the rupture of oocyte envelopes and this process is called ovulation.

As a pisciculturist, one is interested in obtaining the largest number of mature oocytes from a given female so that the fish will yield him the maximum number of fry. An extract of the pituitary gland which contains gonadotropin hormone is injected as a preliminary dose which brings about the preovulatory changes noted above and then a second dose (final dose) is administered which brings about the process of ovulation and spawning.

Mechanism of Ovulation

Ovulation is the process by which fully mature egg *i.e.* an egg that has undergone final maturation (first maturation division) is liberated from the oocyte follicle into the ovocoel (in cystovarian fishes) or peritonial cavity (in gymnovarian fishes). The mechanisms involved in the process of ovulation are not yet clearly understood. It is considered as a complex process involving a number of factors such as eicosanoids, catcholamines, prostaglandins, kinins, angiotensin, histamine, various proteases etc. resulting in:

(a) Separation of follicle wall from oocyte,

(b) Rupture of follicle wall, and

(c) Oocyte expulsion.

A sudden surge in the gonadotropin (GtH II) level is reported prior to ovulation. This sudden elevation of GtH II is supposed to trigger a series of changes which in turn causes the above mentioned steps in ovulation. It is more or less an inflammatory response as main changes that occur during inflammation such as vasodilation, increased vascular permeability, tissue edema, leukocyte infiltration etc. are seen in ovulation.

At the intial stage, a space is formed between follicular epthelium and zona pellucida (chorion). This has been shown to be the result of the withdrawal of microvilli from oocyte surface and follicular cell processes. This results in a disruption of follicle-oocyte connection/communication. The factors causing this disruption are not yet clearly delineated in fish. According to some workers (Oshiro and Hibiya, 1975, 1982) proteases are involved in this. Jalabert (1978) observed that the maturation inducing hormone (17α, 20β dihydroxy 4-pregnene-3one) has stimulatory effect on this process. Strickland and Beers (1979) reported that gonadotropin activates granulosa cells to produce a plasminogen activator which causes the conversion of plasminogen to plasmin. Plasmin has been shown to degrade discreet portion of the follicle causing weakening and consequent rupturing.

Electron microscopic studies have indicated an increase in the number of microfilaments of thecal cells before ovulation. Some scientists (Pendergrass and Schrodev, 1976) observed from the invitro studies that steroids such as hydrocortisone activates the microfilaments. Cytochalacin B, an inhibitor of microfilament contraction has been shown to inhibit hydrocortisone induced ovulation in vitro in medaka (Jalabert and Szollosi, 1975). Smooth muscle cells like cells have been also found in the thecal layer. These studies have pointed towards the role of muscular contraction of follicle wall in oocyte expulsion.

Several other processes, that are suggested to provide a force to expel the oocytes are:

(a) Swelling of the oocyte or production of a layer of gelly by follicle cells (Oshira and Hibiya, 1981).

(b) Alteration in the shape of follicle cells from cuboidal to columnar at a particular region resulting in the constriction of the follicle at that end which may push the oocyte out at the opposite end.

Certain Prostaglandins have been shown to have stimulatory effect on ovulation. Prostaglandin $F_2\alpha$ is reported to induce ovulation invitro of gold fish oocyte provided that these oocytes had undergone maturation invivo after treatment with HCG. Similar results were also obtained in trout (Jalabert *et al.*, 1978). Further, administration of indomethacin an inhibitor of prostaglandin synthesis has been shown to inhibit ovulation in fish and this inhibition was overcome by administering prostaglandin exogenously. But actual level of prostaglandin required for ovulation has not yet been worked out.

Role of catecholamines-the neurotransmitters from the nerve terminals innervating ovary has also been suggested for ovulation. Jalabert (1976) suggested such a role for epinephrine in the ovulation of rainbow trout egg through invitro studies. Similar results have been also obtained in the fishes such as yellow perch, brook trout and carp by various workers (Epler, 1981; Stacey and Goetz, 1982; Goetz, 1983) According to Stacey and Goetz (1987) this stimulatory effect of epinephrine is due to its influence on production of prostaglandin.

ENDOCRINE AND NEUROENDOCRINE REGULATION OF REPRODUCTION AND GROWTH IN CRUSTACEANS

Perhaps, next to fin fishes, crustaceans such as shrimps/prawns, crabs, lobsters etc. are of much fisheries/aquaculture importance. The neuroendocrine system (Fig. 1.13) and mechanisms controlling reproduction in crustaceans are somewhat different from that of fin fishes mentioned above. Clusters of neurosecretory neurons are found in the brain, thoracic ganglia, ganglion of the circum oesophageal connective, tritocerebral commissure etc. In addition, it has a specialised neurosecretory organ called X-organ located in the stalk of eye. The optic peduncle within the stalked eye consists of three ganglionic formations such as medulla externa, medulla interna, and medulla terminalis. The cluster of neurosecretory neurons (X-organ) is located in the medulla terminalis. The axons of these neurosecretory neurons end in neurohaemal organ called sinus gland, which serves as a storage-release centre for these hormones. The sinus gland is at the peripheral junction between medulla externa and medulla interna. As the sinus gland serves for the storage and release of neurohormones secreted by secretory neurons of X-organ it is called 'X-organ-sinus gland complex'. The X-organ secretes several neurohormones such as (i) ovary (gonad) inhibiting hormone (OIH/ GIH), (ii) molt inhibiting horomone (MIH) (iii) light adapting hormone or distal retinal pigment hormone. (iv) hyperglycaemic factor (CHH-crustacean hyperglycaemic hormone) (v) erythrophore concentrating hormone and (vi) neurodepressing hormone. Out of these OIH and MIH are directly involved in reproduction. Secretory neurons from brain and ganglia of circum oesophageal connective are reported to terminate in the post commissural organs located in the tritocerebral commissure. Secretory neurous located in the ganglia of ventral nerve cord send their axon terminals to the pericardial organ located in the pericardial sac. Hence postcommissural organ and pericardial organs are also considered as neurohaemal organs.

Another neuropeptide namely gonad stimulatory hormone (GSH) has been reported to be secreted from the thoracic ganglia. Neurosecretory neurons of supra oesophageal ganglia is believed to secrete a 'gonad stimulatory hormone releasing factor' or GSHRF which stimulates the secretion of GSH from thoracic ganglia. But in Mexican cray fish, *Procambarus bouverri*, GSH has been found in the sinus gland. Till now, this hormone has not been isolated and characterised.

Apart from the above mentioned neuro secretory systems, there are four endocrine glands *viz.* (i) ovary in female (ii) androgenic gland in male (iii) mandibular organ and (iv) Y-organ reported from crustaceans.

Y-organ described by Gabe (1953) is located in the antennary/ maxillary segment of the body anteriorly. This secretes a moulting hormone (ecdysone) which is a steroid

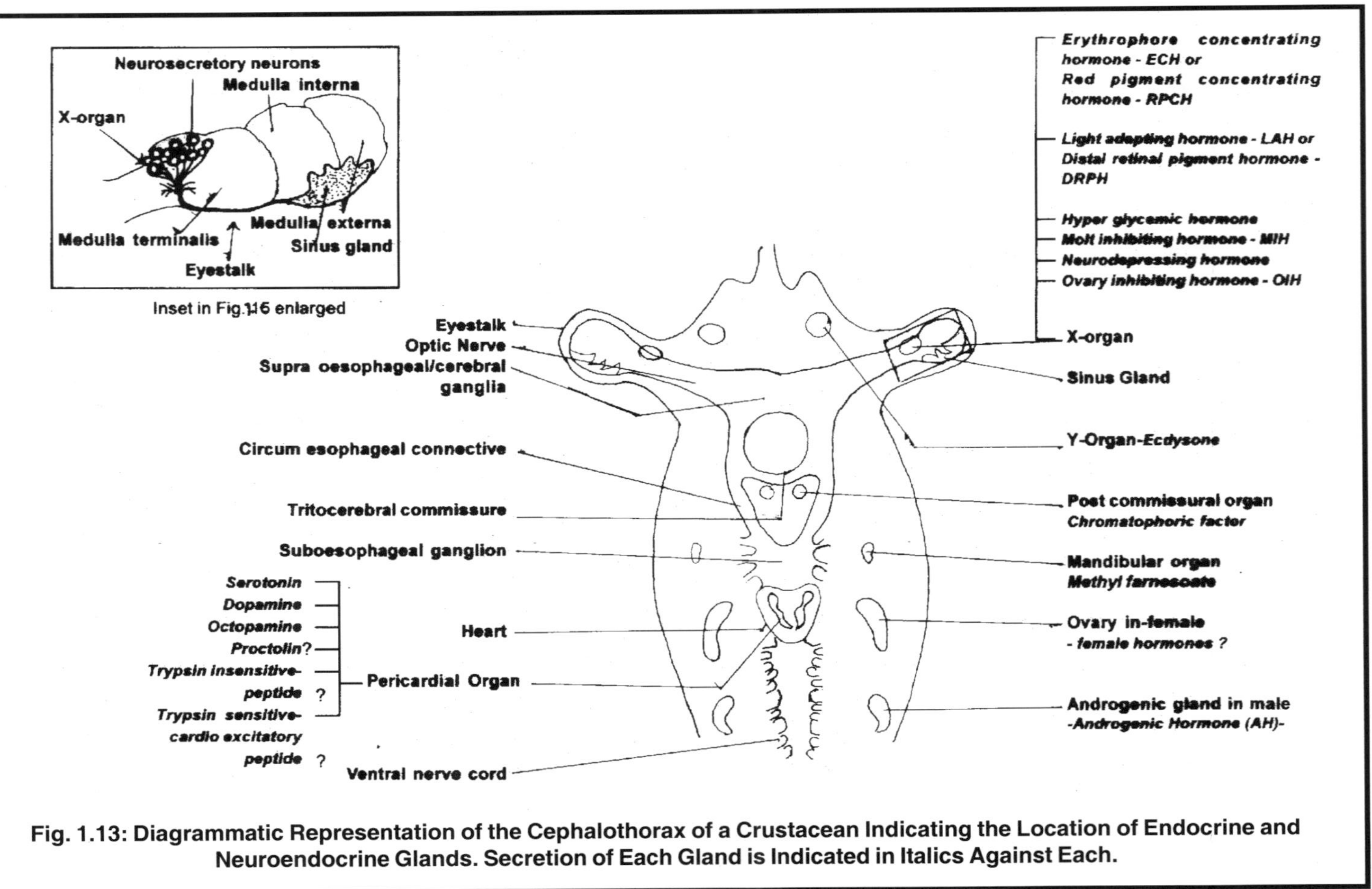

Fig. 1.13: Diagrammatic Representation of the Cephalothorax of a Crustacean Indicating the Location of Endocrine and Neuroendocrine Glands. Secretion of Each Gland is Indicated in Italics Against Each.

hormone (crustecdysteroid) The Secretory activity of this gland is controlled by molt inhibiting hormone secreted by X-organ and secretion of mandibular organ. MIH inhibits Y- organ where as secretion of mandibular organ is stimulatory.

Another endocrine gland, mandibular organ, has been shown to be involved in ovarian growth and vitellogenesis. Mandibular organ synthesises a sesquiterpenoid, methyl farnesoate (MF) and the same has been reported from lobsters, cray fish and shrimps. But in American lobster, *Homarus americanus* MF is shown not to have any role in vitellogenesis as this hormone level is found lowest during vitellogenesis. Despite this, the activity of mandibular organ has been shown to be under the control of eye stalk hormone in many crustaceans as eyestalk ablation led to elevated MF level. As stated earlier MF has been also shown to have stimulatory effect on Y- organ to secrete ecdysone.

Molting (shedding the old exoskeleton) is a regular phenomenon in crustaceans like prawn, crabs, lobsters, etc. as it grows. Molt cycle is devided into four phases such as: (i) premolt (ii) molt (iii) post molt and (iv) inter molt. During premolt stage inorganic constituents of old exoskeleton are reabsorbed and stored in gastroliths (concretions formed in the wall of the stomach) or hepatopancreas. Further glycogen is deposited in the hypodermis and oxygen consumption also increases, indicating accelerated metabolic activity. During molt stage, old cuticle is sloughed off and side by side, there is marked increase in size of the body due to rapid hydration or absorption of water. During post molt stage, a new exoskeleton is formed through the redeposition of chitin and inorganic salts. Inter molt stage is a quiscent phase.

Molting hormone (ecdysone) is secreted by Y-organ during premolt and molt stages, where as during post molt and inter molt stages ecdysone secretion is stopped by the action of molt inhibiting hormone of X-organ. During anecdysis, Y-organ is found greatly reduced indicating there by its dysfunction resulting in lack of molting.

Androgenic glands in males were discovered by Charnianx-cotton (Carpenter *et al.*, 1970). Testis is said to have no endocrine role where as the androgenic gland is shown to be endocrine in function. The secretion of this gland, androgenic hormone controls male secondary sexual characters and spermatogenesis. If androgenic gland is removed from a genetic male, it may develop ovary and change into a female. So also a genetic female can be changed to male by the transplantation of androgenic glands into females.

OIH or GIH of X-organ-sinus gland complex is reported to control the ovarian development. Ovary develops rapidly and undergoes vitellogenesis and maturation if OIH is inhibited. Because of its specific role in vitellogenesis it is also called VIH (vitellogenesis inhibiting hormone) It is for this reason that eye stalk ablation is parctised to induce ovarian maturation in prawns/crabs in hatcheries. During eyestalk ablation X-organ sinus gland complex is removed along with the eyestalk there by the ovary is freed from the inhibitory effect of OIH and MIH. VIH has been isolated and characterised from a few crustaceans and it is found to be a peptide hormone. In Mexican cray fish it has 72 to 74 aminoacid residues.

During recent years, a number of steroid hormones such as 17-β estradiol, progesterone, testosterone etc. have been documented from reproductive and non reproductive organs such as mandibular organ, hepatopancreas, haemolymph, ovary etc. (Subramoniam, T. 1999). Their role is not yet established. Further biogenic amines such as serotonin, dopamine and peptides like met. enkephalin have been implicated in the gonadal function of crustacens. Serotonin stimulates GSH secretion where as dopamine and met. enkephalin have inhibitory effect on ovary (Subramoniam, T. 1999).

EGGS & LARVAE IN FISHES

Among the fishes, diverse types of eggs are noticed. They differ in size, shape, colour and in various features such as buoyancy, presence of appendages etc. The eggs are either pelagic (buoyant) of demersal (benthic). The specific gravity of buyont eggs is almost same as that of surrounding water, whereas specific gravity of demersal eggs shall be more than that of water. Fresh water fishes have mostly demersal eggs. As specific gravity of fresh water is lesser than sea water in order to be buoyant in fresh water egg should have devices to make the specific gravity equal to that of fresh water. Buoyancy of egg is due to hydration of yolk, accumualtion of fats to form oil globule and also due to large perivitelline space as in cyprinids.

The size of the eggs also vary considerably from 0.3 mm to 20 mm in different fishes but mostly it is in the range of 1 mm to 3 mm. Generally, demersal eggs are larger. Most of the marine fishes lay pelagic eggs of 1 to 2 mm diateter. Among the teleost fishes depositing eggs in sea water largest eggs are reported to be that of *Arius commersonii* (family Ariidie), their egg size being 17 to 21 mm dia. Eggs of some selachians may be large upto 90 – 97 m*m (Chlamydoselachus anguineus*). Among the fishes spawning in fresh water, largest eggs reported are of *Gymnarchus niloticus* (10 mm dia), others being Oncorhynchus and Salmo (5-9 mm dia).

The eggs can be classified into sticky (adhesive) eggs and non adhesive aggs based on their ability to adhere or attach themselves to substrates or to each other. Adhesive eggs possess a mucoid deposit on the egg surface, which causes stickiness. Stickiness may be temrporary or permanent. Some of the fishes have moderately sticky eggs while some others have extremely sticky eggs.

Shape of the eggs also vary from the commonly found spherical to elliptical (anchovies) oval, (Rhodeus or European bitterling), cylindrical, pear shaped (gobiidae) or coffin like *(Glossogobius brunneus*).

In a few fishes the eggs are enclosed in a capsule formed by the secretion of oviduct which becomes hard and horny. This can withstand adverse conditions. Such egg capsules are found in *Hydrolagus colliei* and in some selachians. The egg capusle of *H. collei* is 180 mm long where as the elliptical and slightly flattened egg is only 29 × 19 × 13 mm in size. A few fishes lay eggs possessing appendages like tendrils/filamentes. In hag fishes eggs are elongate, ovoid with a number of tendrils at each end. These tendrils are meant to anchor the eggs to vegetations etc. In another fish *Osmerus mordax* egg has a stalk for attachment to stones. In the eggs of many skates tendrils come out from each of four corners. Eggs of some flying fishes have long hair like filaments arising from the surface of spherical eggs.

Many fishes lay eggs singly, but a few fishes such as some species of scorpaenidae and lophiidae lay eggs in a mass. The egg mass may be variously shaped. The egg mass of *Lophius piscatorius* is several meters long and a meter in width and contain several millions of eggs.

Based on the amount of yolk, teleost eggs are classified into (I) oligoplasmic (much yolk and little cytoplasm) and (ii) poly plasmic (rich in cytoplasm). Larger eggs are mostly oligoplasmic eggs Eggs of diameter 1.5 mm and less is either oligoplasmic or poly plasmic. Planktonic eggs are oligoplasmic wehere as demersal eggs may be either oligoplasmic or polyplasmic (Detlaf. T. A., 1972).

The colour of the eggs also vary in different species. Eggs are usually transparent or semi-transparent; coloured (yellowish or greenish or grayish) or colourless. The yellow colour is attributed to the presence of a carotenoid, xanthophyll.

The eggs possess cortial alveoli, at the periphery either in a single layer as in chum salmon or in two to three rows as in gold fish. The yolk granules are mostly smaller at the periphery and larger towards the centre. Sometimes, yolk granule, unite in such a way that it becomes a homogeneous mass. This takes place in most pelagic eggs as it becomes more transparent and less noticable in water column. The yolk may be either protein yolk or fatty yolk rich in lipids. Oil droplets are found scattered over the yolk, below cytoplasmic layer. In some fishes like stickle back, these droplets coalesce to form one large oil drop in course of egg debvelopment. Size of oil drop is larger in fresh water fishes having buoyant eggs than marine. These oil droplets contain glycerides, mixed with phosphatides. They store nutritional material of rich caloric value. Further it provides hydrostatic effect due to its low specific density and ensures buoyancy. In marine fish, oil droplets are in the vegetable pole so that in the developing eggs, blastodisc is directed down and at later stages, oil droplet enters the posterior part of yolk sac as a result it sinks to deeper layers. In fresh water fishes oil droplet is located at the animal pole and later it enters into the anteiror part of the yolk sac during the course of development. As a result of this the larvae floats on the upper surface. Due to heavy yolk content the cleavage which takes place after zytote formation, is meroblastic in most of the fishes but in fishes like lung fishes and chondrostean fishes, it is holoblastic or complete.

According to most of the scientists, embryonic peirod of the fish lasts untill hatching/parturition where as others consider it to last till the young begin to feed themselves. Hubbs (1943) applied the term larvae to the young stages of fish from hatching to juvenile stage when it acquires adult characteristics such as development of fin rays etc. He has further, divide the larval stages into two such as prolarva which retains the yolk and post larva from the yolk sac absorption stage to juvinile stage. Nikolsky (1963) applied the term embryo to the stage upto yolk sac absorption. Hattori (1970) used the term pre larva for the stages from hatching to yolk sac absorption. Balon (1975) identified five stages (1) cleavage phase *i.e.* from zygote stage upto the commencement of organogenesis, (2) embryo -developmental stages from organogenesis till hatching, (3) Eleutheroembryo (Gk. Eleuthero. means free): stages from hatchling to yolk absorption stage, (4) Protopterygio larva from yolk sac absorption to notochord flexure stage and beginning of fin differentiation (5) Pterygiolarva from notochord flexure to differentiation of fin rays. Sette (1945) called the yolk absorption stage to juvenile stage *i.e.* fin ray differentiaiton stage as larva, and post larve from fin ray differentiation stage to the stage with fully adult features. Snyder (1976) differentiated three stages, such as protolarvae from hatching stage to the beginning of notochord flexure stage, mesolarvae from the beginning of notochord flexure to the completion of flexure of ntochord and metalarvae from the previous stage to the fin ray differentiation stage.

Though opinions differ as stated above regarding nomenclature of different stages in the early life history of fishes, a common pattern of life cycle stages can be noticed in the majority of them. But, there are a few fishes of considerable economic importance such as salmonids and eels that exhibit deviations in the early life history stages. A brief review of the same is given below:

Salmonid Fishes

Several species of salmonid fishes are found in both Pacific and Atlantic oceans. These fishes are anadromous in their habit as they move from the ocean to rivers for spawning. Salmonids found in Pacific ocean are grouped under the genus *Oncorhynchus* where as those found in Atlantic are placed under the genus *Salmo.* Five species of Oncorhynchus commonly encountered are *O. nerka* (red salmon), *O. tshawytscha* (chinook salmon). *O. kisutch* (coho salmon), *O. keta* (chum salmon), *O. gorbuscha* (pink salmon) and *O. masou* (masu salmon). The common species of Atlantic salmon is *Salmo salar*.

All species of Oncorhynchus are reported to die soon after spawning. Their natural spawning ground is stream bed with pebbles and stones. They spawn usually from September to January. The eggs develop and hatch out into fry by March/June. Fry of chum salmon and pink salmon migrate to the sea water immediately. But the fry of orther species including that of Atlantic salmon, *Salmo salar*, remain in fresh water for a longer time, even upto one year as in masu salmon or upto 2 years as in Atlantic salmon. At this stage they are called as "parrs". Parrs live mostly at the bottom and are brownish in colour. Darker lateral bars called 'parr marks' are also found in the body. As it reaches an optimum size, to undertake migration to sea, it undergoes pronounced changes in growth, morphology, colourtion, metabolism, osmoregulation and behaviour. At the end of thes changes it developes into a stage called 'smolt' and the process of transformation of smolt from parr is called 'smoltification'. A smolt is slivery in colour and exhibit darkening of fin margins. The body is more slim and stream lined. Changes involved in the process of smoltification are as below:

Changes in Growth

There is a rapid increase in growth rate during smoltification.

Change in Morphology

The body of the fish becomes more slim and streamlined. There is an increase in the proportionate post anal length, decrease in pectroal fin area, change in structure of pelvic fin, jaws, teeth and cloacal folds.

Changes in Colouration

During smoltification there is silvering of the body. This is due to deposition of crystals of purine such as guanine and hypoxanthine in the upper layer of dermis which obscures the melanin pigments which caused the brownish background colouration of parr. Parr marks also disappear. Guanine and hypoxanthine are the end products of degradation of aminoacids such as glycine, glutamine and serine and also intermediate products in the breakdown of nucleotides such as adenosine to urea. Side by side, margin of fins, (dorsal, pectoral and caudal) get darkened with black pigments.

Change in Osmoregulation

Paar living in fresh water has mechanisms to excrete water (increased urine flow due to enhanced reabsorption of water in the uriniferous tubule) and to conserve ions. (preventing efflux of ions and for uptake of ions from water). During smoltification, there is a quickening of mechanisms involved in water conservation (decreased urine flow and increased water absorption in the intestine) and salt excretion (activation of chloride secreting cells of gills to excrete salts by active transport mechanisms with the help of Na^+/K^+ ATPase. So there is an increase in this enzyme in gills during this period.

Changes in Behaviour

The Parr is bottom dwelling (river bed) where as smolt aggregates higher in water column and move downstream in the river towards the sea.

The time taken for developing into smolt varies in different species. Coho salmon showed high ionoregulatory capacity on reaching 10 g but critical size is considered as 20 g. In North American hatcheries, average size of 27-48 g are used for stocking the offshore farms or for ranching. For Atlantic salmon, it takes two years to reach the size range of 14-18 cm but one year smolt of 12.5 – 14 cm are also reported to give good results.

Eels

Eels (fam. *Angullidae*) are catadromous fishes. They live in freshwater for 10 to 15 years but migrate to the sea for spawning. The common species of eel are *Anguilla anguilla, Anguilla japonica* and *A. bengalensis*. *Anguilla japonica* is shown to spawn in ocean at a depth of 400-500 m where the water temperature is 16-17 ^{0}C and salinity is 35ppt and they spawn 1, 000, 000 to 13, 000, 000 eggs. The eggs are buoyant and measures 1 mm. They hatch in the water column within 10 days. The larvae is 6 mm in length and are called leptocephali. They drift in the surface waters, and are caried to the shore where they metamorphose into glass eel. The glass eel enters the river and become pigmented when it is called as elver. The elvers migrate to fresh water and live for a long time. Some of the eels may grow to 180 cm and 20 kg in weight.

Chapter 2

Seed Production of Major Carps

INTRODUCTION

Carps and other cyprinids contribute the largest share in the total global aquaculture production. In the year 1999, out of the total aquaculture production of 33. 31 million tons (excluding aquatic plants), carps and other cyprinids formed 14.90 million tons *i.e.* 44. 7% of the total. Considering the freshwater fish production through farming alone (*i. e.* 18. 5 million tonnes) it formed 80.1% (FAO statistics, 1999). These fishes are cultivated extensively in Asian countries because of their consumer preference and suitable climate prevalent in these areas for its growth. Major carps that are native to the riverine systems of China are called Chinese carps and they include species such as silver carp (*Hypophthalmichthys molitrix*), grass carp (*Ctenopharyngodon idella*), bighead carp (*Aristichthys nobilis*), black carp (*Mylopharyngodon piceus*) and mud carp (*Cirrhinus molitorella*). Catla (*Catla catla*), rohu (*Labeo rohita*) and mrigal (*Cirrhinus mrigala*) which are referred to as Indian major carps are native to the Indo-gangetic riverine systems of India. The common carp (*Cyprinus carpio*), though a native of Eastern Europe enjoys a world wide distribution at present because of its adaptability to both warm and cold climatic conditions. The main native carp species of Thailand is silver barb (*Puntius gonionotus*) which form more than 82% of the total carp production of that country.

Carps are the main stay of aquaculture in India and as a matter of fact, India is called as the 'carp country' with reference to aquaculture because carp flesh is highly relished by the majority of its population and these fishes are cultivated in this country from ancient days. Last three decades have witnessed a phenomenal growth in the farming of these carps in India as a result the market demand for fry for stocking also has increased. According to Gopakumar, K. *et al.* (1999) in order to augment the aquaculture production of India from the present level of 1512000 million per annum to 3312800 mt. per annum, the carp spawn requirement shall be 45000 million per year. In the year 1997-98, India is reported to have produced 15504 million fry.

All major carps mentioned above are seasonal, riverine spawners except common carp, which are biannual spawners that breed also in confined waters. Hence, the technology of seed production of common carp is described separately in this chapter.

INDIAN MAJOR CARPS

Reproductive Biology of Indian Major Carps

Though the three major Indian carps *viz. Catla catla, Labeo rohita* & *Cirrhinus mrigala* (Fig. 2.1) belong to three different genera and differ not only in morphology but also in their habitat and feeding behaviour, they exhibit considerable similarity in several details of reproductive biology such as sexual dimorphism, courtship, mating, spawning, spawning season etc. Hence reproductive biology of three species are collectively described here, mentioning differences if any, whereever required.

Sexuality

All the above mentioned species of I. M. C. are bisexual (hetero- sexual) and sexes can be distinguished only during the breeding season. The identifying features between sexes are related to the length and texture of pectoral fin, condition of genital aperture and size of the belly (Fig. 2.2)

Pectoral Fin

The pectoral fin of male has rough dorsal surface and the same is longer than that of female. In female dorsal side of pectoral fin is smooth.

Genital Aperture

In female it is reddish and swollen where as in male the same is not prominent. Further on applying gentle pressure, milt oozes out through the genital aperture in male, but in female eggs ooze out.

Shape of the Belly

Belly of the female is soft swollen and bulging which is not found in male.

These fishes are polygamous and promiscuous.

Age and Size at First Sexual Maturity

Age and size at first sexual maturity of a given species may vary depending on the temperature and other environmental factors. Generally, all the three major carps attain sexual maturity in the second year and the males mature earlier than female. The female grows faster than male and hence usually, the males are smaller than female in a given population of the same age group. Hence carps of 2+ years and upto 5 years are preferred for breeding. After 5 years senility sets in and hence are not advisable for breeding.

Spawning Season

All the above major carps are seasonal riverine spawners, spawning during the southwest monsoon months (June to August/September). They spawn in inundated shallow areas adjacent to the river during floods. They do not breed usually in the first flood but breeds during middle and latter parts of monsoons. They do not spawn in confined water.

Fig. 2.1: The Indian Major Carps Commonly Used for Aquaculture.
(a) *Catla catla* (catla), (b) *Labeo rohita* (rohu), (c) *Cirrhinus mrigala* (mrigal)
(Courtesy, CIFA, Kausalyaganga)

Courtship and Mating

Natural spawning is noticed in rivers and adjacent shallow inundated terrains during flood times usually during cloudy days accompanied by thunderstorm and rain. Generally, they are found to spawn from early morning hours till evening as reported by several workers.

Sex play is started by males as they chase the females. 'One or several males chase one female and the chasing pair dart about in the water for some time. Then the female is held by the male. The male bends his body around the female, rubbing knocking and nudging. At the climax of this activity the pairs can be seen locked in sexual embrace when their bodies are found twisted around each other with their fins erect and caudal fin quivering. There will be vigorous splashing of water. At this stage the female releases the ova and the male eject out the milt over it. All eggs from the ovary will not be laid at one place, at one time, but at intervals during this time the pair keeps on moving' (Khan, H. A. and Jhingran, V. G. 1975).

Fecundity

Ovarian fecundity in *Catla catla* has been reported to vary from 166773 to 203750 per kg body weight (FAO Fisheries synopsis, No. 32, 1968,) in *Labeo rohita* it is reported to vary from 345500 to 382000 per kg body weight (FAO Fisheries synopsis No. 111, 1975) and in *Cirrhinus mrigala* it is in range of 139056 to 187048 per kg body weight (FAO Fisheries Synopsis, No. 102, 1979.). Thus out of the three carps *Labeo rohita* has greater fecundity than the other two carps. The size of the water hardened eggs also varies in three carps-largest size being that of *C. mrigala* (20000 nos. /lit. of water hardened eggs) and the smallest being that of *L. rohita* (30000 eggs per litre of water hardened eggs). In catla 25000 eggs are usually found in one liter of water hardened eggs.

Embryonic Development and Hatching

Structure of Egg

A freshly laid unfertilized egg consists of cytoplasm filled with yolk. The germinal vesicle or nucleus is located in a narrow, clear layer of cytoplasm in the animal pole. The oocyte is surrounded by a layer of vitelline (plasma) membrane. There is a shell or chorion of varying thickness surrounding the vitelline membrane. A narrow opening called micropyle is present in the oocyte envelope, which permit the entry of sperm into the cytoplasm of egg. Chorion is fragile in an unfertilized egg, which is changed into a tough structure through water hardening. Chorion of fertilized egg consist of tough outer layer and a thick inner layer. This thick inner layer protects the embryo against mechanical, chemical or biological stress.

Soon after fertilization, the micropyle is closed thereby preventing the entry of any other sperm into the oocyte. The fertilized eggs start absorbing water through the oocyte surface into the perivitelline space. Side by side, a colloid derived from the polysaccharides is released from the cortical alveolii of the ooplasm in to the perivitelline space, there by decreasing the volume of ooplasm. Thus the formation of perivitelline space is partly because of the decrease in the volume of ooplasm due to the release of colloid from the alveoli and also partly due to osmotic distension of

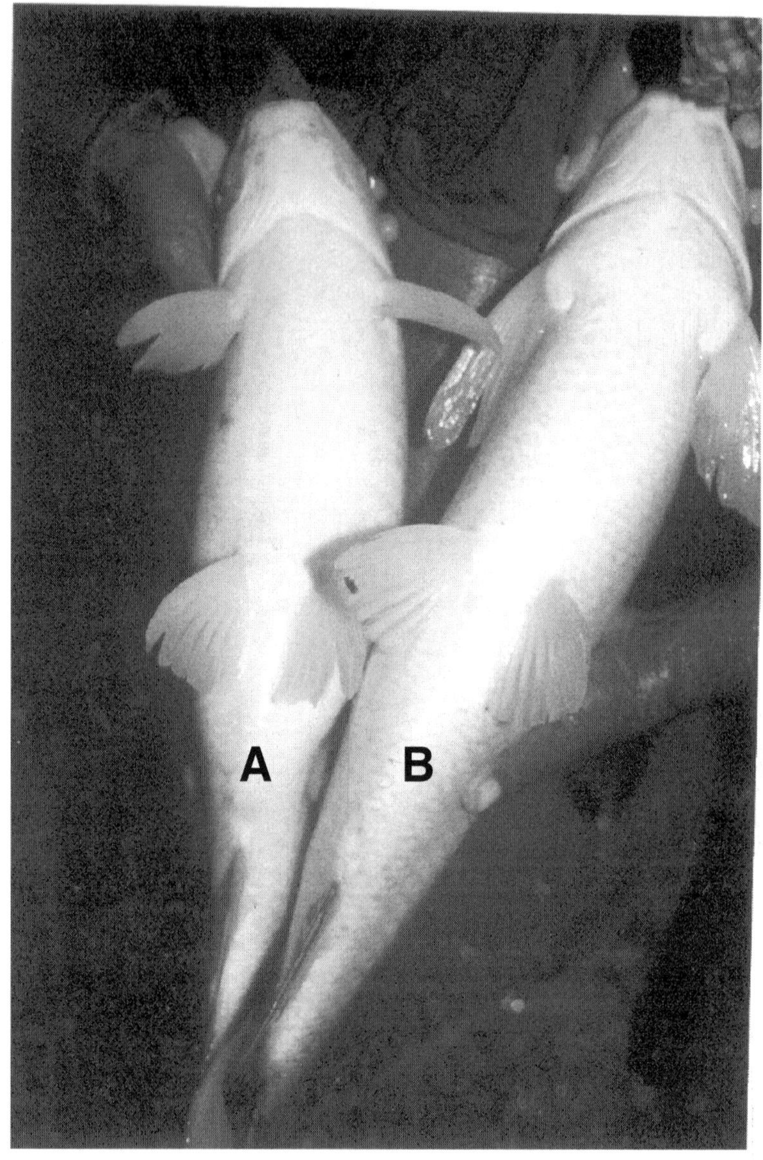

Fig. 2.2: Sexual Dimorphism in *Labeo rohita.*
(a) Female, (b) Male.
Note the Smaller Pectoral Fin, Swollen Belly and Prominent Genital Opening of Female. In Male the Pectoral Fin is Comparatively Larger, Belly is Streamlined and Genital Opening is Less Prominent.

chorion (Blaxter, J. H. S. 1969) As the perivitelline space gets filled with water and colloid, swelling of the oocyte takes place. This phenomenon is called water hardening. At the end of water hardening the egg may attain a size 4 to 6 times their original size. Micropyle usually closes one minute after coming into contact with water, even if fertilization does not take place in case of major carp. This means that time requirement is only a few seconds for the egg to get fertilized after they come in contact with water.

Development of Eggs (Fig. 2.3)

During water hardening the animal pole develops a small hillock of protoplasm on the yolk mass known as egg blastodisc. Cleavage starts about 39-45 minutes after fertilization. Cleavage is not complete or holoblastic. Cleavage furrow is limitted to the cytoplasm of animal pole. Such type of cleavage is called meroblastic cleavage. As the successive cleavages continue, a group of cells are formed and at this stage it is called morula. After further divisions, the cells arrange themselves in the form of a layer of blastoderm. Each cell of the blastoderm is called blastomere. Then blastoderm becomes several layered. Cleavage also results in the formation of another type of cells called periblast which lie between blastoderm and yolk. The embryo develops from the blastoderm. A space develops between the yolk and cell mass which is called segmentation cavity or blastocoel. At this stage the embryo is called balstula. The blastula of teleosts is disc like found over the yolk and hence is called blastodisc.

Even at blastula stage presumptive germ layers (ectoderm, endoderm and mesoderm) can be noted and a fate map of blastula can be demarcated showing the presumptive neural plate and notochord.

Next stage in the development is gastrula. As the time for gastrulation is nearing, the under rim of blastodisc thickens to form the 'randwulst' or marginal ridge. The inner layer of this is called germ ring. The germ ring is thickest at the caudal end, which is named as embryonic shield. The process of formation of gastrula is called gastrulation. Gastrulation occurs by three processes such as invagination, involution and epiboly. An invagination appears at the future endodermal region of blastula at the caudal edge of embryonic shield. The portion that is invaginated grows below the blastoderm and this portion of endoderm is called hypoblast. Due to invagination an opening – blastopore appears at the caudal end. The cells of the endoderm and mesoderm multiply and migrate inward over the lip of blastopore and this process is called involution. During this process the cells of the blastoderm also continue to grow over the yolk which is called epiboly. As a result of this the yolk mass is covered by a layer of cells which is called epiblast, Cells belonging to randwulst and germ ring not involved in involution and future ectodermal cells are involved in this overgrowth. The periblast also grows over the yolk forming an inner layer covering yolk below epiblast. Thus the inner periblast layer and outer epiblast layer enclose the yolk which is now called yolk sac. With the formation of yolk sac, gasturlation is completed. Gastrulation results in an embryo having three germ layers-ectoderm, endoderm and mesoderm.

Side by side with the above changes, primary organ rudiments appear in the anterior part of the embryo. Anterior part of the neural plate from where the brain developes appears first. Notochord gets separated from mesoderm. Mesoderm lying

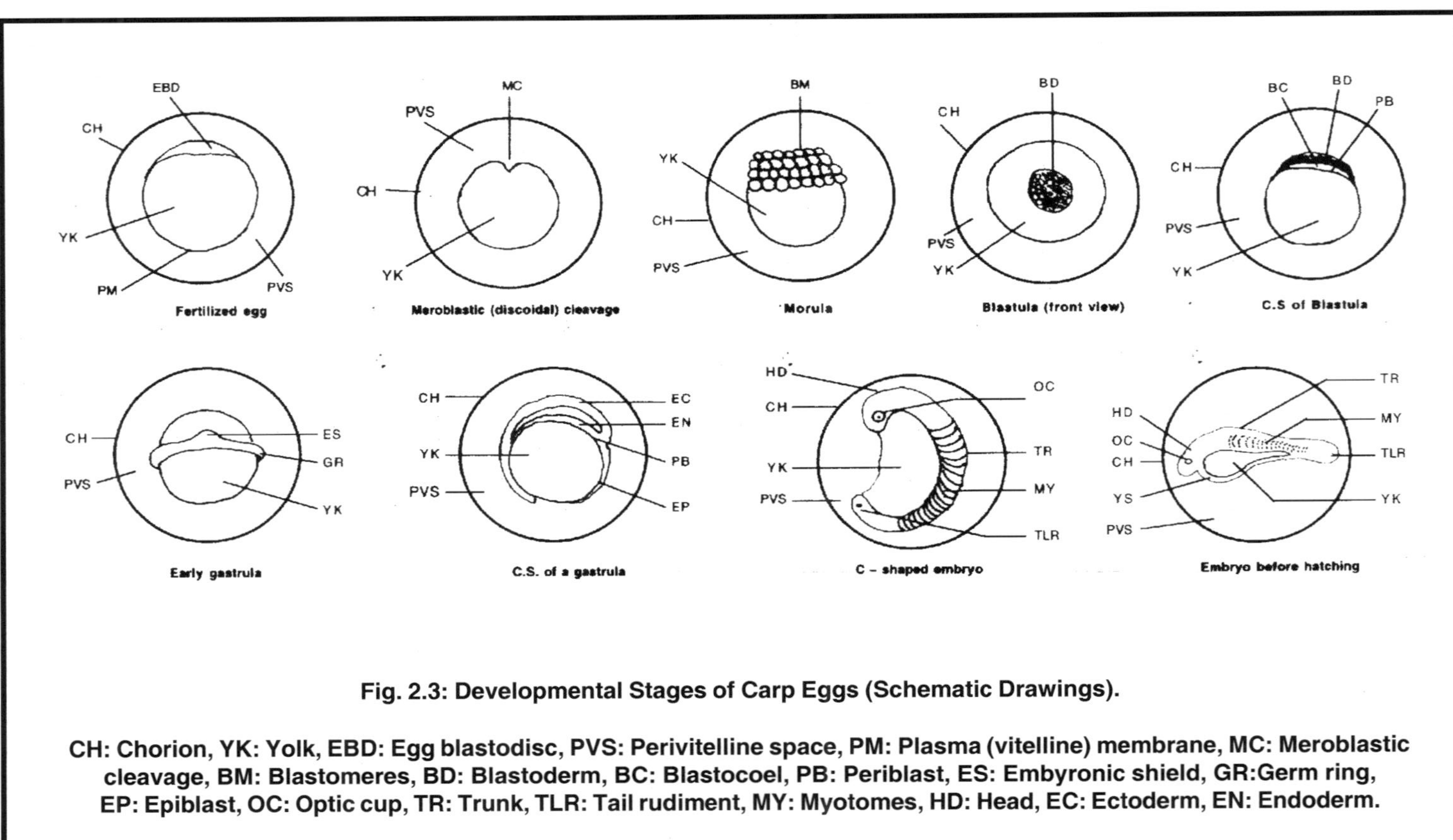

Fig. 2.3: Developmental Stages of Carp Eggs (Schematic Drawings).

CH: Chorion, YK: Yolk, EBD: Egg blastodisc, PVS: Perivitelline space, PM: Plasma (vitelline) membrane, MC: Meroblastic cleavage, BM: Blastomeres, BD: Blastoderm, BC: Blastocoel, PB: Periblast, ES: Embyronic shield, GR:Germ ring, EP: Epiblast, OC: Optic cup, TR: Trunk, TLR: Tail rudiment, MY: Myotomes, HD: Head, EC: Ectoderm, EN: Endoderm.

on both sides of notochord gets segmented. After gastrulation various organs of the body are formed. This results in a small embryo with a cylindrical and bilaterally symmetrical body. The body of the embryo is found elevated from the surface of the yolk as a result the embryo becomes distinct from the yolk sac head projecting anteriorly, tail projecting behind and trunk lying over the yolk. This is seen within seven hours after fertilization. The yolk is digested by periblast and through blood vessels it is supplied to the body of the embryo. As the embryo grows, yolk sac gets reduced in size.

The various organs of the body are derived from the three germ layers *viz.* ectoderm, endoderm and mesoderm. Ectoderm give rise to epidermis and its derivatives like enamel of teeth, olfactory epithelium, lens of the eye, inner ear, brain, spinal cord, retina and optic nerve. Mesoderm divides into three parts, dorsal epimere, and middle mesomere and ventral hypomere. Epimere becomes divided into somites, each somite having three parts such as outer dermatome, middle myotome and inner sclerotome. Dermatome give rise to connective tissue and muscles of dermis of skin, myotome give rise to muscles of trunk, appendicular skeleton, fins and their muscles. Vertebral column is derived from the sclerotome. Kidney, gonads and their ducts are derivaties of mesomere whereas, hypomere develops to somatic and splanchnic layers of mesoderm and enclose coelomic cavity. Mesenchyme cells derived from splanchnic layer develops to involuntary muscles and connective tissue of gut, heart and blood vessels. Mesenchyme of head region gives rise to skeleton and muscles of head and outer layer of the eye.

As the number of myotomes increases, embryo exhibit feeble myotomal movement followed by lashing movement of the tail. Later, as head and tail rudiments are completed, the embyo becomes active and exhibits continuous twitching movement. As the twitching movement increases and as the embryonic development is completed *i.e.* within 16 to 18 hours after fertilization (at temperature between 28 to 30 ^{0}C) embryo hatches out.

Mechanism of Hatching

There are two mechanisms involved in hatching such as (a) mechanical hatching and (b) enzymatic hatching. Mechanical hatching is the process in which egg envelopes are broken down primarily by the mechanical action such as pressure exerted from within, or mastication by the embryo itself. In enzymatic hatching, emergence of young occurs after dissolution or softening of egg envelope by the enzymes secreted by the embryo. This enzyme is called hatching enzyme. These enzymes are secreted into the perivitelline fluid at the time of hatching from the hatching gland cells located in the epidermis of the embryo. Hatching gland cells are unicellular gland cells developed on the surface of the embryo as they reach the hatching stage. Location of these cells differs in different species. In salmonid fishes like rainbowtrout, these cells are distributed on the anterior surface of embryonic body and yolksac and inner surface of pharynx and gill. The hatching enzyme has a choriolytic activity and also proteolytic activity. Because of its choriolytic activity it is called chronionase.

The thick inner layer of chorion that protects the embryo against mechanical, chemical or biological stress effects is also the barrier to the embryo for hatching. Hatching enzymes secreted one hour before hatching, break down the peptide bonds of this layer so that embryo by its movement can break the weakened layer and hatch out. The hatchling with the yolk sac attached to its belly is called sac fry.

Larval Development (Fig. 2.4)

The newly hatched larva of carps has a slender, transparent, non-pigmented body having head, trunk and tail. On the ventral side of trunk is the club shaped yolk sac. It has no mouth and fins though embryonic fin fold may be seen. The eyes are colourless. Pigmented eye is seen from twelve hours onwards. The average length of the hatchling of rohu is 3.78-4. 63mm, of catla is 4.2 to 5.3mm, and of mrigal it is 4.08 to 4.32mm. Soon after hatching, they lie laterally at the bottom occasionally coming up performing irregular movements upto 12 hours. After 12 hours they perform vertical movements. After 24 hours they exhibit slight darting movement covering little distance at a time, upward and horizontal zigzag movements are also seen. By 36 hours feeble horizontal/jerky horizontal movement is seen. By 48 hours they exhibit shooting and darting horizontal movement. Pectoral fin is developed by 24 hours. Dorsal and ventral fin rudiment and caudal fin appears by 72 hours. Caudal fin rays start appearing on fifth day. By sixth day pelvic fin bud, anal fin rays and dorsal fin rays appear. (Chakraborty, R. D. and Murty, D. S. 1972). Mouth, stomach and intestine and gills covered with the operculum are seen by 48 hours. Yolk sac is fully absorbed by 72 hours. The larva starts feeding from fourth day.

Seed Collection from Natural Spawning Grounds

The major carps spawn during monsoon season when the rivers get flooded and the floodwater inundates the adjoining low-lying areas. The inundated terrains are the most favoured spawning grounds for the carps. There are certain advantages for the carp in spawning in the inundated terrains. (i) The fresh flood kills the fauna and flora of the area, the death and decay of the fauna and flora provides the nutrients for the production of microorganisms and zooplakton on which the fish fry feeds. (ii) The area will be also free from natural predators and predatory insects that feed on carp spawn. (iii) The water is also warm and rich in oxygen, which favours rapid development of eggs and larvae.

Collection of floating eggs, larvae and fry of fishes spawning in rivers and inundated terrains is an age-old practice in Asian countries like India and China. It is usually collected by means of funnel shaped nets called shooting nets (Fig. 2.5). These nets are operated along the gently sloping banks of rivers in the path of drifting eggs or larvae. Usually a premonsoon survey is conducted to study the topography of the terrain and riverbank features including the extent of available operational area and likely current pattern during flood. Composition and distribution of fish fauna is also studied to assess the availability of breeders in the area as the spawn production depends mainly on it.

During monsoon season, a number of shooting nets are fixed by means of bamboo poles in the bank of river, the mouth of the net facing the current. Suitable collection

Fig. 2.4(a-d): Larval Development of Indian Major Carps.
(a) *Catla catla,* (b) *Labeo rohita,* (c) *Cirrhinus mrigala*
(*modified from Chakrabarty, R.D. & Murty, D.S 1972*).

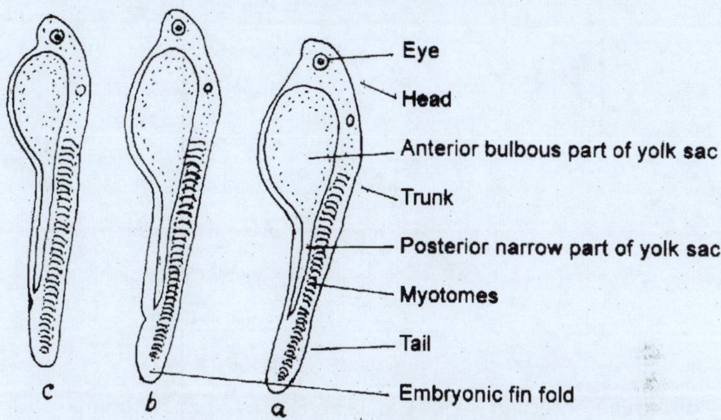

(a) Hatchlings or Prolarvae of Indian Major Carps

Note the length of the posterior narrower part of the yolk sac in relation that of anterior bulbous part. In the hatchlings of mrigal, posterior part is longer than that of anterior part. But in the hatchlings of catla and rohu both parts are approximately of equal length.

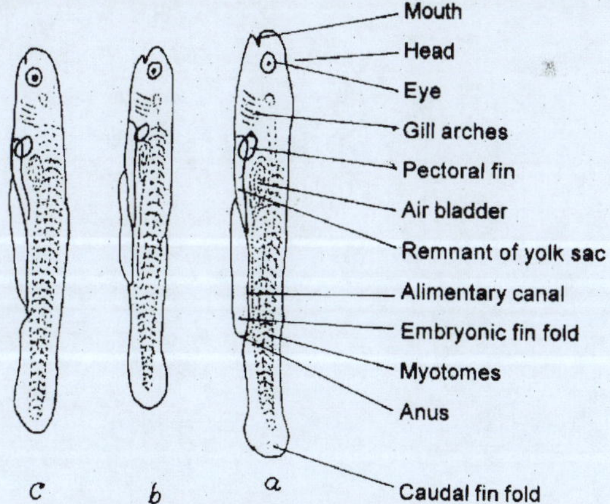

(b) Prolarvae 72 Hours After Hatching

Note the yolk sac which is almost absorbed and the presence of primitive alimentary canal with mouth and anus, pectoral fin rudiments, air bladder and gill arches. In case of catla at this stage a reddish glow is visible in the opercular area which is absent in rohu and mrigal. [Further, a dark triangular spot is seen in the caudal peduncle of catla but in rohu and mrigal, only black pigment spots are seen.]

Fig. 2.4–Contd...

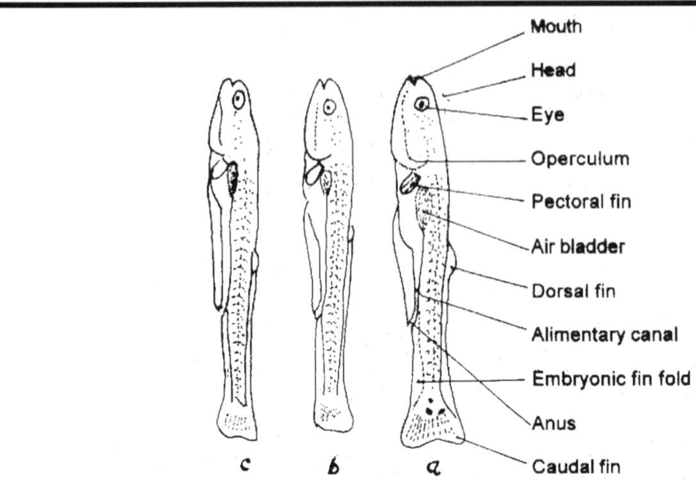

(c) Postlarvae of Indian Major Carps 96 Hours After Hatching (fifth day)

Yolk sac is fully absorved. In case of catla and rohu a reddish hue is seen in the abodminal area above anus which is not found in mrigal. In catla lip margins are thick, in rohu lips are not thick but slightly fimbriated and in mrigal lips are thin. Further, differentiation of dorsal fin from embryonic fin fold is also seen. Black chromatophores in the caudal peduncle is arragned in a semicircle in catla which is not found in rohu and mrigal.

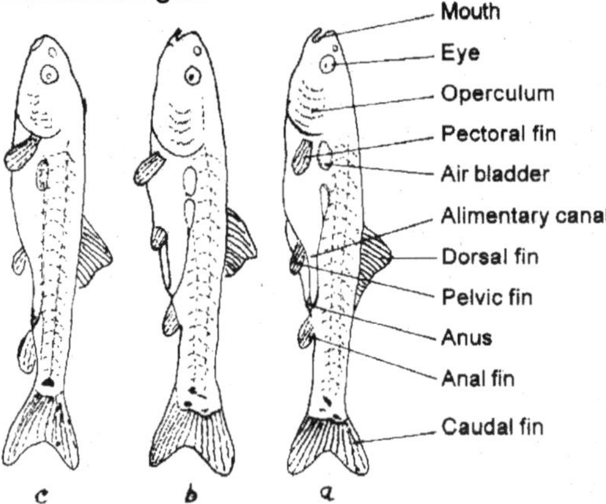

(d) Post Larvae of Indian Major Carps on Tenth Day After Hatching

Embryonic fin fold has dissappeared (except a portion extending from anterior region of pelvic fin to anus which is found in catla and rohu). All fins are with finrays. Slightly upturned mouth with thick lips in catla, thick upper lip over hanging the fimbriated lower lip in rohu, and thin lip and adult like mouth in mrigal can be noted at this stage. Further black chromatophore form a triangular patch in the caudal puduncle of mrigal, but in catla it is less distinct, in rohu two faint black chromatophore concentrations are seen in the caudal base.

ground is the area between two adjacent shadow zone and erosion zone of the river. The net is fixed in such a away that its axis is parallel to the direction of the current. The spawn collection receptacle is tied to the posterior end of spawn collection net around a ring. The drifting eggs/larva/fry enter the net along with current of water and the same is accumulated in the spawn receptacle. The spawn is scooped out from the tailpiece of each net once in ½ hr or 1 hr or 2 hrs depending on the intensity of spawn. It is sieved through a wire mesh net to remove the debris and unwanted organisms. The spawn is measured by measuring cups and then stored in hapas fixed in the water or pits dug in the river bank.

Some workers have studied the behaviour and availability of spawn in rivers in relation to hydrological and hydrobiological features. Usually during monsoon season, the water level in a river gradually rises, reaches a peak and then recedes. This is the pattern of a flood and this is repeated twice or thrice during a season. Usually the first flood contains spawn of undesirable species of fish such as predatory fishes and minor carps like *Puntius* sp. where as the successive floods will have spawn of major carps. ['Earlier breeding' of predatory fish is an adaptation for its survival because by the time its fry hatches out and starts feeding, the spawn of major carps will form the food for the fry of predatory fishes.] Usually, spawn of major carps congregate at the water surface even upto 3 feet depth. At 3'-6' feet depth stray specimens and below 6' dead ones are usually found. Current velocity conducive to spawn catches is shown to be 0.5 – 3 kms/hour and the current direction should be parallel to the bank of the river. During heavy rain and wind water current becomes disorderly and hence the spawn collection becomes erratic. Overcast sky with gentle breeze and with or without drizzle is ideal for spawn collection. Availability and abundance of spawn was not related to DO turbidity or pH of the water.

Till a few years back, spawn collected from natural spawning ground such as rivers as described above during the monsoon season formed the only source for seed used in aquafarming. This had some inherent disadvantages. (i) There was uncertainty in the availability of seed as collection from natural spawning ground depended on the vagaries of nature. (ii) Further, it was impossible to get pure stock of seed of a given species as seed procured from rivers had always a mixture of seeds of different species of fish including that of predatory fish. (iii) Further, the process of procurement of seed from riverine spawning ground is labour intensive and (iv) heavy mortality of the collected seed also was observed.

Bundh Breeding

The first attempt to breed the major carps in confined waters was by bundh breeding technique. Bundh is a seasonal or perennial tank located at the slope of a catchment area. During rainy season, water from the catchment area is allowed to flow through narrow canals into the pond and the provision for water outlet is also provided which is usually guarded by a screen made of split bamboo and straw or wire mesh net of desirable mesh size. The flow of water during rainy season in this system simulates the riverine condition. Breeders of major carps are released into the tank and they breed during rainy season. There are two categories of bundhs such as dry bundh and wet bundh. Dry bundh is seasonal, as it remains dry during summer

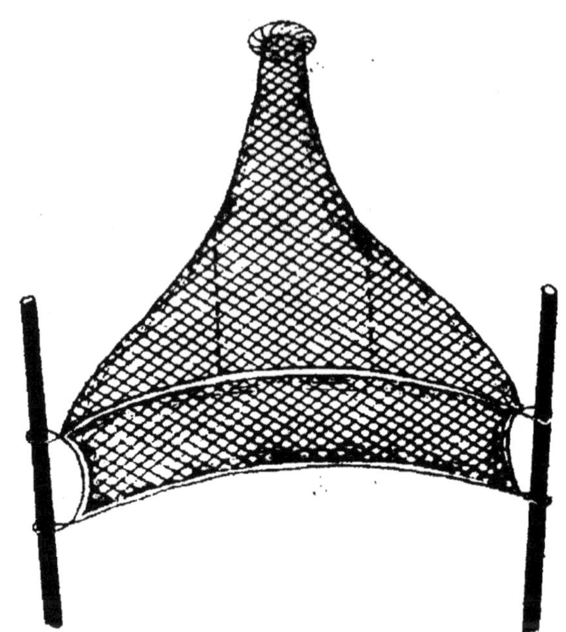

Fig. 2.5: A Shooting Net Used for Riverine Spawn Collection.

Fig. 2.6: A Sample of Pituitary Gland of Carps Preserved in Alcohol, Used for Hypophysation.

seasons. Wet bundhs are perennial. In wet bundhs brood stock can be reared in the same tank where as in dry bundh brood stock has to be released into the pond during rainy season as the ponds gets filled with the rainwater.

In one season as many as four crops can be taken from a given bundh and cost of production of spawn has been shown to be minimum in dry bundh breeding. In a few bundhs 'sympathetic breeding technique' was adopted to produce larger quantity of spawn. In this system a few pairs of mature fish after injection of pituitary extract are released into a bundh along with several other uninjected fishes. The injected brood fish breed under the influence of pituitary extract. This spawning act influences the breeding behaviour of other uninjected fishes and they start courtship and spawning even in the absence of rain, water current etc. This was initially done by fish farmers of Bankura and Midnapur district of West Bengal. Another version of dry bundh is called 'Bangala Bundh' originally designed and operated by an enterprising fish farmer of Mogra village in West Bengal. This had also gained popularity during nineties. It consisted of a pond preferably of rectangular shape of 75′ × 25′ or 50′x20′ with cemented side walls and pucca bottom. The slope of the bottom is adjusted in such a way that when water is filled in the pond, first part of 50′ length posses about 1 meter depth and the rest 25′ about 1.4 to 2 meter depth. The bottom of the first part is provided with 6″ to 13″ of fine river sand. A bump separates the shallower zone from the deeper zone to prevent slipping of sand. About 50-60 Kg of male and female breeders are released after giving low doses of pituitary extract or ovaprim. All other conditions such as water current, flow and showering systems are maintained like a dry bundh. Showering effect is produced by G. I. pipes with pores provided on both sides Water current is maintained with inlet and outlet pipes. From this about 75 lakhs of spawn is produced at a time. Successive breeding has been done, changing the bottom sand after every breeding after completely draining out the water. (Dubey, G. P. 2000)

The concept of bundh breeding of major carps is reported to have originated after a private fish farmer named Manu Teli, who observed breeding of carps in bundhs in the year 1882 (Dubey, G. P. 2000). He collected spawn from Sorabati bundh in Bankura district. Spawn collection from bundh on a commercial scale was started only in the year 1902 at Simlapal belt of Bankura district, which became a major practice in these areas by 1926.

As spawn production through bundh breeding became an easy and economically viable activity, more and more bundhs were created later in these districts. By the end of seventies, Bankura district alone had 320 dry bundhs and 30 wet bundhs and the Midnapur district had 90 bundhs. During nineties, these districts operated about 1400 bundhs producing about 4000 million spawn out of which 65% was from Bankura and 28% from Midnapore. These bundhs are operated from March to August and 20% of the brooders are injected with pituitary or other inducing agents. In M. P. the first dry bundh was established in the year 1958 at Sonar Talaiya in Nowgoan in Chatterpur district. By mid nineties 30 dry bundhs, 50 wet bundhs and 7 Bangla bundhs were operated in M. P. Following this, other states like Rajasthan and Maharastra also constructed dry bundhs. At present Rajasthan is reported to have 257 dry bundhs contributing to 65-70% of the total spawn production of the state.

Maharastra is reported to have 25 dry bundhs and 12 wet bundhs producing about 975. 45 lakhs eggs (Dubey, G. P. 2000).

Artificial Propagation Technique

The most important discovery that revolutionised aquaculture is perhaps, the technique of inducing major carp to breed in confined waters through the injection of pituitary gland extract. These carps grow well in confined waters such as ponds and lakes, their gonads also undergo growth, vitellogenesis and maturation as in other fishes that live in natural environment but they do not breed naturally in confined water. This was a serious handicap in the development of freshwater carp culture and this hurdle was over come by the hypophysation technique. Various steps involved in the artifical propagation technique are outlined below:

(a) Brood husbandry,

(b) Hypophysation,

(c) Egg incubation,

(d) Larval care and nursery rearing.

Brood Husbandry

Brood husbandry comprises of

(a) Brood raising, and

(b) Brood rearing.

Brood Raising

It is a very important aspect of induced breeding operation. Care taken in raising a brood stock helps in the recruitment of healthy prospective brood fish that prevents inbreeding depression and genetic drift in the offspring which is very often encountered in commercial carp hatcheries. For this purpose healthy yearlings of desired species of carps are collected from natural grounds or from farm-reared stock from different hatcheries. This prevents breeding between offsprings of the same parents or between parents and offsprings. Care is also taken to avoid the collection of fingerlings from wastewater, industrial effluents or sewage culture system. These fingerlings are initially kept under quarantine conditions for 2-3 months to avoid the entry of unhealthy or diseased fish to the hatchery system.

The size of ponds for brood fish raising ranged from 0.2 ha to 0.5 ha, preferably rectangular in shape having a water depth of 1.5 meter. Stocking rate can be @ 1500 kg/ha to avoid crowding. Stocking ratio of different species are maintained as 3:2:2:2:1 of catla : rohu : mrigal : grass carp : silver carp respectively. ' Drainable ponds or ponds having water replenishment facilities are considered better for the purpose.

The ponds are prepared by eradicating aquatic weeds, predatory and weed fishes. The effective way to eradicate predatory and unwanted fishes is by the application of bleaching powder and urea. Bleaching powder ($CaOCl_2$) and urea, if used separately or combinedly can kill the fishes. When used separately urea releases unionised ammonia (NH_3) which is highly toxic to fish. When bleaching powder is used separately it produces free residual chlorine ($HOCl+OCl^-$) which are toxic to the fishes. When both are used combinedly, combined residual chlorines (chloramines) are produced such as NH_2Cl, $NHCl_2$, NCl_3. These chloramines are highly toxic to fishes. Hence it is recommended to apply urea @ 100 to 250 Kg per ha. first. After 12 to 24 hours, bleaching powder containing 30% chlorine is applied @ 150 to 250 kg/ha in the pond. This kills the fish within 15-25 minutes. This has been proved to be one of the best methods as all fishes are killed and toxicity also lasts only for a shorter duration

After the pond is made free of aquatic weeds, predatory fishes and weed fishes, organic manure and inorganic fertilizers are added in the pond to obtain the desirable level of plankton. Application of lime is essential to maintain the water quality of the brood pond. The dose of lime is decided based on the pH of water as shown Table 2.1.

Table 2.1: Dose of Lime Based on the pH of the Water

pH	*Dose (Kg/ha)*
4. 0-4. 5	1000
4. 5-5. 5	700
5. 5-6. 5	500
6. 5-7. 5	200

The common organic manure applied in the pond shall be raw cowdung @ 5-10 tons/ha. as base manure followed by fortnightly application of 1-2 tons/ha based on the plankton population of water. The inorganic fertilizer requirement is calculated based on the content of nitrogen and phosphorous of the pond soil. The pond soil is classified into low, medium and high as shown below with regard to nitrogen and phosphorous content and the recommended dose of fertilizer is given Table 2.2.

Among the different fertilizers, phosphate fertilizer is considered most important in controlling the productivity of fishpond. For most of the phytoplankton species nitrogen: phosphorous:carbon (NPC) ratio required is 16:1:106 which indicates that even it trace quantities phosphorous influences the primary productivity in the presence of other two elements in sufficient concentrations. Phosphorous added to the water declines rapidly due to dilution, assimilation by macrophytes, algae, bacteria, fungi, etc. and adsorption to bottom sediments. Hence fortnightly application of phosphate fertilizer is recommended to sustain the productivity. In addition Potassium fertilizer @ 40 Kg per ha. in split doses is also recommended. According to S. K. Das and Jana, B. B. (1996) concept of pond fertilization should be based on NPC rather than NPK, as availability of potassium is high in the pond sediment.

Optimum desirable water quality parameters for the pond are temperature 25-35°C, colour greenish, turbidity 8-12 cm, pH 7.8 to 8.4, total alkalinity 80-150 (mg/litre) ammoniacal nitrogen (NH_4N) 0.2-1. 2mg/l, nitrate nitrogen (NO_3N) 0.08 to 0.5 mg/litre, nitrite nitrogen less than 0.01 mg/l, ammonia (NH_3) less than 0.01mg/l, phosphorus (P_2O_5) 0.1-1. O mg/l. Iron less than 0.5 mg/l and manganese less than 0.3 mg/l.

Feeding can be done @ 2-3 % of the body weight once daily.

Table 2.2: Fertilization Schedule Developed by CICFRI Based on the Soil Nutrient Status *(S. K. Das & B. B. Jana, 1996)*

Pond Types	*pH*	*Soil Condition Organic Carbon (%)*	*Available Nutrients (mg/100g)*	*Nutrient requirements (Kg/ha)*	*Total Quantity of Fertilizer or Manure for ponds (Kg/ha/yr.)*
Low	below 5.5	below 0.5	N-below 25	N-200-250*	Calcium ammonium nitrate 500-600 or urea 225-290
			P_2O_5 below 3	P_2O_5-100-125**	Single super phosphate 315-405 Or Triple superphosphate 110-145
				Org. carbon 600-720	Cattle dung 10000-12000 Or Gobar gas slurry 20000-30000
Medium	5. 5-6. 5	0.5-1. 5	N 25-50	N 150-200	Calcium ammonium nitrate 350-500 or Urea 156-225
			P_2O_5 3-6	P_2O_5 75-100	Single superphosphate 219-315 or Triple super phosphate 75-110
				Organic carbon 480-600	Cattle dung 8000-10000 or Gobar gas slurry 16000-20000
High	6. 5-7. 5	1. 5-2	N above 50	N 100-150	Urea 112-156 or Ammonium sulphate 225-330
			P_2O_5 above 6	P_2O_5 50-75	Single super phosphate 156-219 or Triple super phosphate 54-76
				Org. carb. 300-480	Cow dung 5000-8000 or Gobar gas slurry 10000-15000

* The easily available source of nitrogen or nitrogen fertilizer is urea which contains about 46% nitrogen. So if nitrogen requirement of a pond is 200 kg/ha/year amount of urea to be added shall be 200 × 100/46 = 434. 8 or 435 kg/ha/year. Though this is the optimum dose to be applied, recommended doses are only 225-290 kg/ha/year.

** The easily available source of phosphorous is single super phosphate (SSP). Commercial grade of SSP contains only about 16% phosphorous. So, in order to provide 100 kg of phosphorous to pond ecosystem, the quantity of SSP to be added shall be 100 × 100/16 = 625 kg/ha/year. But the recommended dose is only 350-405 kg/ha/year.

Brood Stock Rearing

In order to develop potential brood stock for hatchery use usually carps that are of 2 to 3 years are collected from the above pond and reared. Over- aged fishes (above 5 years) are not advisable to recruit. Usually, fish that have bred at least once in the earlier year have been proved better for breeding work. Such fishes are named 'professional brood stock' (Gupta, S. D. *et al*, 1995). Stocking density is maintained at 1000 kg/ha. 25-30% of the water of the pond has to be replenished at least once in a month (from January to March). Propsective spawners are selected and reared atleast for 5-6 months before the breeding season. The stocking pattern is "main species" as professional brood stock 60% and the rest four species 10% each {for example if catla is the main brood stock it will be catla 60%, rohu 10%, mrigal 10%, silver 10% and grass carp 10%. The fish are to be fed on a formulated diet containing 30% protein (Gupta, S. D. *et al*, 1995).

Formulated feed for Carp brood fish

Groundnut oil cake	70 kg
Rice bran	28.4 kg
Common salt	1.5 kg
Trace element	0.1 kg
Vitamin C	10 g
Vitamin E	3 g

(Trace elements: ferrous sulphate 50 g, copper sulphate, 8 g, zinc oxide 6.7 g, manganese sulphate 15.4 g, potassium iodide 4.2 g, cobalt chloride 2 g, calcium carbonate 13.7 g).

For a pond having grass carp as dominant group soybean cake 50 kg, ground nut oilcake –25 kg, ricebran 20 kg, fishmeal 5 kg, is recommended (Rath, S, C. *et al.*, 1999). In catla dominated pond broadcasting of powdered feed is preferred. For rohu, semi soaked feed is suspended in a perforated bag in the water column. For mrigal and grass carp, soaked feed ball is recommended.

If one comes across any case of infection it is better to isolate the fish from the rest of the stock. Routine prophylactic measures are to be followed during management practices.

Induction of Spawning

The technique of induction of breeding by administration of pituitary extract is called hypophysation technique.

It was B. A. Houssay of Argentina in 1930 who injected the extract of a fresh pituitary gland collected from a fish, *Prochilodus platensis* in to a viviparous fish *Cresterodon decemmaculatus* that resulted in premature birth of young. Following this Brazil was the first country to develop hypophysation technique on a commercial scale. In the year 1934, Von Ihering *et al.*, of Brazil succeeded in inducing fishes to spawn through administration of pituitary gland. This was followed by Russia in 1937 where sturgeons *(Acipencer stellatus)* were induced to spawn. In India, the first attempt to induce *C. mrigala* to spawn by the injection of mammalian pituitary extract was done by Khan, H. (1937). Later Chaudhuri, H. (1955) succeeded in inducing *Esomus dandricus* to spawn with pituitary gland of catla, and *Pseudotropius* with the pituitary of *C. reba.* Chaudhuri and Alikunhi (1957) successfully induced *Labeo rohita, C. mrigala, C. reba L. bata* and *Puntius sarana* to spawn with carp pituitary.

The different steps in the hypophysation technique are as follows.

(a) Collection of pituitary gland,

(b) Preservation of pituitary gland ,
(c) Preparation of pituitary gland extract, and
(d) Administration of the pituitary extract to the brood fish.

Collection of Pituitary Gland

The pituitary gland is usually collected from a freshly killed or ice preserved mature fish. The donor fish for pituitary can be carp of either sex. Among carps common carp is the most preferred donor fish due to availability of mature fish round the year. Pituitary gland of a marine fish, Tachysurus also has been tried and found effective. The brain is dissected out by cutting open the dorsal side of skull and then the pituitary is picked up after the removal of the brain from the skull. A few workers suggest to use a cork screw like borer to take out the pituitary along with the corresponding area of ventral side of skull and the portion of the brain into a watch glass and then taking out the gland from it for preservation. Some prefer to collect the pituitary by opening the brain cavity through the foramen magnum

Preservation of Pituitary

(a) *Alcohol preservaton*: The freshly collected pituitary is preserved in absolute alcohol in amber coloured phial. Two to three changes of alcohol at three to four hours intervals helps in defattening the gland and keeps the gland preserved in better conditions. This can be kept in a refrigerator or in a cool, shady place. Refrigeration increases the shelf life of the gland to two years. The phial is to be closed tightly to avoid entry of moisture. Alcohol is to be changed from time to time.

(b) *Acetone preservation*: The fresh gland is put in fresh acetone or a in dry ice chilled acetone and kept in a refrigerator for 36-48 hours. During this period acetone is changed 2-3 times at 8-12 hour intervals. Then glands are taken out of acetone, put on a filter paper and allowed to dry at room temperature for one hour. Weight of the dry gland is taken. The weighed glands are transferred to separate sterile phials and stored in a refrigerator. It can be preserved for 6-12 months. Such dehydrated and defatted pituitary glands can be lyophilized or stored frozen at -20^0C or below for a number of years.

Preparation of Pituitary Gland Extract

The extract of the gland is usually prepared just before injection. The gland (Fig. 2.6) is weighed and homogenised in distilled water or 0.3% saline. Add distilled water in such a way that the final volume shall be 0.2 ml/kg of the breeder. Usually the volume may not exceed 1 to 1.5 ml for one fish.

After homogenising the gland, it is centrifuged in a hand centrifuge. The supernatant fluid is used for injection

Preservation of Pituitary Gland Extract

(a) Ibrahim, H. and Chaudhuri, H. (1966) devised a method for preserving pituitary gland extract in glycerine. After homogenising in distilled water it is diluted with 1/3 of total volume calculated at desired concentration and kept in refrigerator for 24 hours. pure glycerin is added to make up the total volume. This mixture is kept in refrigerator for 24 hours after which it is centrifuged and supernatant fluid is transferred to ampoules for future use. This practice ensured more uniform hormone potency per unit volume of extract and also saved the time for the preparation of injection dose at the time of breeding operations.

(b) Pituitary extract in propyleneglycol. The extract is preserved in prophylne glycol and can be kept in refrigerator for 30 days.

(c) The pituitary gland is immersed in 1.5% TCA for 6-12 hours TCA extracts the gonadotropin into it. TCA containing gonadotropin is injected to the fish. TCA extract can be kept in refrigerator for 10 days. If 2.5% TCA is used, the gland may be kept only for 3 hours. The dose should be 14 mg/kg body wt. for 1.5% extract and 22-26 mg/kg body weight for 2.5% TCA extract.

Administration of Hormones

Dose of the Pituitary Extract

Usually, the female is given 2 doses of pituitary gland injection. The first is called initial dose or preparatory dose or priming dose and the second is called final dose or decisive dose or resolving dose. The male is usually given only one dose at the time of the second dose given to female. Dosage required for successfully spawning vary from species to species and from fish to fish of even the same species. The dosage depends on readiness of the brood their age, size and sensitivity. When spawning is induced with the administration of a single high dose it is called as 'knock out dose'. Knock out dose is given when the breeder is in a mature state for a long time.

The first dose of injection is given to advance the final maturation of oocytes. (See the chapter on reproduction). The interval between initial dose and final dose varies from species to species or from place to place depending on the temperature. For females of Indian major carps one initial (preparatory) dose and after 5-6 hours another final (decisive dose) is administered. But in subtropical or temperate climates sometimes one preparatory dose followed after 18-22 hours by 2 decisive doses at an interval of 6-8 hours between them is given. In some cases such as *Orinoco cachama* (*Collosoma oculus*) several preparatory doses at 24 hr intervals and 2 decisive doses at 6 hr interval is given. More hormone is required for ovulation if the ovary is bulky.

Solvent used is usually physiological saline (0.3% or 0.6%). Quantity of the solvent is not of much importance, if it is too little loss of one drop would mean loss of considerable quantity of hormone. If volume is too much, administration of the fluid will be a problem. So the quantity at the range of 1-1. 5 ml is generally used.

The injection of pituitary gland extract is done either intramuscularly or intrperitioneally (Fig. 2.7). Intramuscular injection is done either on the dorsal part of caudal peduncle or in the dorsal muscle above lateral line and below the anterior part of dorsal fin. While injecting care must be taken not to prick through the scale. Intraperitioneal injections are done through the inner side of the base of pectoral fin.

The time between last decisive dose and ovulation depends mainly on temperature. This is expressed in terms of hour grade. This is arrived at by multiplying

Fig. 2.7: Injection of Hormones into a breeder.
(Courtesy, CIFA, Kausalyaganga)

Fig. 2.8: A Breeding Hapa in Operation.

the time taken for ovulation after decisive dose with water temperature. For example if a fish takes 10 hours for ovulation after the last decisive dose at 20°C it is equivalent to 200 degree hours. The degree hours depends on the species of fish, size and age of the female, type of treatment etc. in case of Chinese Carps such as grass carp, silver carp and big head it is 200-220 degree hours at 21 to 22° whereas, for common carp it is 240 to 260 degree hours at 21-22°. In a given species hour grade varies with the type of treatment. For example, in common carp, when only one decisive dose is administered degree hours is 340 to 360.If one preparatory dose is given 24 hrs prior to decisive dose it will be only 240-260.If decisive doses are given 6-8 hours interval, it will be 200-220.Further, if breeders are of small size, hour grade is only 130 to 150.

Spawning

Spawning is usually carried out in traditional breeding hapas or in spawning pools of a Chinese circular hatchery or ecohatchery

Male-female Ratio for Spawning

To achieve successful induced spawning, it is advisable to put together one female with 2 males when breeding is done in breeding hapas. But in ecohatchery male to female ratio can be 1:1 provided males are of equal size and strength.

About 3 to 3.5 kg brood/m^3 can be kept in the spawning pool, having good water quality and 5-6 ppm DO.

Spawning in Breeding Hapas

Breeding hapas of varied size have been used in India for breeding major carps since last five decades. Still it is practiced widely for small scale production of carp seed.

A breeding hapa is a rectangular box shaped structure stitched out of fine meshed mosquito net cloth or nylon net cloth or fine meshed markin cloth. The size of the hapa may vary depending on the size of the brood fish to be used from 3.5 × 1.5 × 1 m to 1.8 × 0.9 × 0.9 m. At four upper and lower corners of the hapa, laces are fixed so that the net can be tied to the 4 bamboo poles fixed in the water column at both upper and bottom corners. The bamboo poles are to be fixed in such a way that on tying to it, the hapa shall remain in a fully stretched out condition. While fixing the hape in the water, care is to be taken to keep at least 10-20 cm of the upper surface above water and to keep the lower horizontal surface not touching the muddy bottom. On the upper surface, at one end is the opening which can be closed by a flap of net. This opeining is meant for introduction and removal of brood stock and also for collection of egg. Usually after hypophysation, the breeding pair are released into the breeding hapa fixed in the water (Fig. 2.8) during evening hours and the pair under the influence of hormones does courtship and mating. Fertilized eggs are collected during morning hours.

Spawning in the Spawning Pool of Chinese Circular Hatchery

A spawning pool used in Chinese hatchery system (Fig. 2.9) consists of a circular tank, the diameter of which vary from 5 to 8 meters having a depth of about 1 to 1.5

Fig. 2.9: A View of a Chinese Circular Hatchery.

Fig. 2.10: A Hatching Hapa in Operation.

meter. The tank may be made of brick work or RCC or FRP. The bottom of the pool shall slope towards the centre where the opening of the outlet pipe is located. The outlet pipe leads to the egg collection chamber. Inlet pipe of 2″ or 3″ diameters is arranged diagonally at 60^0 through the wall of the tank to create a circular water flow inside during operation. Over the spawning pool, two numbers of water showers of 6″ diameter are fixed at a height of one meter. In certain hatcheries, the egg delivery pipe from the centre of spawning pool are made to open directly into the hatching or incubation pool. In such cases it is advisable to have separate egg delivery pipes from the centre to different egg incubation units instead of taking several connections from one pipe. Injected breeders are released into the tank usually in the evening hours. During recent years, injected well-reared breeders have been found to respond and spawn at any time of day night cycle.

The females are given the initial dose of 2-3 mg/kg body weight and after 5-6 hours males are given pituitary gland extract injection @ 4-6 mg/kg and females 10-12 mg/kg. Quantity of the gland to be administered depends on the readiness of the female - their age, size and sensitivity. Water current in the circular pole is created at a speed of 0.2 to 0.5 m/second. Under the influence of hormones and the water current that simulates riverine flow, male and female engage themselves in courtship behaviour finally resulting in spawning. When the female releases the egg, the male ejects the milt and fertilization takes place in water.

Disturbance free environment is favourable for better spawning.

Following are the main points to be considered in the operation of a spawning pool:

(a) About 3.5 kg brood fish/m^3 can be stocked in a spawning pool provided with good quality water having DO of 5-6 ppm.

(b Water depth in spawning pool may be maintained at 0.6 m to 1 m depending on the density of brood stock.

(c) Brood in the pool is to be kept under shower before and after hormone administration.

(d) No water current is required before or just after hormone administration. Water current is allowed one hour before the calculated spawning time. This initiates the excitement of spawner and spawning.

(e) Speed of water current in the pool should be between 3-5 m/second or two litre per second.

(f) The spawning of each female is usually completed within 1 to 1.5 hrs of initiation. Water flow can be stopped soon after the spawning is over. Brood fish are to be collected and released into the brood stock tank.

(g) Before and after each operation, spawning pool is cleaned and disinfected with strong formaldehyde and potassium permanganate.

Assessment of Egg Quantity and Fertilization Rate

The next step in the hatchery operation is assessment of egg quantity and fertilization rate before it is taken for incubation. Measurement of egg is done by volumetric or gravimetric methods.

Volumetric method

The eggs are measured in a cup or beaker. The number of eggs in one cup is counted in triplicate and the average of three estimations is taken to know the numbers of eggs per cup. Multiplying this with the number of cups measured, total number of eggs is calculated.

Gravimetric method

In this method, a sample of eggs is weighed and the number of eggs in the known weight is counted. This can be repeated 3 times to arrive at the average figure. After knowing the number of eggs per g weight, weight of the total quantity of eggs is found out. By multiplying the number of eggs per g weight with the total weight of eggs, the total number of eggs can be obtained.

Calculation of fertilization rate: whereas the latter are opaque (A simple field identification can also help the hatchery operator to distinguish the fertilized and unfertilized eggs accurately by dipping the eggs in a dilute solution of $KMnO_4$ taken in a petridish. The unfertilized eggs get coloured where as the fertilized eggs do not take the colour). By counting the number of fertilized eggs from the total number in a given sample, percentage of fertilization is calculated.

Incubation of Eggs

In a commercial hatchery, facilities are built up for incubating the fertilized eggs so that requirements of developing fish eggs are fully met under controlled conditions. The requiremnets for developing fish eggs are (i) Optimum temperature (ii) Constant flow of water (iii) Disturbance free environment (iv) clear plankton and pollutant free water (v) prophylactic measures against bacterial/fungal diseases and (vi) maintenance of optimum water quality parameters

Optimum Temperature

For Indian carps such as catla, mrigal and rohu optimum temperature is 28° – 30°C and incubation period is 14-20 hours. For Chinese carps such as grass carp and silver carp it is 22°C to 25°C and the incubation period is 1 to 1.5 days. For common carp incubation period is 7½ days at 12°C, 3 to 4 days at 20°C to 22°C and 2 days at 28-32°C.

Constant Flow of Water

The metabolic wastes from the developing embryo such as ammonia and carbon dioxide are liberated into the water and the accumulation of the same is highly toxic to the developing embryo. Hence in a carp hatchery, provision for maintaining constant flow of water is ensured so that developing eggs are not exposed to lethal doses of metabolic toxicants.

Disturbance Free Environment

The fishes exhibit better courtship behaviour in the absence of human/other mechanical interference and hence a stress free environment is essential for better spawning efficiency.

Clean, Plankton and Pollution Free Water

As running water from rivers or canals or reservoirs are utilised generally for hatchery use, it is necessary that the water should be made free of silt and plankton before it is let in to the hatching pool. The water may be allowed to stand for 24 hours to allow the settling of silt and debris and then can be passed through filters before use in the hatchery.

Prophylactic Measures Against Bacterial and Fungal Diseases

Saprolegnia is a common fungus that affects the eggs. This affects the dead eggs first and later it causes heavy mortality of the the developing eggs. Hence, as a prophylactic measure eggs are dipped in 0.1 to 0.2 ppm malachite green. In case of disease occurrence, eggs are to be treated in 5ppm malachite green by dip treatment Some workers use Tannin solution instead of malachite green. Preparation of tannin solution. Dissolve 5-8 g tannin in 10 litre of water. 1 to 2 litre of this solution is added to 30-50 litre of incubation media and stirred for 10-20 seconds. Then wash it. Growth of bacteria may be due to use of water which is dirty and containing heavy organic load. Plankton like cyclops hurt the egg shell. Chironomid larvae bite the eggs with mandibes and damage the eggs

Maintenance of Optimum Water Quality Parameters

Optimum water quality parameters as noted below are to be maintained for the spawning and incubation pools.

Turbidity	Clear to 20 cm.
PH	7, o to 7.5
Total alkalinity	40 to 60 mg. /litre
P_2O_5	Trace to 0.01 mg/litre
NH_4 N	Trace to 0.02 mg. / Litre
NO_{3-} N	Trace to 0.01 mg/litre
Iron	Trace to less than 0.4 mg. /litre
Manganese	Trace to less than 0.1 mg/litre

If total alkalinity is above 100 mg. per litre it is harmful. In such cases water has to be treated with alum to reduce alkalinity. To generalise 1 ppm alum can reduce 1 ppm alkalinity

Incubation Devises of Carp Eggs

Hatching Hapas (Fig. 2.10)

Two enclosures or hapas are used one in side the other (outer hapa and inner hapa) for incubation. The hapas are made of nylon net the outer hapa has a mesh size of 0.5 mm and inner hapa has the mesh size of 2 to 2.5 mm. The hapas are fixed in the pond by means of bamboo poles. The eggs after prophylactic treatment are placed at the bottom of inner hapa. After hatching, the hatchlings move out of the inner hapa through its large size mesh and are collected in the outer hapa. The empty egg shells are left in the inner hapa which are removed along with the inner hapa after the hatching is completed

Funnel Type Incubators (Fig. 2.11)

The incubator is funnel shaped. This type was first used in the Zoug lake area of Switzerland and hence it is named as Zoug jars. It was originally made of glass each having 6 to 16 litres capacity. At present, such jars are made up of fibre glass of 40-80 litre capacity and of even 200 litre capacity. If jars of metal are used, it is to be coated with neutral paint. The broad end of the funnel may be closed with rubber head or plastic head and the same is provided with an out- let. The eggs are placed in the funnel after the water is let in through the inlet to funnel connected below. The upward moving water causes the egg to move and roll and in this process the eggs get sufficient oxygen and in turn the carbon dioxide diffuses out. The metabolic wastes and carbon dioxide are removed along with outlet water in the constant flowing system.

Incubation Pool (Hatching Pool) of Chinese Circular Hatchery

The hatching pool or incubation pool (Fig. 2.12) is circular in shape having two chambers outer chamber and inner chamber. The outer chamber may be of 3 m to 6m diameter and the inner chamber of 0.8 –1. 5 m and depth of 1 to 1.5m. The diameter of the outer chamber may be 3 to 4 times the diameter of inner chamber. The circular wall separating the outer chamber from the inner chamber is provided with windows which are fitted with fine meshed net (mesh size 1/60″ to 1/80″) that allows the water to flow from outer chamber to inner chamber and not the eggs or spawn. At the centre of the inner chamber is the out let pipe. The water inlet pipes are fixed at the bottom of the outer chamber. These are 'duck-mouth inlets' and are placed at an angle so that inlet water move in a circle in the outer chamber. The vertically errected pipe at the centre maintains the water depth in the incubation chamber at the desired level. The eggs are collected from the egg collection chamber and released into the outer chamber of hatching pool after prophylactic treatments. As a constant flow of water is maintained in the pool, eggs released in to the outer chamber gets sufficient oxygen as they are drifted along with water current in a circular motion. Duck mouth inlets are arranged in a row, 6-12 numbers equidistant from each other and from both inner and outer wall of the chamber. This causes a unidirectional water flow during operation which results in movement of eggs without touching the screen or wall of the chamber. The pool can hold about 7 lakhs fertilized eggs per m^3. The speed of the water flow is maintained at 0.4 to 0.5 m. /second in first twelve hours and at 0.1 to 0.2 m. per second in the next six hours which is then increased to 0.3 to 0.4 m. per second. Good water circulation inside the pool is very important for the success of the hatching pool. If the circulation is not proper eggs settle at the bottom which results in decomposition. This causes the growth of pathogenic organism and saprophytes which in turn cause mortality of developing egs. There has to be a spawn delivery pipe to collect the spawn after the hatching is over.

In certain hatcheries, the water from the spawning pool is directly allowed to flow into the outer chamber of hatching pool so that eggs, which are laid in the

Fig. 2.11: Glass Jar Hatchery in Operation.
(Courtesy, CIFA, Kausalyaganga)

Fig. 2.12: Incubation Pool (Hatching Pool) of a Chinese Circular Hatchery in operation.

spawning pool directly reaches the outer chamber without going through the egg collection chamber.

To summarise, the following are the points to be considered while operating an egg incubation unit of an ecohatchery:

(a) Where the egg delivery pipe from the spawning pool opens directly into the outer chamber of hatching pool, care is to be taken to see that eggs are received on a water cushion to avoid injury to eggs.

(b) It is better to mud plug the openings of egg delivery pipe and spawn delivery pipe so that no blind space exists in the unit. This prevents the loss of eggs and spawn due to formation of the 'low oxygen pockets' during operation.

(c) Direction of duck-mouth inlets and speed of the water are maintained in such a way that it keeps the developing eggs away from the screen and wall of incubation chamber.

(d) The speed of water current is to be maintained at 0.4 to 0.5 m/second for first twelve hours and then 0.1 to 0.2 m/second for next six hours and then 0.3 m-0.4 m/second for rest of the operation.

(e) Dissolved oxygen in the pool should not be less than 4ppm.

(f) Cleaning the pool of floating, suspended or settled debris including dead eggs and spawn is necessary for increasing the percentage of survival and recovery of spawn. The cleaning is done as follows:

Surface cleaners

A wooden or bamboo stick of 2 cm dia meters is kept on water surface across the outer chamber and in between walls. This helps in accumulating the foam, debris and insects efficiently

Sub surface cleaners

A perforated wooden plank of four to five cm. width is fixed to surface half submerged. It helps in cleaning the subsurface debris of water of the chamber.

Column cleaners

5 to 6 pieces of coir rope (1 –1. 5 m long) with smooth bristles are tied to a stick of 2 cm dia. equidistant from each other. The stick with rope is fixed to the water surface of incubation chamber. Column debris containing dead spawn and egg shell stick to the rope. This is removed periodically and washed in $KMnO_4$ and refixed again.

Bottom cleaners

From the second day of operation, dead spawn, eggshell and fungal mat that settles to the floor is siphoned out by means of 2 cm diameter stiff polythene pipe.

Cleaning the screen

Inner surface of the screen is cleaned using a brush from the inner chamber.

The incubation pool (hatching pool) of the ecohatchery is the most popular and extensively used incubation device in recent years.

Larval Care and Nursery Rearing

Early larval stages are the most crucial and vulnerable stages in the life cycle of a fish. During this period, the young ones are defenseless against predators, susceptible to microbial attacks, prone to diseases and sensitive to fluctuations in the environmental factors such as dissolved oxygen, temperature, alkalinity etc. and to handling stress. Hence the rate of survival of these stages depend on the maintenance of optimum water quality parameters, availability of adequate choice food and a predator free aquatic environment. At present, the national average survival rate from hatchling to fry stage is only 14% which is indicative of the lacunae that exists in our present day larval care protocol followed in the seed productuon centres.

The freshly hatched larvae are called spawn or prolarvae. A prolarva measures 4.6 mm long and 1 to 1.5 mg in weight. They do not feed on exogenous food for about 72 hours. During this period they utilize the yolk present in the yolk sac. By the time the yolk sac is fully absorbed i. e on 4th day it grows to 6.5 to 7 mm long and weighs 1.6 to 2mg. At this stage it is referred to as postlarva. The postlarvae start feeding on exogenous food, its choice food being live feed organisms – zooplankton. They feed on ciliates, rotifers, cladocerans and copepods in succession. During the initial post larval stages, they may also feed on microalgae *viz. Pediastrum boryanum*, *Coelastrum acicularis*, *Cryptomonas morsoni* and N*itszchia* sp. etc. Usually from 7 mm to 10 mm size they prefer small size rotifers and after 10 mm size they may feed on larger rotifers, cladocerans and copepods.

In view of this in a carp hatchery complex, there should be facilities not only for brood stock raising, spawning and hatching but also for spawn conditioning and nursry rearing of spawn to fry stage or even upto fingerling stage.

Generally carp seed is classified in to three categories *viz.* Spawn, fry and fingerling. The term spawn is applicable to hatchlings upto 8 mm size *i.e.* till yolk sac absorption. The fry stage itself can be distinguished in to two substages such as early fry stage (8 mm to 25 mm size) and advanced fry stage (from 25 mm to 40 mm size). The fingerling stage refers to those from 40 mm size to 150 mm size (early fingerling stage from 40 mm to 100 mm and advanced fingerling from 100 mm to 150 mm size). The rearing of postlarvae to fry stage i. e up to about 20 mm size (one crop cycle) usually takes 15 days which can be reduced to ten to twelve days under intensive nursery rearing practices.

Spawn Conditioning

The spawn can be kept in the incubation chamber till the yolk sac is absorbed and then it is be shifted to separate spawn holding tanks after assessing the quantity of spawn by volumetric method. Usually the spawn is allowed to pass from incubation chamber to a spawn collection chamber by gravity flow. The spawn collected from this chamber is measured by means of measuring cups. The number of spawn in a cup is calculated first and this is multiplied with the total number of cups of spawn to obtain the total number of spawn. Usually the spawn count varies from 600 to 650 numbers per ml. In the spawn holding tank facilities for aeration and water exchange are provided to avoid mortality of spawn due to oxygen scarcity or water pollution because of accumulation of metabolic wastes. Some workers prefer to use race ways with flow-through system for this purpose.

During early years, the hatching was done in hatching hapas. In such cases the inner hapa is removed along with the remains of the egg shell etc. and the spawn will be retained in the outer hapa till yolk sac absorption.

Nursery Rearing

By fourth day after hatching the spawn is to be released in to a well prepared nusery tank for growing the post larva to fry stage. (Carp spawn of multiple spawning takes only 60 hours for yolk sac absorption where as in the traditional method it takes seventy two hours for complete yolk sac absorption). If the spawn is released into a tank which is not well prepared, there may be heavy mortality of the spawn which may be due to the following reasons.

1. Wide variations in the physicochemical conditions between spawn holding chamber or outer hatching hapas and nursery tank

While releasing the spawn the temperature pH, DO alkalanity etc. of the nursery tank are to be checked. The ideal water quality parmeters shall be water temperature; 28^0C to 32^0c, turbidity; 8 to 15 cm, pH ; 7.5 to 8.0, dissolved oxygen; 5mg/ltr or above, free carbon dioxide below 5mg/ltr, total alkalinity; 60 to 80 mg/ltre P_2O_5 0.04 to O. 1mg/litre, NH_4N 0.08 to 0.2 mg/litre, NO_3N 0.04 to 0.08 mg/litre, NH_3 less than 0.01 mg/litre, nitrite less than 0.01 mg/litre, iron less than 0.3 mg/litr and manganese less than 0.15 mg/litre.

Usually polythene bag containing spawn is left floating in the nursery tank for about half an hour to equalise the water temperature of polythene bag and tank gradually so that spawn is acclimatised gradually to any temperature variation.

2. Dissolved oxygen

In a pond ecosystem, depletion of dissolved oxygen is noticed during early morning hours. During night oxygen producing process (photosynthesis) is arrested where as oxygen consuming process (respiration) continues. Hence facilities for aeration should be provided to enhance dissolved oxygen level during critical hours.

3. Accumulation of metabolic wastes

When a high density of spawn is stocked in the nursery tank, nitrogenous metabolic wastes such as ammonia and respiratory wastes such as carbon dioxide get accumulated in the ecosystem beyond the tolerable limit. This becomes more pronounced when there is heavy loss of water due to seepage and or evaporation. This loss can be met by the addition of water. In order to overcome this problem partial water replenishment is suggested daily.

4. Dearth of choice live feed organisms for the spawn

Carps spawn with the yolk sac fully absorbed when released into the nursery tank feed on the available live feed organisms voraciously. Thus the natural plankton fauna get exhausted within 1 – 2 days. Hence sustained production of live feed organisms is to be ensurred through repeated manuring and fertilizing. Further, artificial feeding is also required for healthy growth of spawn.

5. The presence of predatory insects or predatory fishes

Predatory insects and predatory fishes are to be eradicated before the stocking of spawn to avoid loss of spawn due to predation by these organisms.

Preparation and Management of Nursery Tank

Nursery ponds of small size (0.02 to 0.05ha area) having a water depth of 0.5 to 1 meter are ideal for spawn stocking. The ponds can be either earthern ponds or brick lined with cement plastering. In the latter case a soil layer of 15 to 30 cm is to be provided at the pond bottom. Drainable ponds are desirable as it facilitates (i) complete eradication of predatory and weed fishes before spawn stocking and (ii) drying the pond bottm for ploughing and manuring prior to the release of water. The latter is practiced in many countries as a routine procedure for nursery tank prepartion. Ploughing the pond bottom enhances aeration of the soil there by the process of oxidation sets in – sulphides are changed to sulphates, ferrous iron is changed to ferric form and ammonia to nitrites and nitrates. Further mineralisation of humus is also enhanced. Nutrients remain adsorbed to the soil under such conditions. As the water is filled in the pond again reduction process sets in at the pond bottom -ferric form changing to ferrous, sulphate to sulphides, nitrates and nitrites to ammonia- there by losing its adsorption capacity causing release of the nutrients to the water phase gradually. This encourages the sustained production of plankton essential for spawn survival and growth. The different steps in the preparation and management of nursery tanks are (i) weed clearance (ii) eradication of predatory fishes and weed fishes (iii) manuring and fertilizing (iv) control of aquatic insects and (v) spawn stockig and supplementary feeding.

Weed Clearance

Excess growth of aquatic weeds is usually found in perennial nursery tanks. If there is excessive growth of aquatic weeds in a nursery tank, the weeds are to be eradicated from the pond before the spawn is released. Presence of aquatic weed is harmful as noted below:

(a) Aquatic weeds utilize the available nutrients in the water as a result plankton production is reduced.

(b) It causes oxygen depletion in water, particularly during night hours.

(c) It limits the living space of fish and hinders their free movement.

(d) It provides shelter to predatory fishes, predatory insect and weed fishes which are harmful to spawn and fry.

(e) It promotes siltation in pond.

(f) It obstructs netting operation and hence makes the harvesting difficult.

The weeds can be removed manually by hand picking and uprooting. Rooted submerged weeds can be removed mechnically by dragging log weeders fitted with spikes and barbed wire. Apart from this weedicides can be applied to destroy the weeds. The common weedicide used is 2, 4-D (2, 4-Dichlorophenoxy acetic acid) @ 4.5 to 6.7 kg/ha. Apart from this other chemicals such as Taficide (2, 4 – D sodum salt) 4 – 6 kg/ha or simazine, 6kg/ha are also recommended. For fully submerged weeds such as hydrilla aquathol (Disodium 3, 6 endosohexa hydrophthalate) is recommended as 2, 4-D is not very effective.

Eradication of Predatory Fishes and Weed Fishes

Presence of predatory fishes in a nursery pond is harmful as they directly feed on carp spawn/fry. Further, as most of the predatory fishes breed earlier than carps, their fry shall be present in the pond by the time the carp spawn is released and hence they shall feed voraciously on carp spawn. Weed fishes shall compete with carp spawn/fry for food. Hence the pond is to be made free of predatory as well as weed fishes before the release of spawn.

In case of drainable ponds complete dewatering can be done to eliminate the predatory and weed fishes. Where dewatering is not possible repeated drag netting coupled with use of hooks and lines are recommended. Even this may not result in eliminating all the harmful fishes. Hence some workers have suggested use of fish poisons to kill all the fishes. The fish poisons are of different categories viz (i) plant derivative poisons (ii) chlorinated hydrocarbons (iii) Organophosphates and (iv) urea and bleaching powder.

Plant derivative poisons

The common fish poisons derived from plants are (a) mahua oil cake and (b) derris powder. Mahua oilcake contains 4 – 6% saponin which is toxic to fish by causing haemolysis of red blood corpuscles. It may be added in water at a dose of 75 ppm but for complete removal of all fishes 200 – 250 ppm is recommended at least 2 weeks before stocking to allow complete detoxification in the aquatic environment to avoid mortality of spawn due to its toxicity. Derris powder is prepared from the dry root of derris plant, *Derris trifoliata.* It contains rotenone (5%) which damages the respiratory organs of fish as it works like a contact poison. Recommended dose is 4ppm for weed fishes, 6-10 ppm for air breathing fishes. Toxicity lasts from 4 to 12 days depending on the dosage used.

Chlorinated hydrocarbons

This include poisons such as endrin, dieldrin aldrin etc. it is not advisable to apply them in water because of their high residual toxicity and consequent biological magnification in the ecosytem.

Organophosphates

Poisons such as malathion, phosphamidon, DDVP (nuvan) monocrotophos etc. belong to this category. The toxicity is due to their property of inhibiting acetylcholin -esterase thereby causing dysfunction of nervous system resulting in the death. These chemicals are easily degradable in aquatic environment and hence has very less residual toxicity. But its use is also limited because of its harmful effects on cattle and humans.

Application of bleaching powder and urea

Bleaching powder (with 30% chlorine) is applied at a dose of 150 to 250 kg/ha in the ponds twelve to twenty four hours after the application of urea @ 100 to 250 kg/ha. This has been proved as one of the best method as all fishes are killed and the toxicity also lasts for a shorter duration.

Manuring and Fertilizing the Pond

Next step in the prepartion of nursery pond is the application of manures and fertilizers, which is aimed at a sustained primary production in water. To start with liming is done. Dose of lime has to be decided based on the pH of water (see table 1.of brood husbandry section) Application of lime has several advantages *viz* (i) It has a buffering action in water (ii). It can neutralise the harmful effects of acids such as humic acid and sulfuric acid (iii). It speeds up decomposition of organic matter (iv) It counteracts the toxicity of some metallic ions such as K, Na, Mg (v) It can aggregate soil colloidal particles and make it release phosphorous into overlying water (vi). It is necessary for the formation of cell wall of some algae, molluscan shell, and exoskeleton of arthropods (vii) It works as a disinfectant killing harmful bacteria and fish parasites present in water and (viii) it is also required to meet the calcium requirement of the growing fish

Five to seven days after the application of lime, organic manure is to be applied. The best organic manure is raw cowdung. It may be added @10-15 tons/ha, 5 – 7 days before stocking. If mahua oil cake is used to eradicate predatory fishes, cowdung may be reduced to 5 tons/ha. Cowdung promotes nitrification, formation of good colloidal structure and in the formation of good pond mud. Next to this inorganic fertilizers are to be added. (See table 2 for fertilization schedule based on soil nutrient status). Some workers have recommended applications of a mixture of ground nut oilcake (GOC) @ 750 kg per ha. Raw cowdung 250 kg per ha. and SSP @ 40 kg per ha. in liquid form. Dense population of zooplakton is reported to develop within 24-96 hours.

Another system of phased manuring of nursery ponds has been practiced in certain areas like Maharashtra since a few years. The schedule is as noted below. In this system a crop of fry is raised from spawn within a period of 10 days and hence a crop cycle consists of ten days.

On the day preceding stocking

The pond is fertilized with a mixture of single superphosphate of lime (150 kg), triple super phosphate of lime (50 kg), cow dung (700 kg), oil cake (700 kg) per ha.

Then the pond is inoculated with 30-50 ml of Daphnia/Moina.

Day 1 of Crop cycle	:	Stock carp spawn @6-10 million/ha
Day 2 of crop cycle	:	Apply oil cake @ 350 kg and Cattle dung @ 87. 5 kg/ha.
Day 3 of crop cycle	:	Apply oil cake @ 175 kg & cattle dung @ 43. 75 kg/ha.
Day 4 to 9 of crop cycle	:	Apply oil cake @ 87. 5 kg, cattle dung 22kg and rice bran @ 100 kg/ha.
Day 10 of crop cycle	:	Harvesting

If catla spawn is to be stocked quantity of manures may be doubled from 4-9 days. In a well manured and fertilized pond, in order to stock the spawn, there should be at least 1.5 to 2 ml plankton in about 50 lit of water.

Control of Predatory Insects

The pond environment provides an ideal habitat for aquatic insects mainly because of the stability of the water in the pond unlike water current that exists in streams or wave action in ocean. Most of the insects live in shallow water areas where there is abundant marginal weeds and grasses. Plants provide hiding places, material for cases and cocoons, place to lay eggs, anchorage, food and oxygen for insects. Plant materials such as roots, stem and leave of aquatic plants are eaten by a number of insects whereas a number of them feed on other insects, fishes, their eggs and larvae and other animals. It is these predatory forms that are harmful to carp spawn and larvae. Predatory insects usually have prehensile forelegs or strong lower lips that help in seizing and holding its pray.

About 3% of the total insect population is found in water. Many of the aquatic insects are harmful to carp spawn as they attack and kill them. Majority of aquatic insects belong to three orders such as (i) Coleoptera, (ii) Hemiptera, (iii) Odonata though several species of other groups such as Plecoptera, Trichoptera, Ephemeroptera, Lepidoptera, Hymenoptera etc. are also found in aquatic ecosystem. (Fig. 2.13).

Order Coleoptera

These insects are commonly called as beetles. 3 famlies of coleoptera *viz.* dytiscidae, hydrophilidae and gyrinidae are mainly found in fresh water ponds. Common dytiscids are *(I) Cybister confusus, (ii) C. tripunctatus and (iii) Eretes sticticus*. Both adults and larval stages of dytiscids attack the spawn/fry. A newly hatched larva of *C. confusus* can devour 20 numbers of advance stage carp fry of 24-40 mm length within a period of 24 hours. Larvae of *Sternolophus rufipes* of family hydrophilidae (commonly called water scavenger beetles) kill fish spawn. Larvae of *Dineutes* of family Gyrinidae feed on even small fish.

Order Hemiptera

The three families of hemiptera *viz.* notonectidae belostomatidae and nepidae are found in fresh water ponds inflicting heavy casuality to carp spawn. Notonectids are commonly called back swimmers. The most common back swimmer in nursery pond is *Anisops bouvieri*. According to Gorai and Roychoudhuri (1962) its larvae feed on fish spawn, from fifth instar onwards. Even carp fry of 10-13 mm long are killed by the giant water bug (Belostomatidae). They secrete a toxic substance through saliva which can kill the fish within a very short time. Water scorpion (Nepidae) such as *Laccotrephes griseus* attack advanced carp fry and feeds on them. Water stick insects *(Ranatra elongata)* are also predators of carp spawn and fry.

Order Odonata

The common insects of this order found in fresh water are dragon fly nymphs which are highly predatory on carp spawn.

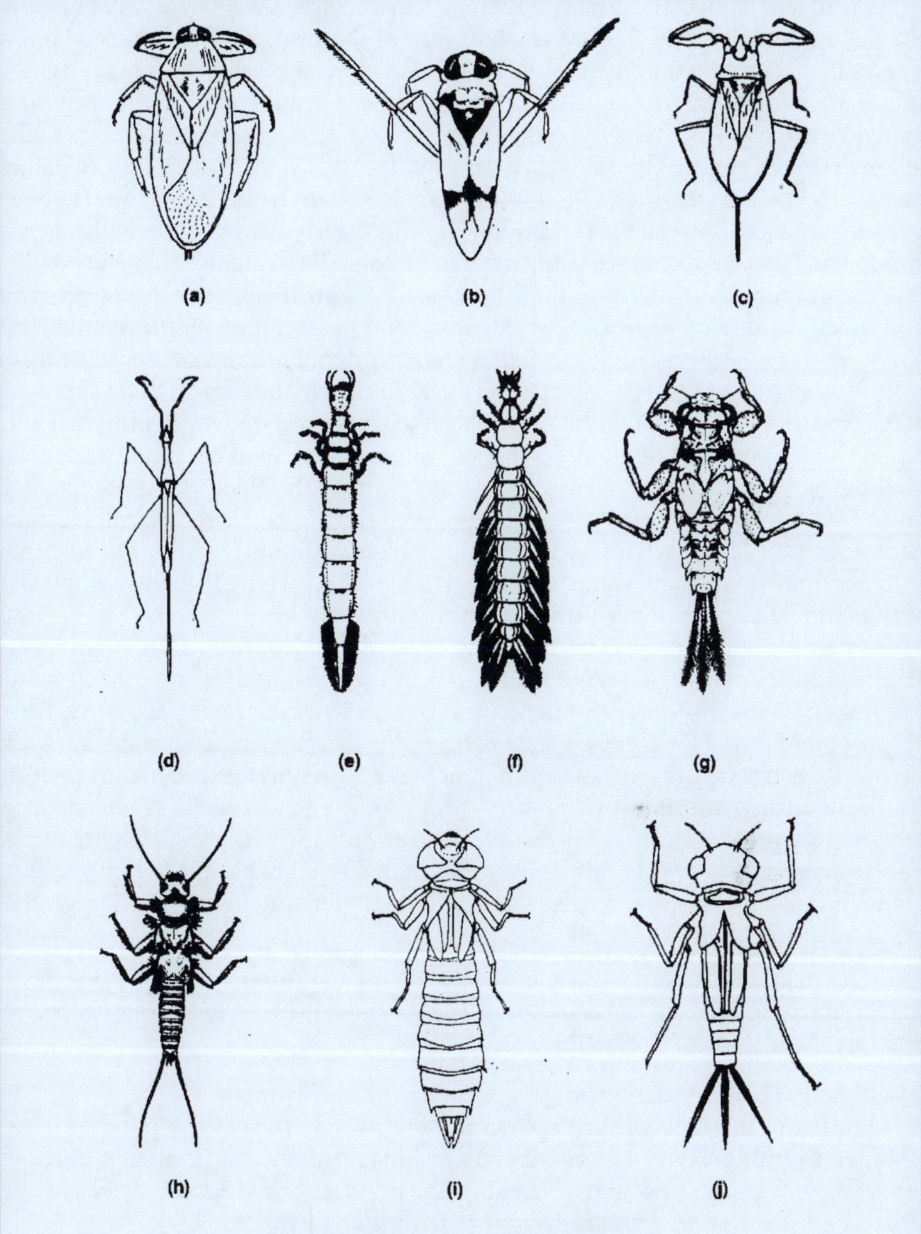

Fig. 2.13: Common Aquatic Insects.

(a) *Belostoma* sp. (Giant water bug), (b) *Notonecta* sp. (Back swimmer), (c) *Nepa* sp. (Water scorpion), (d) *Ranatra* sp. (Water stick insect), (e) *Cybister* sp. (Diving beetle) larva, (f) *Dineutes* sp. (g) *Perla* sp. (Stone fly) nymph, (h) *Ephemerella* sp. (May fly) nymph, (i) *Aeschna* sp. (Dragon fly) nymph, (j) *Argia* sp. (Damsel fly) nymph.

In order to control the insect population in a nursery tank it is necessary to know the biology, particularly the respiratory habit of the insects. There are only a few insects that spend their entire life cycle in water. In most cases, the larval or nymph stages are completed in water where as the adults are not found in the water (eg. Dragon fly, damsel fly, stone fly). But in some cases, both larvae and adults are found in water (eg. Gyrinide, Dytiscidae, Hydrophilidae) Based on the respiratory habits, aquatic insects are classified into (I) semi aquatic and (ii) truly aquatic insects. Semi aquatic species perform aerial respiration while living in water where as truly aquatic species perform gas exchange using oxygen dissolved in water through specialized tracheal gills or blood gills or through body wall. Tracheal gills are expansions of the body wall containing extension of tracheal system. This may be present beneath abdomen as in the larvae of caddis flies, or head and thorax as in larva of stone fly or rectum as in dragon fly larvae, In damsel fly larvae it is in the form of three flat plates at the posterior end of the body. Blood gills are expansion of body wall provided with blood, which contains hemoglobin as in Chironomids. Semi aquatic insects have well developed tracheal system with spiracles through which atmospheric air is taken into the body for respiration. Those remaining just below the water surface have respiratory tubes which can pierce through the water surface film. For example, water scorpion (*Nepa*) and water stick insect (*Ranatra*) have long filaments at the tip of abdomen which form tubes through which air can be taken into the body. These forms can remain fully submerged only for a short duration. Diving insects of Hydrophilidae and Dytiscidae exhibit an unique adaptation for under water respiration. These insects, carry air bubbles either between abdomen and elytra (as in Hydrophillidae and Dytiscidae) or at the tip of the abdomen (as in Gyrinidae). While diving, these insects use oxygen from air bubbles and in turn the carbon dioxide from the body diffuse into the surrounding water. As oxygen from air bubble is used, oxygen pressure within the bubble is decreased as a result oxygen from water diffuses into the bubble. Side by side, nitrogen escapes from the air bubble into the surrounding water. It is calculated that oxygen in the bubble under water shall be 13 times that originally found in it.

The principle behind the measures taken for killing predatory insects is prevention of aerial (tracheal respiration) of these insects by adding an emulsion that will form a uniform layer over the water.

Procedure for Killing Predatory Insects

For better survival of carp spawn it is necessary to clear the pond of insect population immediately before spawn stocking. If there is a long gap between eradication of insects and spawn stocking, the pond may get repopulated as insects can migrate from nearby ponds because of their flying habit.

The common method to kill predatory insects is to spray an emulsion of mustard or coconut oil with cheap washing soap at a proportion of 56:18 kg/ha, 12-24 hours before stocking. As the emulsion forms a layer over water; it prevents the tracheal respiration of insects, there by resulting in their death. Dragnet is operated to remove all the dead insects. Instead of soap, teepol B-300 (560 ml of teepol in 56Kg mustard oil) has been used by a few workers. Other combinations of emulsions tried are as follows (I) 50 lit of diesel oil + 25-30% (by weight) of cheap washing soap per ha.

(2) 1 liter of light speed diesel oil, 0.75 ml Hyoxid 1011 and water 40 ml at a rate of 1040 ml per 200 m^2 of water surface (3) kerosine @ 75 lit/ha or high speed diesel @ 50 lit/ha. Application of such emulsion has to be done on calm days as wind will break up the oil film on water surface.

Spawn Stocking and Supplementary Feeding

In a well prepared pond carp spawn can be stocked at a rate of 6 million to 10 million/ha. Stocking is done either in the morning or evening hours to avoid mortality of spawn due to abrupt change in temperature. It is better to limit the water depth at the time of stocking to 40 cm. Later at 4 days interval, water depth can be increased in a phased manner. Low initial depth is advantageous to carp spawn. Initially, the mobility of spawn is limited and hence if the water is less, it can reach the live feed organism easily. Phased increase of water helps in improving the water quality and also provide ample space for proliferating plankton.

Soon after stocking, spawn feeds on zooplankton voraciously. The starter feed for spawn is rotifers such as *Brachionus* of 100-150 mm size. Hence at this stage, water should have a rotifer density of 2000-5000/liter. This is followed by microcladocerans like *Daphnia*, *Moina* and *Bosmina*. Larger copepods like cyclops are harmful to spawn.

Supplementary Feeding

As the freshly stocked spawn start feeding on natural live feed organisms of the pond, these planktonic organisms may get exhausted within 2-3 days. Hence the natural feed is supplemented with artificial feed. Finely powdered mustard oil cake or groundnut oil cake and rice polish (1:1) is given 4 times the weight of the spawn for first 5 days. Then from 6th to 12 day, the quantity can be doubled. For optimum growth and maximum survival protein in the diet has to be 40-45%, carbohydrate 26-30% and the feed has to be fortified with vitamins and minerals and micronutrients like cobalt chloride.

CHINESE MAJOR CARPS

Grass carp *(Ctenopharyngodon idella)* and silver carp *(Hypophthalmichthys molitrix)* (Fig. 2.14) are the two Chinese major carps introduced in India for polyculture along with Indian major carps. Grass carps are voracious feeders of aquatic plants and exhibit fast growth rate *i.e.* 2.5 kg in six months rearing (Sinha *et al*, 1973) in a polyculture system and can attain a weight of upto 50 kg. Silver carp, a phyto-planktophagous fish is also an easy and fast growing fish that has been reported to grow upto 20 kg (Huet, M. 1986). Considering the species wise global production of cyprinids, silver carp contributed the largest share (3380861 mt) followed by grass carp (3160066 mt) in the year 1999 (FAO Fisheris statistics). It is because of the large-scale culture of these carps in several countries such as China, Taiwan, Thailand, Malaysia, Japan, Ceylon, India, Pakistan, Nepal, Philippines, USSR, Burma, HongKong, Singapore, U. A. R., Israel, Vietnam, Hungary, Poland, Romania, Australia etc. In India fingerlings of silver carp and grass carp were brought to the pond culture division of CIFRI, Cuttack in the year 1959 from Japan and Hongkong respectively. Ever since this, through rearing and seed production of the above stock, these fishes

Fig. 2.14: The Two Chinese Major Carps Used Extensively for Culture.
(a) Siver Carp (*Hypophthalmichthys molitrix*),
(b) Grass Carp (*Ctenopharyngodon idella*)
(Courtesy, CIFA, Kausalyaganga)

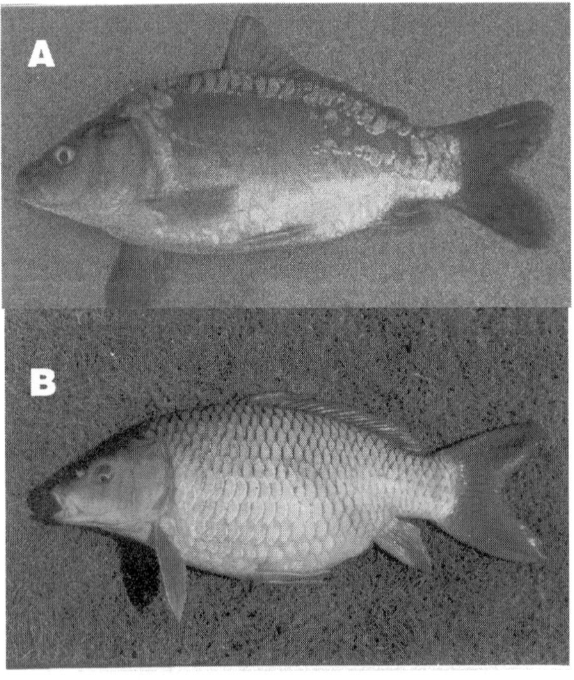

Fig. 2.15: Common Carp (*Cyprinus carpio*).
(a) Mirror Carp, (b) Scale Carp

are distributed to all parts of this country by this time, Another commonly cultured Chinese carp is the big head *Aristichthys nobilis.* Like Indian major carps, Chinese craps are also seasonal, riverine spawners. They do not breed in ponds.

Reproductive Biology

Age and size at first sexual maturity varies from place to place depending on the temperature and the climatic conditions. Female silver carp attains sexual maturity on attaining 2-3 years (2-6 kg) in South China, 4-5 years (2-6 Kg) in Central China, 5-6 years (2-6 Kg) in north east China, 2-3 years (> 2 kg) in Taiwan, 6-9 years (6-8 kg) in Rumania. Males usually become mature one year earlier than female. (Kuronema, 1968). Similarly, grass carps also attain sexual maturity on attaining 4-5 years (6-8 kg) in south and central China, 6-7 years in north east china, 4-5 years in Taiwan, 4-7 years in Europe, 8-9 years in Siberia and Ukraine and 10 years in Moscow region of USSR. In India, the original consignment of silver carp brought from Japan attained sexual maturity on attaining 2 years, but their progeny became mature on attaining 11 months. In case of grass carps females of the first stock brought from Hongkong matured within 3 years where as male matured one year earlier.

Fecundity varied from place to place and depending on the age and size of the fish. The ova measures 1.2 to 1.3 mm in diameter and fecundity has been estimated at 82 ova per g body weight in grass carp and 161 to 170 ova per g body weight in case of silver carps.

Breeding season also varies from one geographic region to another. Breeding season of grass carp is April to August in central China, May to July in Taiwan, June-July in USSR, March to August in India and March to October in Malaysia. For silver carp, it is April to July in China, June to July in Japan, and May-July in India.

Breeding Techniques

Prior to the popularisation of hypophysation technique riverine spawn collection was extensively practiced in the mainland China. By hypophysation technique, silver carp was first bred in China in the year 1955, and grass carp was bred in 1958. Later these fishes were bred in countries like USSR, Taiwan, Japan in 1956 and in India during 1962 (Alikunhi *et. al.*, 1962 and Chaudhuri *et. al.*, 1966). The basic principles and practices for hypophysation of Chinese carps are similar to that of Indian major carps described earlier. The dose of hormone (pitutary gland extract) administered and interval between initial and resolving dose varied in different places. In China, 3 to 3 1/2 common carp pituitary glands per Kg body weight or 2-3mg acetone dried pituitary per Kg body wt is given either as a single dose or in 2 doses. If in 2 doses, the first dose was only 1/3 or 1/4 of the total and the balance is given as second dose. HCG was found ineffective in grass carp. Courtship was noticed 6-8 hours after the resolving dose. Usually, the injected breeders were stripped and artificial insemination was done by mixing the milt from the male with the eggs taken from female. In USSR, the initial dose administried was 2-4 mg of dry pituitary followed after 24 hours a resolving dose of 20-35 mg (Konardt, 1968). According to Vinogradov (1968), dose may be adjusted according to the weight of female – a female weighing 6-8 kg. is given initial dose of 2.5 to 3mg. and resolving dose of 17 to 21mg. In females of weight 10-12 Kg, resolving dose is increased to 25 to 35 mg carp pituitary extract. Males are injected only at the time when female receive resolving dose. Here also the artificial insemination by hand stripping was practiced. In India also, various doses have been tried by different workers from the year 1962 onwards. On an average 10-14 mg of pituitary gland per kg of female was administered in female where as the male received only 2-5mg/kg when the female was given the resolving dose. Artifical insemination by stripping was also carried out about 6-8 hours after the resolving dose.

The procedure for stripping and artificial fertilization shall be same as that of common carp described else where. The ova of silver carp are bluish in colour and measures 1.2 to 1.3 mm in dia. and water hardening is completed within 30 minutes. The water hardened egg measures 4.2 to 4.7 mm in dia incubation period is 18 to 20 hours at 27-29^{0}C. The eggs of grass carps are yellow to deep golden brown in colour and measures 1.2 to 1.3 mm in diameter. The water hardening is completed within half hour and the same measures 4.5 mm in dia. The incubation period may be 20 hours at 27-29^{0}C.

Though, invariably, stripping and artificial insemination was practiced for the seed production of grass carp and silver carp in India and else where (Bhimachar, B. S. and Tripathy, S. D. 1972) a few workers have reported partial natural spawning in grass carp after hypophysation (Choudhury *et al*, 1966; Choudhuri *et al.*, 1967). Further, hand stripping was not always successful as sometimes stripping was done at an improper time. There was also poor survival of spent brood due to stripping stress. During recent years, a few workers (Rath, S. C. *et al*, 1999) have succeeded in breeding grass carp in ecohatchery after hypophysation of breeders reared on' foliage - free diet' which is described below:

Brood Husbandry

Two ponds of 0.1 ha. size were stocked with grass carp yearlings of 0.4 to 0.5 Kg weight along with four other carp species such as catla, rohu, mrigal and silver carp in a ratio 70:10:10:7:3 at a stocking density of 1000 Kg per ha. These ponds were completely free from all types of aquatic weed. These fishes were fed on a formulated feed prepared from soy bean, GOC, rice bran and fish meal (10:5:4:1) containing 36. 75% crude protein, 6.8% crude fat, 14.25% crude fibre 14.25 total ash 13%. The fish were fed 2-3% of the body weight per day. The soaked food was kept in trays suspended in the water column. 20-30% of water was replenished once in a month from January to March. Another two ponds of same size were stocked with the fish of the composition and density as given in pond I and II (Experimental group - NFD group). In these ponds grass carp was fed on the aquatic weed *Hydrilla verticillata* adlibitum. (FD group) The proximate composition of foliage was crude protein 14.1%, crude fat 6.5%, ash 19. 3%, crude fiber 6.9%. Water was also replenished once in a month from January to March.

The breeders of 2+ years age were selected from the above ponds after observing the maturity status of the breeders using catheter. Males with the oozing milt having spermatocrit value of more than 90% were selected. 35 sets from control group and 17 sets from experimental group were used for induced breeding. Transportation of brooders were done using hammocks with water to avoid stress. Ovaprim was administered to breeders by intraperitoneal injection at the recommended dose and released into the spawning pool. The breeding was noticed 9 hours after the ovaprim administration. The eggs were collected in the egg collection chamber and after prophylactic treatment were incubated in the jar hatchery.

The above experimental study indicated that though the growth rate of grass carp fed on foliage diet was double that of non foliage diet group (average weight of FD group was 5.02 ±0.48 Kg whereas that of NFD group was 2.45±0.4 Kg) the rate of breeding responce was 88. 2% in NFD group whereas in FD group it was only 20%. Spawning fecundity of FD group was 0.47±0.13 × 10^5 per kg body weight whereas in NFD group it was 0.9±0.15 × 10^5 per kg body weight. The rate of fertilization was also more in NFD group (89. 6±5. 33) where as in FD group it was only 68. 75±23. 1.Spawn recovery rate was also better in NFD group (0.76±0.13 × 10^5 per kg body weight) compared to 0.25±0.08 × 10^5 per kg body weight in FD group (table 3). These workers have further observed that grass carp preferred slow water current *i.e.* 2 to 3 meter/sec and less water depth (0.5 to 0.6m) for spawning.

Table 2.3: Induced Spawning (with ovaprim @ 0.5 ml per kg Body Weight) of Grass Carp *(Ctenopharyngodon idella)* Reared on Foliage Diet (FD) and Nonfoliage Diet (NFD) in the Spawning Pool of Chinese Circular Hatchery (*modified from Rath S. C. et. al., 1999*)

Diet	*Breeding*	*Number of Females Used*	*Number of Females Bred*	*Percentage of Breeding Response*	*Spawning Fecundity × 10^5/kg Body Wt. ± SD*	*Percentage of Fertiliza-tion ± SD*	*Spawn Recovery × 10^5/kg Body Wt. ± SD*
FD	First	35	7	20	0.47±0.13	68. 75±23. 1	0.25±0.08
					(0.3-0.7)	(33-90)	(0.15-0.4)
	Second*	9	–	–	–	–	–
NFD	First	17	15	88. 2	0.9±0.15	89. 6±5. 33	0.76±0.13
					(0.6-1. 1)	(74-96)	(0.5-0.950)
	Second*	7	4	57. 1	0.66±0.15	82. 75±6. 8	0.45±0.11
					(0.5-0.85)	(76-92)	(0.35-0.95)

* Second breeding 70 days after the first breeding.

Figures in the bracket indicate range.

BREEDING OF COMMON CARP

Introduction

Common carp, (*Cyprinus carpio*), is a major carp cultivated extensively in tropical, sub-tropical and temperate regions of the world because of its high adaptability to varying climatic and environmental conditions as well as to divergent feed availability. It grows in cold as well as in warm waters and tolerates salinity up to 7ppt. It is omnivorous and feeds on food items as per its availability. It prefers bottom fauna and flora - it stirs up vigorously bottom of pond for worms and other benthic fauna. Common carp alone contributed to 7.98% of the total aquaculture production and 14.2% of the total fresh water fishes cultured in the year 1998. Though opinions differ among scientists regarding the original habitat, it is assumed that Central Asia within latitude 35°-50°N and longitudes 30°-135°E is the original home of this species. It is a hardy fish that can also tolerate varying oxygen levels in the water. It can withstand injuries and has inherent capacity to regenerate injured tissue. Apart from this, unlike other major carps, the common carp breeds in confinement, as a result of which breeding and seed production becomes easier and cheaper. In view of these features, common carp is cultured world wide

There are three varieties of common carp based on the pattern of scales *Cyprinus carpio Communis* (scale carp) Body is completely covered by regularly arranged scales (Fig. 2.15). *Cyprinus carpio specularis* (mirror carp) Body is unevenly covered with a few large and bright scales leaving a large area of the body nacked. (iii) *Cyprinus carpio nudus* (leather carp) Body is completely devoid of scales, a row of degenerate scales may be present along the base of dorsal fin or at the base of other fins.

Reproductive Biology

Age and size at first sexual maturity depends on the temperature and other climatic factors of the environment. In the plains of tropical countries like India, where temperature varies from 18°-35°C the females become mature within 6-8 months where as males mature about 2 months earlier. But in the hilly areas of India it takes one year to attain sexual maturity. In Europe where temperature varies from 15°-18°C, it takes 3-4 years, In USSR at 17-19°C, it takes 2-5 years, Indonesia at 19°-30°C it takes 1 to 1½ years. In India, the common carp has 2 peak breeding seasons such as (1) January to March (2) June to August, though mature males and females may be found throughout the year. If they are provided with adequate space and food, males can be found oozing throughout the year. Another interesting feature of this fish is the large size of testis in male, right testis is usually smaller than the left. Males are usually smaller in size than the female. A mature testis may form 20-32% of the total body weight.

During breeding season, the females are easily distinguishable by the swollen and soft abdomen, and a prominent protruding fleshy vent, the distal margin of which is gently cleft at the middle. The males also may exhibit bulging abdomen because of the large size of testis, but pectoral fin and tubercles on the scales can be the distinguishing features. The fecundity also varies with the age and size of the female. Size of the oocyte varies from 1.0 to 1.4 mm and fecundity has been estimated to be 180000 per kg. of fish.

Natural spawning is done in confined waters, spawning in shallow, marginal weed infested areas. It lays sticky eggs and the same are attached to floating weeds. Perivitelline space is very narrow. Yolk is yellow to light brown in colour.

The incubation period depends on water temperature. It is 7½ days at 12°C, 3½ days at 20°C and 2 days at 28°-32°C. The newly hatched larvae remain attached to weeds for 2-6 days till the yolk sac is fully absorbed. After the yolk sac is absorbed it starts feeding on zooplankton.

Artificial Propagation of Common Carp

There are three methods to produce seed of common carp:

(a) Collection of eggs by providing egg collectors in the spawning tank (seminatural breeding).

(b) Stripping the female to collect the egg and fertilizing artificially by the milt collected from adult male.

(c) Breeding in eco -hatchery system.

Collection of Eggs by Providing Egg Collectors in Spawning Tank (Seminatural Breeding)

This is adaptations of the natural spawning habit of the fish. These are of 4 types: (a) Dubisch method (b) Sundanese method (c) Chinese method (d) Indian method.

Dubish method

This is named after the fish culturist Dubisch and it is practiced in some parts of Europe and Indonesia. In this method, grass is grown on the bottom of a dry pond to a height of about 40cm. Then the pond is filled till the water reaches the top of the grass. After prophylactic treatment, by dipping in 2.5% salt solution, breeders are released into the pond - one female along with two males. Usually in Europe Commoncarp breeds during the period from middle of May to July. When the carp breeds the eggs are attached to the grass. After the spawning is over, the breeders are caught and removed. The eggs are allowed to hatch. When the hatchlings reach post larval stage, they are removed and released into the prepared nurseries.

Sundanese method

In the Dubisch method described above, netting out the adult carp from spawning tank may cause disturbances and damage to the eggs that are attached to the grass. Hence in this method, practised widely in Indonesia, spawning tank is provided with egg collectors, called kakabans (Fig. 2.16). During spawning, the eggs get attached to the kakabans and after spawning is over, kakabans containing eggs are removed to a hatching tank.

Spawning ponds are elongate having an area of 25 to 30 m^2 in area and with a hard bottom devoid of mud and silt. If the pond is muddy it is dried for a few days before filling. Brood fish is released in the afternoon after4 kakabans are placed in position. Thus in this method practised extensively in Java and Indonesia maintenance of brood fish breeding and hatching are done in separate tanks. Breeders are segregated sex wise and kept in separate ponds on artificial diets.

Kakabans are made of horse hair like fibres of Indjuk plant *(Arenga pinnata/A. saccharifera)*. The indjuk fibres are soaked in water for 5 days. Then the fibres are arranged into thin layer (1. 2 to 1.5cm long) and it is pressed longitudinally between two split bamboos (4-5cm wide). Margins of the fibre is trimmed to produce even end. Kakabans are placed transversely on a long bamboo pole. The long bamboo pole is held in place between 2 pairs of shorter poles, fixed at the bottom at both ends of the pond. The kakabans are placed in such a way that fibres of adjacent kakabans just

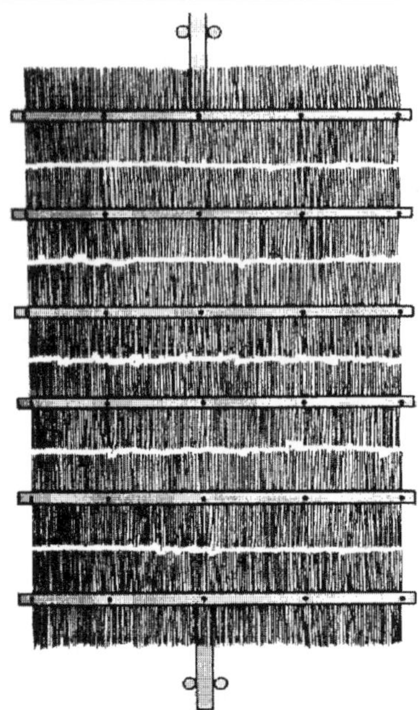

Fig. 2.16: Kakabans Used for Common Carp Breeding.

Fig. 2.17: Hapa Incubation System of Common Carp.
(a) A Series of Hatcing Hapas with Hydrilla Twigs as Egg Collector, (b) Developing Eggs on Hydrilla Twig.

touch. As the bamboo pole is floating, the whole structure moves freely with changes in water level.

While spawning takes place, a gentle flow of water is maintained. As the freshly laid eggs get attached to lower surface, kakaban can be turned over. When both sides are full, kakaban is placed in hatching ponds, which are usually 20 times the area of spawning tank. Five to eight pieces of kakabans are required for a female of 1 kg weight.

Before transfer to the hatching pond they are gently washed to ensure removal of mud or debris that may be attached to the eggs. In hatching ponds, kakabans are placed transversely on floating bamboo pole leaving a gap of 5-8cm between fibre of adjoining kakabans. Kakabans with eggs attached on both surfaces are pressed at least 8 cm below water surface by bamboo poles placed across the ends of kakabans parallel to central pole, held in place by bamboo pieces kept at each end of the pond parallel to kakabans. The pond has a number of outlets at different levels to regulate outflow of water. Eggs hatch out and on 7th day, artificial feeding is started. When fry becomes three weeks old, they are collected

In certain parts of Indonesia such as central Sumatra, indjuk fibres are scattered in the spawning ponds and the eggs are attached to this. The breedes are removed from this and eggs are allowed to hatch, spawn is collected when 5 days old.

In some parts of Indonesia, spawning enclosure is a compartment in a larger pond, separated from the pond proper by a temporary, earthern dyke. When spawning is completed, the dyke is cut opened and spawners are allowed to escape and then the dyke is closed again. After seven days, the dyke is opened again and larvae are allowed to escape into the pond. This is called Tjimindi method.

Indian method

Breeding is done in hapas and *Hydrilla* is used as egg collectors. One female brood fish along with 2-3 males are introduced into the breeding hapa, during evening hour. Aquatic plants such as *Hydrilla* is thoroughly washed and then released into the breeding hapa. 2 Kg of *Hydrilla* is required for 1kg weight of female. Within 6 - 10 hours after being released into the hapa, fish spawns in instalments and the fertilized eggs get attached to *Hydrilla* (Fig. 2.17). In the morning, the egg collector (*Hydrilla*) is removed from breeding hapa and released into hatching hapa. One hatching hapa can hold atleast one kg of egg collectors and it will have 40000-100000 eggs. After hatching, spawn is collected and measured and then released into nursery tank for rearing to fry stage.

Chinese method

Chinese practiced a modification of Dubisch method. Plants like Ceratophyllum, Eichornia etc are used as egg collectors. The plants are placed inside a square bamboo frame so that weeds do not stray away from spawning area.

Stripping and Artificial Fertilization

Though common carps breed naturally in ponds and seminatural breeding techniques utilising egg collectors are practiced for seed production in several parts of the world, stripping the male and female for collection of gametes and artificial fertilization by mixing male and female gametes and thereafter incubation in incubation units has been practiced by several workers like Woynarovich (1969) of Hungary.

Female carp breeders are anaesthetised using MS 222 and then a suture is made in the vent. Then they are injected with 2.5 to 3.5mg of pituitary/kg body weight. They are released into hapa fixed in the water. Then the males are also given a lower dose of pituitary extract and released along with female. It takes 8 to 26 hours for the females to become ripe. When the females are ripe, they are stripped by pressing the belly and eggs are collected into a plastic tray. The male is then stripped and the collected milt is mixed with the eggs for fertilization. It is necessary to remove the stickiness of the eggs to enable proper handling and successful hatching. If stickiness is not removed eggs will stick to each other and get clumped up. To start with, the eggs are washed with increasing quantities of salt-carbamide solution after which they are washed in weak tannin solutions. Tannin stops the swelling of eggs, so it is applied only when water hardening is completed.

Salt-carbamide (fertilization) solution is prepared from 30g. of carbamide and 40gm of NaCl in 10 litres of clean pond water. The solution is poured over the egg mixed with milt and the mixture is stirred with a plastic spoon or feather for 3-5 minutes. Subsequent stirring can be done with hand. The eggs absorb the solution and begin to swell. Add fresh quantities of salt-carbamide solution intermittently. The eggs swell ten times of its original size during swelling.

After 1 to 1½ hours, swelling is completed. A part of the salt-carbamide solution with dissolved sticky materal is drained off during the course. If at this stage, the eggs are transferred to fresh water, eggs may tend to stick together again in clumps. Hence in order to get rid of the sticky material completely, it is necessary to wash the eggs 2-3 times with fertilization solution. The eggs are then transferred to tannin solution having 5-8g. tannin in 10 litters of water. This solution is freshly prepared every time. Add 2-3 litters of swollen eggs to 2-4 litters of tannin solution kept in a plastic bucket. After stirring for 3-5 seconds, pour clean water into the bucket. After the eggs settle down, water is drained out, preferably by using a strainer. A small quantity of tannin solution is poured into the bucket and after stirring briefly, clean water is to be added followed by draining. Tannin is very harmful if it remains in contact with eggs for longer duration. Therefore eggs must be washed in clean water repeatedly.

Salt-carbamide solution not only helps in removing the stickiness but also increases the viability or life span of sperm. Sperm remains active for 20-25 minutes in carbamide solution where as in fresh water, sperms remain active/viable only for a maximum of two minutes. It also dissolves the sticky material closing the micropyle where as in eggs with the sticky material, micropyle is closed within one minute.

Cream-milk Method for Removing Stickiness of Eggs

Apart from the salt-carbamide solution described above, a solution of milk powder has also been used for removing the stickiness of common carp eggs. 20g. of full cream milk powder (fat content 26-28%) is dissolved in one liter of water and this solution is used for the treatment of fertilized eggs. This is equivalent to a solution of

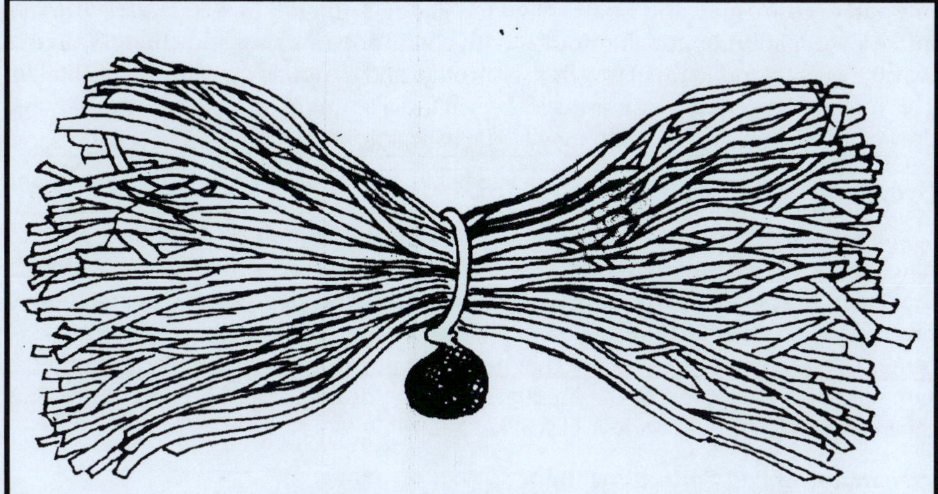

Fig. 2.18: A Bundle of Synthetic Fibres Tied in the Middle Used as a Egg Collectors of Common Carp.

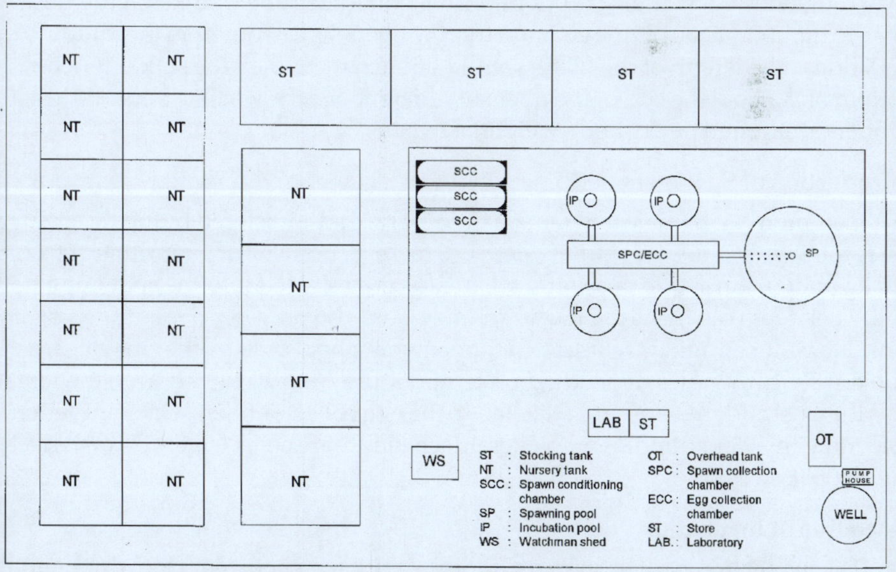

Fig. 2.19: Layout Plan of a Model Carp Hatchery Complex (Not Drawn to Scale).

one part of cow's milk and 9 parts of water. The egg-milt mixture is treated with this milk powder solution for 45 minutes with continuous stirring and changes. In this way the eggs are found to get free from each other and are found suitable for incubation. This method is preferred by many workers in India as this treatment does not damage the fertilized eggs and it is also easy for farm practice.

Breeding in Ecohatchery System

During recent years some workers have used the Chinese circular hatchery (eco hatchery) for the breeding of common carps also. Breeding procedure is similar to that of Indian major carps but with some additional devices and management practices. In this method bundles of synthetic fibers peeled from sacs used for packing cement, fertilizer etc. are provided in the spawning pool for egg collection and the same are also incubated in the incubation pools for hatching. The procedure is as given below (Rath, S. C. and Sarkar, S. K., 2002).

Preparation of Egg Collection Fibers

Synthetic fibers are peeled out from sacs and cut into uniform size of 3 to 4 feet length and are washed thoroughly. They are grouped into bunches of approximately 50gm wt. and tied at the center, keeping both ends of the fiber free. A sinker of desirable size is fixed to the central knot before putting the same in the spawning pool (Fig. 2.18).

Preparation of Spawning Pool

Several bunches of synthetic fibers prepared as noted above are put in the spawning pool. Soon after it is put in the water because of the sinker at the center it sinks to the bottom, but as the ends are free the fibers spread out from the middle to all directions. The length of the fibers should be such that when the sinker touches the bottom of the pool the end of the fiber should be above the water. About 300 to 400 g of fibers is required per kg body weight of female.

Introduction of Spawners

Common carps (male and female ratio 1:1) are administered inducing agents at the recommended doses (pituitary extract 2 to 3 mg/body wt or ovaprim @. 1 to. 2 ml per kg body wt in single dose only to female) and released into the spawning pool. The water depth can be 1.5 to 2 ft. and sprinklers are also provided above the spawning tank. No water current is required in the spawning pool unlike other major carps. In due time (which varies depending on temperature from 10 to 15 hours) the female lays the eggs and the same are attached to the egg collector fibers. On completion of spawning this fiber bundles with eggs attached to it are collected and transferred to the incubation pool.

Operation of Incubation Pool

The incubation pool is operated as usual with water current from duck-mouth taps located at the floor and also sprinklers from above. The eggs hatch out after 30 to 48 hours depending on temperature. After hatching fibers with the dead eggs and egg shells of hatched out ones are moved leaving the hatchling in the hatching pool. Spawn recovery has been reported to be as high as 90-92% in this system.

Using synthetic fibers of sacs as egg collectors have been reported to be advantages as these fibers are easily available at nominal cost and there shall be no oxygen depletion in the system. Other egg collectors such as *Hydrilla* consume oxygen and deplete oxygen of the water which may affect the egg development. Further, these fibers do not undergo disintegration and putrefication in water easily and it can be used again and again cleaning each time after operation.

ADVANCES IN CARP BREEDING

As mentioned earlier the prime requirement for producing quality fish seed in the hatchery is the availability of good quality parental stock. A good quality seed is one which will have (I) faster growth rate (ii) better feed conversion efficiency (iii) high disease resistance and (iv) better adaptability to wide fluctuations in environmental factors such as temperature, dissolved oxygen and other stress conditions. So a fish breeder should carefully select the male and female brood fish for breeding from different stocks to avoid mating between offsprings of the same parent and also between parents and offsprings. This prevents 'inbreeding depression' which is the cause for production of weak and defective offsprings. Genetic drift is to be avoided by chosing effective number of breeders of different age groups. Mixed breeding i. e breeding between different species or genera is to be prevented by releasing only fish of one genus into the spawning pool at a given time for breeding. This prevents any inadvertent cross breeding and consequent genome pollution. Further, in a hatchery, economy measures have to be practiced to reduce the cost of production. Utilizing the same brood fish repeatedly in the same year to get maximum number of progeny is one way of achieving this goal. This is called as multiple breeding technique. Another method is to cryopreserve the gametes and to utilize the same for artificial fertilization as and when the seed is required. Another breakthrough in this field was the formulation of synthetic compounds and their use to induce spawning in carps and other fishes.

Multiple Breeding of Major Carps

Though hypophysation technique and subsequent advances in inducing spawning through administration of synthetic compound have revolutionalised the carp seed production technique, its applicability only during the monsoon season remained as a major constraint in the expansion of aquaculture as the availability of seed for stocking was limited to the monsoon months. By this time scientists have successfully induced the same fish to breed four times in an year-twice in premonsoon months (March-May) at an interval of 40 to 45 days, and once in late monsoon months (August and September) in addition to monsoon months *i.e.* June-July (Gupta S, D. *et al.*, 1995). Thus, a given brood fish can be bred 4 times twice during pre- monsoon, once during monsoon and another time during late monsoon times. This was achieved through careful pond management practices, selection of brood stock and scientific feeding practices.

Selection of Brood Stock for Multiple Spawning

Breeders that have bred atleast once during the previous season are selected for undertaking breeding works as 'virgin breeder' is not found suitable. Such breeders are named as 'professional breeders'. (Gupta *et al.* 1995). Age of the fish should be 2 to 5 years as overaged fish do not respond well to inducing agents. They are stocked @1000 kg/ha and fed an a formulated diet fortified with minerals and vitamin @ 1 to 2% of the body weight (for further details on brood stock raising and rearing please see the desription under brood husbandry in artificial propagation technique). 25% to 30% of the water is replenished at least once in a month from February to May with fresh canal or reservoir water. Other management practices can be of routine nature. Such breeders are found to mature at least three months earlier.

Handling of breeders used in multiple spawning has to be done in such a way as to cause least stress to the animal as noted below.

(a) Breeders are to be transported in the canvas bags (hammocks) in water.

(b) Intraperitoneal injection is preferred as it provides least injection stress when compared to intra muscular route.

(c) In the spawning pool, only the required flow of water should be maintained and duration of water flow also should be controlled to the required time. In case of administration of pituitary gland extract, water flow may commence 4 hours after final dose and in case of ovaprim given as single dose, water flow may start five hours after the injection. The flow of water has to be continued for two to three hours as within this period effective spawning is found to be over.

(d) As soon as the spawning is over, the spent breeders must be removed from the spawning pool. They are dipped in 5 ppm potassium permanganate before releasing in pond. Further at regular intervals they are checked and treated with the said solution to avoid secondary infection or for quick recovery.

(e) As they are used for subsequent maturity of ovum, feeding and other management practices are to be done meticulously.

Premonsoon Breeding

This commences as early as March. The spawn yield is reported between 0.5 to 0.6 lakh spawn/kg body weight of fish. The spent breeders are maintained for 40-45 days and used again for induced spawning when the production of spawn was 1 to 1.5 lakh spawn/kg body weight.

Monsoon Spawning

After a lapse of 40-45 days of second spawning, a third crop can be obtained during June-July. As this is the natural spawning season of major carp the spawn yield has been seen to increase to 1.5 to 2 lakh spawn/kg body weight.

Late Monsoon Spawning

Though the fish breeds during the August - September period, if maintained in a stress free environment and following the precautionary measures mentioned earlier, the yield of spawn will be the least (0.4 to 0.5 lakh/kg). It has been also observed that the percentage of rematuration of individual for 2nd, 3rd and 4th breeding was 100%, 79-100%, 11-30% respectively in *Catla catla.*

Care of Larvae in Multiple Breeding

Special care is recommended for spawn produced through multiple breeding. It is found that yolk of spawn get absorbed within sixty hours after hatching but in case of traditional spawn it take 72 hours for complete absorption of yolk. So it is necessary to provide feed after sixty hours. Intermittent sprinkling of prophylactics is also recommended in spawnery to maintain hygeinic condition.

Thus, through multiple breeding technique, the yield of spawn per kg body weight of breeder is enhanced 3-4 fold than single breeding and cost of production of seed is also reduced. This also provides seed for stocking during season other than monsoon and even for multiple cropping in the culture practices of a given water body.

Cryo Preservation of Carp Milt

Though cryopreservation of milt of several species of fishes has been successfully practiced, the same was not standardised for indian major carps till recently. (for details on the principles of cryopreservation please refer to the chapter on cryopreservation of fish gametes).

Cryopreservation Protocol for Carp Milt

The healthy male breeders are selected during peak breeding season by gently pressing the belly to see the oozing milt through the vent. Such males with oozing milt are called milters. Then the males are subjected to hypophysation. Within three to four hours of hormone administration, the milt from the male is collected by stripping into an ice cold sterilised tube. Yield of milt after hypophysation is 6-10 ml/kg body weight. But without hormone treatment yield is only 0.5 to 1 ml per kg body weight. Care is taken to avoid contamination of the milt with urine, mucus, faecal matter etc. The milt is then kept in a refrigerator. The quality of the milt should be assessed for sperm count and sperm motility by microscopic examination and through spermatocrit value. If more than 80% of the spermatozoa are motile showing jumping forward movement, sperm count is between 2×10^7-2. 5×10^7/cu mm and spermatocrit value is 70 it is considered as good sample. Then the selected milt is diluted by a mixture of extender and cryoprotectant.

Preparation of Extender

Out of the several extenders cited in literature three are used for carp milt which are as follows:

A	:	NaCl-400mg, $CaCl_2$-23 mg, KCl-38mg, $NaHCO_3$-100mg, NaH_2PO_4-41mg Mg SO_4-23mg in 100ml double distilled water.
B	:	NaCl-600mg, $CaCl_2$-23mg KCl-38mg, $NaHCO_3$-100mg, NaH_2PO_4-41mg Mg SO_4-23mg in 100ml double dist. water.
C	:	NaCl-750mg, $CaCl_2$-20mg, KCl-20mg, $NaHCO_3$-20mg, in 100ml double distilled water.

The above solution can be kept in a refrigerator for one month. Diluents are prepared by mixing cryoprotectants such as Dimelthyl sulphoxide (DMSO)/Glycerol with extender as shown below:

Diluent I

DMSO 5ml

Glycerol 10ml

Extender A 90ml

Diluent II

Glycerol 10ml

DMSO 8ml

Extender B 90ml

Diluent III

DMSO 15ml

Extender C 65ml

The mixing of cryoprotectant with extender is an exothermic reaction and hence the diluant is kept in a refrigerator to maintain isothermal condition. Any one of the above diluents can be used for mixing the milt. Usually the ratio of milt and diluent is maintained at 1:4.

Next step is the equilibration of milt - diluent mixture. This mixture is kept at 8-10°C for 45 minutes for equilibration. The equilibrated milt-diluent mixture is filled in 'french straw' (0.5ml) or visotubes (5ml) and sealed. In case of french straw, it is fumed over liquid nitrogen for ten minutes and then stored in liquid nitrogen at -196°C. If visotubes are used, it can be cooled at 15°C/minute by a programmable cooling chamber. The success of cryopreservation depends on the optimum concentration of cryoprotectant, equilibration time and freezing rate.

Whenever required, reactivation of the sperm can be done by thawing. In case of carp milt, fast thawing is preferred as slow thawing cause crystallisation in the sperm cells. The french straws are made to slush swrirling them in a water bath at 38 $\pm$ 2°C for 7-9 seconds. Then put a few drops of fresh water into the milt. This reactivates and restores the mobility of the sperm. Visotubes may need 65-70 seconds for slush formation. It has been calculated that a visotube containing 3ml thawed diluted milt is sufficient to fertilize 1 to 1.5 lakh eggs. The fertilized eggs are to be incubated in the hatching pool maintaining flowing water.

Cryopreservation of carp milt has made the gamete exchange programme possible for evolving superior stook for breeding and culture practices by hybridising different strains irrespective of distances. Application of this technique might help in overcoming 'inbreeding depression' very often noted in the present day aquaculture practices. This technique also helps in making the male gamete available for fertilization in case of fishes where male and female mature at different time. It also reduces the cost of male brood stock maintenance by exploitation of limited stock.

Synthetic Compounds Used for Induced Spawning

Though Hypophysation technique has revolutionised the aquacultural practices, ever since its adoption on a commercial scale, pisciculturists faced several problems in course of time. (i) Good quality pituitary glands were not available in sufficient quality in time (ii) The procedure of preparation of crude pituitary gland extract (CPGE) and administration to the fish in field conditions particularly during monsoon season involves cumbersome procedure (iii) Crude pituitary extract being a mixture of hormones may have adverse side effects on gametogenesis and other functions (iv) The pituitary extract is variable in potency and has short storage life (v) Species specificity has been shown for fish gonadotropin, for example salmon gonadotropin is ineffective on sea bream. (vi) CPGE has less percentage of successful spawning, less number of eggs per kg body weight and less fertilization rate and hatching rate. This led to a search for alternate synthetic hormone preparations that could be preserved and used as and when required by pisciulturist.

Synthetics compounds used for induced breeding of fishes under captive conditions are of four categories (1) Gonadotropins (2) Gonadotropin releasing hormones (GnRH) or (Lutenizing hormone releasing hormone-LHRH or LRF) and their analogues (GnRH-A/LHRH-A/LRF-A) (3) Steroids and (4) Other drugs.

Gonadotropins

Gonadotropins are hormones secreted by the gonadotrophs of pituitory gland. Though, two distinct gonadotropins such as FSH (Follicle Stimulating Hormone)and LH (Luteinizing Hormone) are reported in higher animals, many scientists believed that there is only one gonadotropin in fish performing the function of both FSH and LH. Recently, several studies have revealed the presenceof two gonadotropins GtH I, and GtH II in fishes. GtH I is concerned with follicle development and vitellogenesis whereas, GtH II is concerned with final oocyte maturation, ovulation and spawning. As a surge in Gonadotropin II (GtH II) has been shown to cause ovulation and spawning in fish, synthetic gonadotropins were the first to be tested for induced breeding. Gonadotropins tested for its efficacy in induced spawning are (1) mammalian luteinizing hormone (LH) (2) Human chorionic gonadotropin (HCG) and (3) Pregnant mare serum gonadotropin (PMSG). All gonadotropins are glycoproteins, each molecule consisting of two subwrits such as a and b-subnits. a-subunits of all molecules are similar whereas, differences exist between b-subunits of different gonadotropin molecules.

Partially purified or purified form of gonodotropins have been prepared from the pituitary glands of teleosts such as common carp, pink salmon, rainbow trout, tilapias, winter flounder, pike eal etc through extraction and chromato graphic procedure. But this has been produced in limited quantities and hence were used only on an experimental scale, except perhaps partially purified gonadotropin of pacific salmon (SG – G1000) which was produced in larger quantities.

Luteinizing Hormone (LH)

Ovine LH has a molecular weight of 28500 and has 15.5% carbohydrate b subunit has 119 aminoacids. But use of gonadotropins II, though effective in inducing spawning is not cost effective and hence not practised in fish culture.

Human Chorionic Gonadtropin (HCG)

HCG is secreted by the Langerhans cells of the chorionic villi of human placentra and it appears in the blood and urine during early pregnancy shortly after the implantation of blastocyst in the wall of uterus and reaching a peak between 50th and 60th day of pregnancy. After this peak blood and urinary titres of the hormone drop to low levels and at the level it remains constant until parturition.

Chemically, HCG molecule resembles LH having an a and b subunit. It has a MW of 36700 and carbohydrate containing is 30-35%. The b subunit has 30 additional residues at carhoxyl terminus. Though primary function of HCG is maintenance of pregnancy because of its similarity with LH molecule, it has also a gonad stimulating effect if injected to other animals. Hence HCG under different brand names (Prolan (CIBA, Organon, Bayer) Antuitrin –S (Parke-Davis) Gonatropin (Teikokyzoki Co) APL (Ayerst Lales) Sumaach (Infar) have been used for induced breeding of fishes. Effective dose is 100-2000 IU/kg fish for carps.

Synahorin, trade name for a commercial preparation, a mixture of HCG and mammalian hypophysial extract, has been used for inducing ovulation and spawning in fish.

Pregnant Mare Serum Gonadotropin (PMSG)

This is another type of gonadotropin like molecule secreted by the endometrial cups of the uterus of a pregnant mare. It appears in the blood on the 40th day of pregnancy, remains high until 120th day and then it drops and disappears after 180 days. As the molecular configuration is similar to that of LH, having a and b subunits, it functions as a gonadotropin, if injected to other animals. Hence, administration of PMSG has been found useful in inducing ovulation and spawning in fish. Sunderraj and Goswami, S. V. (1966) successfully induced spawning in *Heteropneustes fossilis* with 500 IU of PMSG.

Puberogen

It contains 63% FSH, and 34% LH. Dosage is 50-200 IU/kg body weight for female and 20-25 IU/kg body weight for male.

Gonadotropin Releasing Hormone (GnRH)/Luteinising Hormone Releasing Hormone/Factor (LHRH/ LRF)

Gonadotropin releasing hormone (GnRH) or Luteinising hormone releasing factor (LHRH/LRF) was first isolated from porcine hyopthalamus by Schally and coworkers in the year (1971). Later, in the year 1972, Guillmin and associates isolated it form ovine hypopthalmus. It has been shown to be a decapeptide P glu-His-Tryp-Ser-Tyr-Gly-Leu-Arg-Pro-Gly NH_2 This neuro hormone secreted by the GnRH/LRF neurons of hypophysiotropic area of hypothalamus is regulating the secretion of gonadotropins (GtH) from the pituitary gland.

Out of the nine numbers of GnRH molecules synthesised so far, 4 are from fish (sea bream, catfish, salmon and dog fish). The first fish GnRH to be sequenced was that of Chum salmon, *O. keta* (Sherwood *etal*, 1983)

Sea Bream

1	2	3	4	5	6	7	8	9	10
Pglu	His	Trp	Ser	Tyr	Gly	Leu	Ser	Pro	$GlyNH_2$

Salmon

Pglu His Trp Ser Tyr Gly Trp Leu Pro $GlyNH_2$

Catfish

Pglu His Trp Ser His Gly Leu Asn Pro $GlyNH_2$

Dog Fish

Pglu His Trp Ser His Gly Trp Leu Pro $GlyNH_2$

GnRH Analogs (GnRH-A/LRF-A)

Scientists were successful in synthesizing analogs of GnRH substituting amino acids at position 6, 7 or 10.If glycine at position 6 is replaced by D-Alanine (D-Alanine^6 LRF) the molecule becomes 4 fold more potent whereas a-Alanine ^{6} LRF is only only 0.01 times more potent. D Phe^6 LRF or D Tyr ^{6} LRF are 15 fold more potent, D Tryp^6 LRF is 36 fold more active than original. Des-Gly^{10} (Pro^{9}N-Et)LRF and des Gly^{10} (Pro^9 –N-CH_2 CF_3)-LRF are 4 and 9 times more potent than LRF Des – Gly^{10} (DALa^6, Pro^9-N-Et) LRF is 15 times and des-Gly^{10} (DTrp^6 Pro^9 -NEt) LRF is 144 times more potent.

The high potency of LHRH analogs that are modified at 6th, 7th and 10th positions has been attributed to (1) Higher receptor affinity of the molecule due to changes in overall conflrmation, hydrophobicity or charge characteristics (2) relative resistance of peptide to degradation in the various biological test systems. For example LHRH (GnRH)is slowly degraded by plasma or semen and tissue extracts of Testis, anterior pituitary and and brain rapidly inactivates both biological immunological properties. The bond between Gly^6 and Leu^7 was cleaved by brain extracts. But LHRH analogs with D-ala^{9}Net modification were resistant to cleavage. They also exhibit prolonged activity *in vivo*. For example, 10 mg of D-Trp^6 LRF causes two fold higher plasma gonadotropin than 10 µg of LRF. Thus GnRH analogs are resistant to enzymatic degradation and hence has prolonged life span in circulation.

GtH II release is under dual control. It is stimulated by GnRH and inhibited by Dopamine which is a Gonadotropin release inhibiting factor (GRIF). Dopaminergic neurons inervating the GnRH neurons of hypothalamus inhibits the release of GnRH through dopamine released from nerve terminals. Dopamine may also act directly on the pituitary gland to modulate the action of GnRH as well as release of GtH. Chang and Peter (1983) observed marked potentiation of GtH II release and ovulation in response to GnRH-A along with pimozide, a dopamine receptor antagonist. This finding led to the use of GnRH-A along with dopamine antagonist for induced ovulation and spawning in cultured fish. This procedure is called Linpe method-named so after the scientists who invented this (H. R. Lin and R. E. Peter, 1996).

Further, studies on the relative potency of various analogs of GnRH and various dopamine antagonists have shown that salmon GnRH and domperidone respectively are more effective in most of the fishes. But in a few fishes like Atlantic Salmon *(Salmo salar)* and gill head seabream mGnRH – A is more potent than sGnRH-A. In a few fishes like Atlantic croaker *(Micropogonius undulatus)*, gill head sea bream *(Sparus aurata)* and striped bass *(Morone saxatilis)*, dopamine has been shown to be having no inhibitory effect on GtH II release. In such species GnRH alone could induce spawning. Comparative advantages of analogs are (i) it has longer stability (II) it is effective in various species of fish (iii) the time from injection to spawning is predictable and (iv) there is reduced stress on brood stock as only one injection is required.

Ovaprim

It is a synthetic compound manufactured by Syndel laboratories inc. Vancouver, British Columbia, Canada and marketed in India by Agri vet farma of Glaxo India Ltd., Bombay.

It consists of Salmon gonadotropin releasing hormone analogue (sGnRH-A)-D.Arg^{6}Pro^{9}NEt- and domperidone which is a known dopamine antagonist. Recommended single dose of ovaprim is 0.5 mg/kg of fish. The ovaprim has been shown to be highly successful and advantageous when compared to the traditional method of CPGE administration. Use of ovaprim has resulted in higher percentage of spawning success, higher number of eggs per kg body weight of brood, higher fertilization rate and higher percentage of hatching.

Studies based on 3 year field trial with Indian major carps from 1988 to 1990 have indicated that average number of eggs obtained per kg body weight of brood fish was 11400 with ovaprim as compared to 85000 with CPGE. Average number of fry obtained per kg body weight of brooder was 72000 with ovaprim as against 43000 with CPGE. Thus 40% increase was obtained using ovaprin than CPGE (Nandesha *etal* 1991)

Ovatide

It is a synthetic compound launched by Hemmopharma, Bombay. It is also a combination of GnRH analogue with dopamine antagonist pimozide. Recommended dose of ovatide for major carp is given below

	Male (ml/kg body wt.)	*Female (ml/kg body wt.)*
Catla	0.2-0.3	0.4-0.5
Rohu	0.1-0.2	0.2-0.4
Mrigal	0.1-0.2	0.2-0.4
Silver carp	0.2-0.25	0.4-0.5
Grass carp	0.2-0.25	0.4-0.5

Ovapel

It is developed by university of Godollo in Hungary. It is combination of mammalian GnRH analog D-Ala^6, Pro^9 Net-mGnRzH and water soluble dopamine raceptor antagonist metaclopramide. It is prepared in pellet from, each pellet contains 18 to 20 microgram of mGnRH-A and 8 to 10 mg of metaclopramide. Recommended dose is 1-2 pellet/kg of fish in rohu and mrigal. Pellet can be dissolved in water and injected.

WOVA – FH

It is a synthetic gonadotropin releasing hormone analogue prepared and marketed by WOCKHARDT (Biostat Agrisciences, a division of Wockhardt life sciences Ltd, Mumbai India.) The recommended dose is 0.5 ml per kg body weight of fish.

All the synthetic compounds listed above are found most effective in inducing spawning in fish. Attempts have been made by scientists to deliver the above mentioned combination of drugs (GnRHA + Domperidone) orally along with food to the gravid fish in order to avoid stress to the fish and also to reduce the labour and involvement of scientist/technician in each breeding trial. Thomas and Boyd (1989) found that oral administration of mammalian GnRH analogue in the diet of spotted sea trout *(Cynoscion nebulosus)* induced spawning after 32-38 hrs of feeding. McLean and coworkers in 1991observed that mammalian GnRH and its analogs are absorbed by the gut in coho Salmon after oral administration. Oral delivery of sGnRH- A and domperidone to *Puntius gonionotus* induced ovulation in 88 to 1005 within 48 hours (Sukumasavin *et al.*, 1992) In spite of these observation, till now, a commercially viable formulation for oral delivery of the inducing agent along with feed to induce ovulation and spawning in fish has not been manufactured.

Steroid Hormones

Both cortico steroids such as 11-deoxycortico sterone acetate (DOCA) and progestins such as 17a-hydroxy –20b -dihydro progresterone have been used by some workers to induce spawning in fish.

The first successful use of corticosteroid to induce spawning in teleost was in *Misgurnus fossilis* with a single dose of 1mg DOCA per fish (Kirshenblatt, 1952, 1959) later, Sunderraj and Goswami (1966, 1972) induced ovulation in *Heteropneustes fossilis* with 5mg DOCA per fish. These authors found other steroids such as cortisol and cortisone also effective at a dose of 25 mg per fish. Other works in the use of corticosteroids are that of Hogendorn (1979) in *Clarias lazera* (50mg DOCA per kg fish) and Khoo (1974) in gold fish (100 mg DOCA per kg). Jalabert, *et al.* (1977) found that DOC at 1 mg per kg was ineffective in carps.

Several studies have been conducted to know the efficacy of progestins (17α hydroxy progesterone and 17α 20β dihydroxy progesterone) in inducing ovulation in trouts, (Jalabert e*t el* 1978), common carp (Jalabert, *et al.* (1976) and northern pike (De Montalembert, *et al.* 1978). These studies have indicated that while progesterone derivatives induce final maturation, ovulation occurs only when the endogenous gonadotropin concentration is sufficient or when a minimum dose of gonadotropin is supplied exogenously.

At present neither corticosteroids nor progestins are utilized commercially for inducing spawning in fish.

Other Drugs

Prostaglandins

Prostaglandins have been shown to be involved in the follicular rupture during ovulatian (Jalabert and Szollosi, 1975, Kagawa and Nagahama, 1981) of the different prostaglandin tested, Prostaglandin F2α has been shown most effective. Injection of indomethacin, an inhibitor of prostaglandin has inhibited ovulation in gold fish and administration of prostaglandin to such fish had an overriding effect on such inhibition (Stacey and Pandey, 1975), Further, injection of gonadotropin that induce ovulation has been shown to cause increase in the endogenous prostaglandin in several fishes. In gold fish, injection of PGF 2α (10mg/kg body weight) in to the third ventricle induced female spawning behaviour (Stacey and Peter, 1979). Jalabert *et al.* (1978) observed PGF 2α induced ovulation *in vitro* in rainbow trout in which final maturation was induced *in vivo.* These studies have indicated that prostaglandins may be used in future to stimulate natural spawning in culture condition.

Antiestrogens

Antiestrogens are synthetic compounds that can compete with estrogen for binding on estrogen receptors. There is a decrease in estradiol level during final maturation and ovulation. Estradiol has been shown to have negative feed back inhibition on gonadotropin release. The lowering of estradiol is hence found to enhance gonadotropin secretion. Two compounds belonging to this category are (i) Clomiphene citrate and (ii). Tamoxifen. Clomiphene citrate was first used to induce ovulation in woman as a fertility drug (Greenblatt, *et al.*, 1961). Following this this drug has been tested in teleosts by several workers on an experimental basis (Pandey *et al.* (1973) in gold fish, Pullin (1975) in *Pleuoronectes platessa*, Singh and Singh (1976) in *H. fossilis*, Ueda and Takahashi in 1976 in loach, *Misgurnus anguilicaudatus*, Breton *et al.* (1975) in *Cyprinus carpio*. In spite of success in some of the experimental studies this drug is not used commercially for inducing ovulation in fish.

Tamoxifen, another antiestrogenic compound has been also used to induce ovulation in Coho salmon (Donaldson *et al.*, 1978b, 1981b) Coho salmon primed with low dose of partially purified salmon gonadotropin (0.1 mg/kg) was injected with tamoxifen at a rate of 1mg/kg in one group and 10 mg/kg in another group. There was 100% GVBD and 85% ovulation when compared to 84. 6% GVBD and 77% ovulation in control fish. This compound has been also used to induce ovulation in loach *(Rutilus rutilus)* by Worthington *et al.* (1981) by injection of the same at a dose of 1mg/kg two times at four days interval. In spite of this, tamoxifen is also not used in large scale breeding works.

DESIGN AND CONSTRUCTION OF A MODEL CARP HATCHERY (Chinese circular hatchery/eco- hatchery)

Introduction

All relevant details on the technology of major carp seed production have been presented earlier. This section deals with the planning, design and construction of a model carp hatchery of Chinese circular hatchery or ecohatchery type. A carp hatchery complex should have facilities not only for spawning and incubation of eggs but also for raising brood stock, rearing spawn to fry or even up to fingerling stage. To facilitate marketing there should be facilities for packing and transporting the spawn/fry/ fingerlings. Success of a hatchery depends on proper planning, design and site selection.

Site Selection

Suitability of the site is the first criterion to be investigated before establishing a carp hatchery. Suitability of the site depends on various factors such as availability of a perennial source of good pollution free water, topography of the area, soil type, proper road communication and availability of electricity. Hence a survey has to the conducted to study the above parameters in order to select the suitable site.

The area should be plain or gently sloping and should not be prone to floods during rainy season. Adequate, good quality water should be available either as surface water or ground water. An open well or a tube well can be dug and it should have high water yielding capacity of atleast 20000 liter per hour to meet the heavy water requirement of hatchery. Existence of a river or stream or canal or reservoir in the vicinity of the site is most desirable. The soil for the construction of rearing tank for fry and brood stock has to be of impervious type such as clay or silty clay or clay loam etc. The area should have optimum temperature which shall facilitate good and healthy growth of larvae and adults and also for incubation of eggs, embryonic development and hatching. A study of the rainfall pattern, soil and water quality parameters etc. is to be undertaken before finalising the site.

After the site is selected, area required for different units of hatchery such as brood stock raising and rearing units, spawning units, incubation unit, spawn conditioning and spawn rearing units, packing and marketing units are to be apportioned based on the actual requirements to achieve the target (Fig. 2.19).

Logistic dimensions of the spawning pool, hatching or incubation pool and area of a brood rearing and nursery tanks of a hatchery complex is illustrated below

Calculation of Brood Fish Requirement

The total area of the brood stock rearing and raising ponds shall depend on the number of brood fish required which in turn depends on the target of seed production.

Suppose that an entrepreneur sets the target for a production as 50 million spawn per year. Then the quantity of brood fish required shall be calculated as shown below.

The standardized spawn yield of a major carp is, on an average, one lakh per kg body weight.

So the female brood fish required for producing 50 million spawn shall be 50 million/1lakh = 500 kg.

Assuming that an equal number of adult males are required for fertilizing the female, total brood fish required shall be 1000 kg.

To cover risk factor, add 30% to the calculated figure.

So the total brood stock required shall be 1300 kg.

To generalise, the formula for calculating the brood requirement is:

$$\text{BRR} = \frac{\text{SPT}}{\text{SRR}}$$

Where,

BRR = Brood fish required (in Kg)

SPT = Spawn production target

SRR = Spawn recovery rate (expressed in terms of number of spawn per kg body weight)

Calculation of the Area Required for Brood Fish Rearing Tanks

The optimum stocking density for rearing brood stock is 1000 kg per hectare. So to rear 1300 kg of fish, area required shall be (1 × 1300)/1000 = 1.3 ha

Or the formula for finding out the rearing area of brood fish shall be as follows.

$$ABRP = BR \times \frac{1}{SD}$$

Where,

ABRP = Area of brood fish production pond,

BR = Weight of brood fish expressed in kg,

SD = Stocking density of brood fish per ha.

In addition, another 1.3 ha. shall be required for raising the brood stock so that continuous supply of brood fish is ensured every year. Further, pond space shall be required for releasing the spent brood fish after each spawning. So the total area for brood husbandry shall be 2.6 ha.

Calculation of Spawning Pool Dimensions

The stocking of brood fish in breeding pool is at the rate of 3.5 kg per cubic meter of water. Suppose that in one spawning cycle, the hatchery manager intends to produce 50 lakhs eggs, the mass of female required shall be 50 lakhs/1. 5 lakhs = 33. 33 kg or say 34 kg (assuming spawning fecundity of major carp as 1.5 lakhs per kg body weight). For 34 kg females equal mass of males are also required to effectively fertilize the eggs.

So the total mass of brood fish to be released into the spawning tank shall be

$$34 \times 2 = 68 \text{ kg}$$

From this data the volume of water required can be calculated as one cubic meter of water is required for 3.5 kg of fish.

So the volume of water required is 68/3. 5 = 19. 4 cubic meter. From the volume of water estimated, it will be possible to calculate the diameter of the spawning pool (from the formula for calculating the volume of a cylinder

$$i.e.\ V = \pi r^2 h$$

V = Volume

r = Radius

h = Height

The ideal height of the water column in the spawning pool shall be 1 meter.

So substituting the known volume of water in the equation value of 'r' can be calculated

$V = 19.4\ m^3$

$h = 1$ meter

$\pi = 22/7$

$19.4 = 22/7 \times r^2 \times 1$

$19.4 \times 7 = 22 \times r^2 \times 1$

$r^2 = 19.4 \times 7/22 = 135.8/22 = 6.17$ or say 6.2

So, r (radius) = $\sqrt{6.2} = 2.489$ or say 2.5

So, diameter shall be $2.5 \times 2 = 5.0$ meter.

Thus, diameter of the spawning pool shall be 5 meter, height shall be 1.2 meter *i.e.* one meter for water column and 0.2 meter as free board.

Suitable range of spawning pool diameter shall be 5 to 8 meters. (Fig. 2.20)

Calculations of Dimensions of Incubation Pool

The quantity of eggs that can be incubated in one cubic meter or 1000 litre of water is 7 lakhs. So the water required for incubating 50 lakhs eggs shall be 50/7 or 7.14 cubic meters.

As the eggs are incubated in the outer chamber of the incubation pool, the volume of outer chamber refers to the total volume of incubation pool – the volume of the inner chamber. The volume of outer chamber inclusive of inner chamber is $V = \pi R^2 h$ where 'R' is the radius of the outer chamber including the inner one, 'h' is the height.

Volume of inner chamber $V = \pi r^2 h$, where 'r' is the radius of the inner chamber and 'h' is the height.

So the volume of the outer chamber alone shall be

$$\pi R^2 h - \pi r^2 h \text{ or } \pi \times (R^2 - r^2)$$

The value of 'h' is one meter.

The ratio of 'R' to 'r' is 4:1.

So $R/r = 4/1$

Or, $R = 4r$

Substituing 'R' with '4r' the equation shall be

$$V = \pi[(4r)^2 - r^2] = \pi(16r^2 - r^2) = \pi \times 15 r^2$$

As 'V' is 7.14 cubic meter equation shall be

$$7.14 = \pi \times 15r^2$$

So, $r^2 = 7.14/\pi \times 15 = 0.151$

$r = \sqrt{0.151} = 0.388$ or say 0.4 meter.

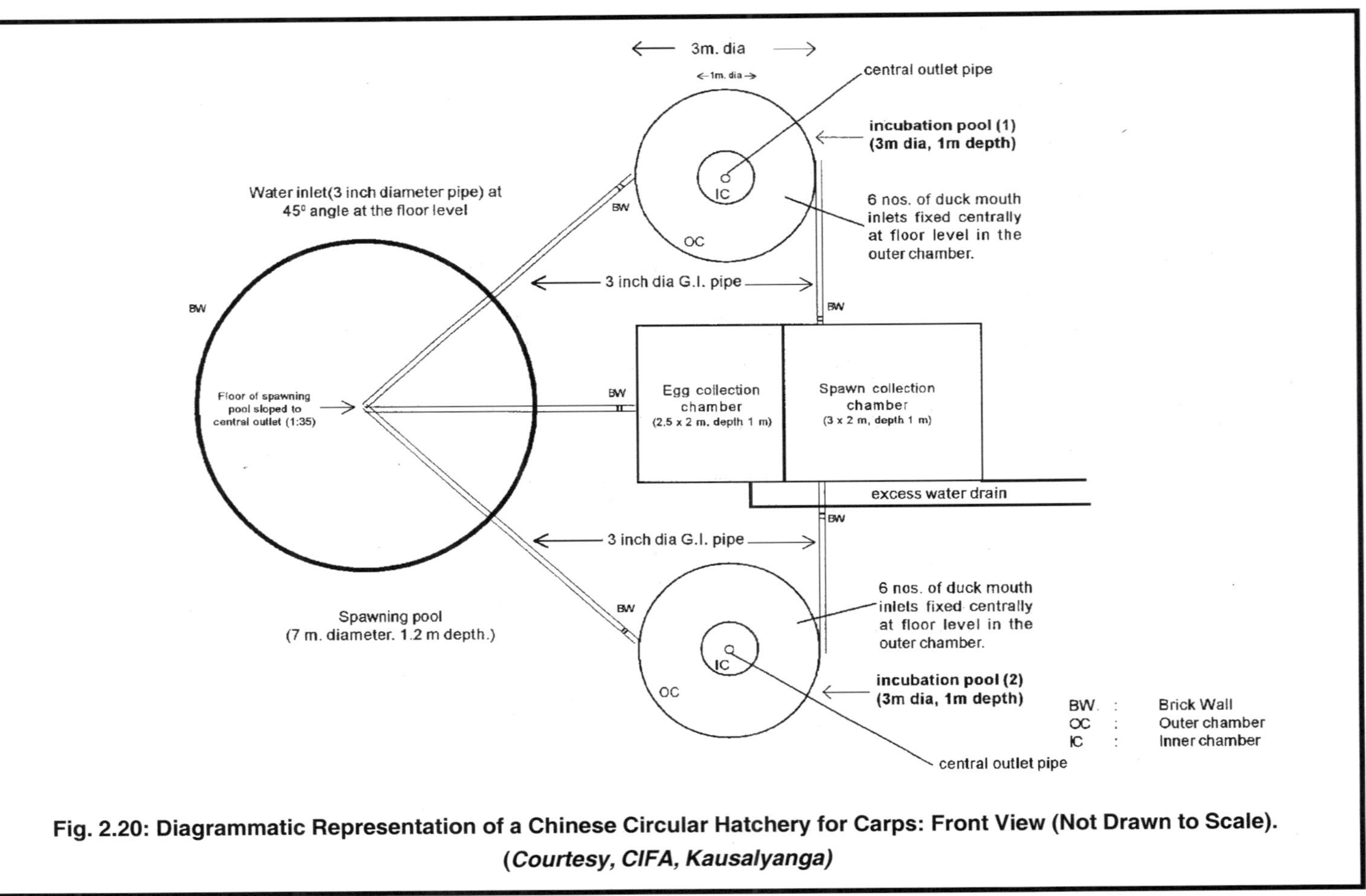

Fig. 2.20: Diagrammatic Representation of a Chinese Circular Hatchery for Carps: Front View (Not Drawn to Scale).
(Courtesy, CIFA, Kausalyanga)

So R = 1.6 meter (radius of the outer chamber including the inner chamber shall be 1.6 meter).

So diameter shall be 3.2 meters and height shall be 1 meter (Fig. 2.21)

Estimation of the Area Required for Nursery Tank

The average stocking density of spawn for nursery rearing can be taken as 5 million per hectare.

So for rearing 50 million spawn, area required shall be 50/5 = 10 ha. In one season, a given pond can be used 3-4 times for spawn rearing. The duration of a crop cycle is 15 to 20 days. If four crops are raised in one season, for rearing 50 million spawn area required shall be 10/4 = 2.5 ha.

To generalise, the area for a nursery tank shall be

$$SPT/SSD \times NCC$$

Where,

SPT = Spawn production target,

SSD = Spawn stocking density,

NCC = Number of crop cycle.

Estimation of the Water Requirement and Size of the Over Head Tank

The water requirement for one incubation pool has been calculated as 4 litre per second or 14.4 cubic meter per hour (1 cubic meter = 1000 litres). In case of breeding pool the requirement is 2 litre per second or 7.2 cubic meter per hour. When both breeding and hatching pools are operated side by side total water requirement shall be 21. 6 cubic meter per hour. This is in addition to the water required for initial filling of the pool upto the desired level of 1 meter. As the number of hatching pools shall be more than one, water requirement can be increased depending on the number of incubation pools. To meet the requirement of 21. 6 cubic meter per hour, size of the over head tank shall be 3m × 3m × 3m. As the water may be stored only upto 2.5 meters, the water storing capacity of such a tank shall be 3 × 3 × 2.5 cubic meter or 22. 5 cubic meters. Depending on the number of incubation pools to be operated, the size of the overhead tank can be altered.

There should be at least three pumps, two pumps for pumping water to the overhead tank and another for management of stocking/rearing tanks. One generator also shall be required to meet urgent emergency requirements.

Packing and Marketing Unit

This includes a spawn conditioning unit and other facilities such as oxygen cylinder, polythene bags etc. If the spawn is sold, it has to be acclimated in a spawn conditioning tank prior to packing. Tanks with running water facilities are preferred for spawn conditioning to avoid mortality. The spawn or fry is measured in cups and then transferred to polythene bags with water and the same is sealed after filling with

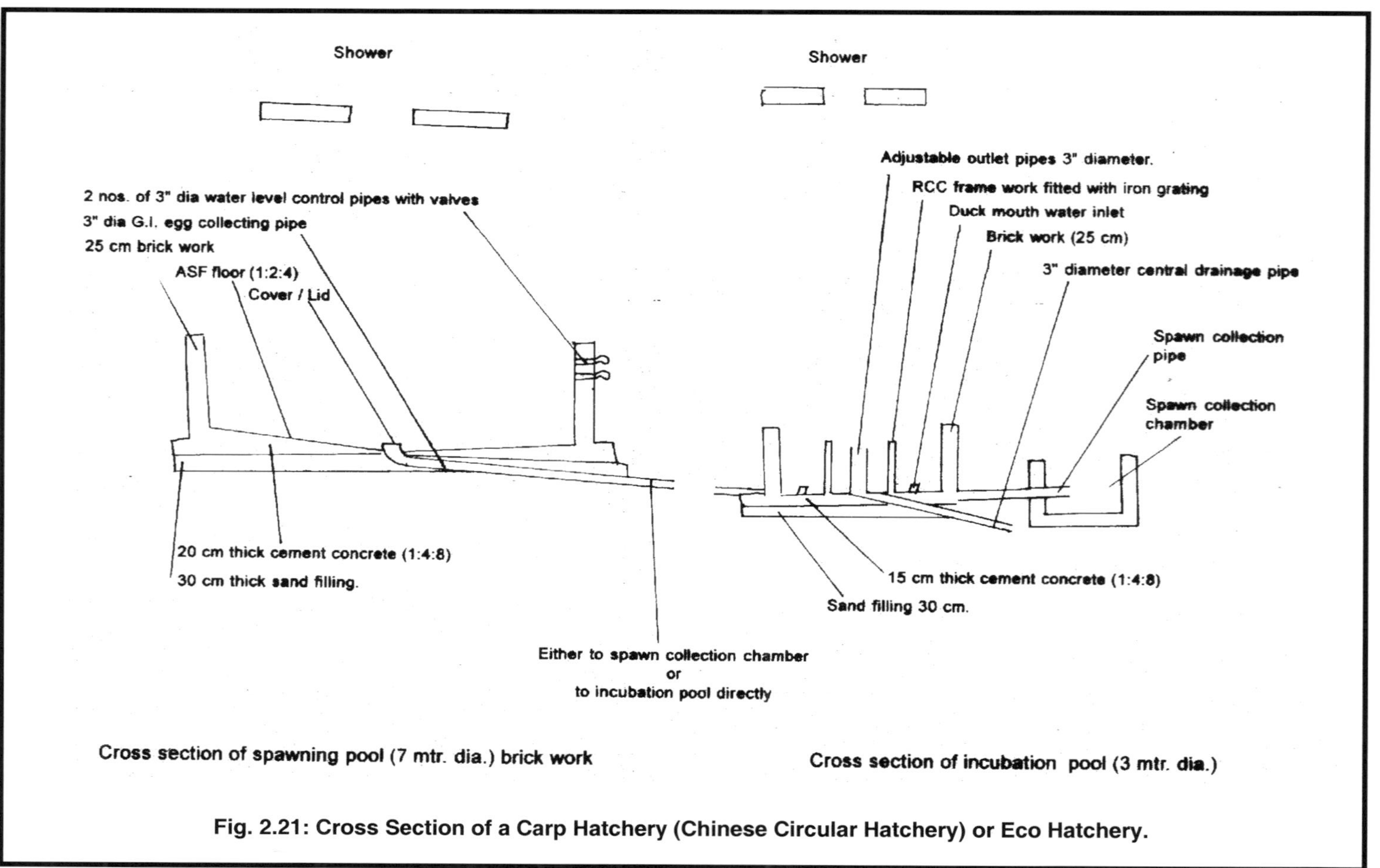

Fig. 2.21: Cross Section of a Carp Hatchery (Chinese Circular Hatchery) or Eco Hatchery.

oxygen. To prevent damage and consequent leakage and loss of spawn, during trasport, polythene bags with spawn are kept in empty tins or plastic cans.

Construction and Operation: Some Important Hints

Spawning pool, incubation pool, spawn/egg collection chambers, spawn conditioning chamber etc. are to be brick work with cement concrete base. The overhead tank should be RCC built on RCC columns and brick wall. The space below the overhead tank can be used as pump house or store. For water supply connections and drainage, P. V. C. pipes are better. Water supply installations are to be planned carefully as water flow has to be controlled at every unit of of the hatchery.

Usually surface water contains organisms which may attack the eggs and spawn in a hatchery. Further, in some places surface water may contain suspended solids which may form a coating over eggs or clog the gills of fry. Such waters are to be filtered prior to use in the hatchery. Such water may be stored in a sedimentation tank to allow sedimentation of larger particles and then filtered through a sand filter before pumping to the overhead tank for supply to the spawning/hatching pools. A sand filter may filter particles larger than 0.025 mm size. Ground water is usually free from suspended solids or pathogen. But it will be dificient in dissolved oxygen and may contain harmful chemicals such as ferrous salts. Hence aeration is recommended for such water before it is released to settling tanks for sedimentation of ferrous salts. For hatchery operation alkalinity should not exceed 100mg/litre (ideal range being 40 to 60) and iron should not exceed 0.4 mg/liter. pH of the water may be 7-7. 5, P_2O_5 from trace to 0.01 mg/litre, NH_4N from trace to 0.02 mg/litre, NO_3 –N from trace to 0.01 mg/l, nitrite and unionised ammonia should be less than 0.01 mg/l

Spawning pool may have a single inlet with 3 mm diameter pipe kept at an angle of 45^0 to create circular movement of water. In case of incubation chamber a series of duck mouth inlets are provided in the outer chamber in the floor which facilitates circular movement. (For further details on spawning pool and hatching pool refer to the description of spawning pool and incubation devices given under Indian major carps). In case of a system where eggs are received directly from spawning pool to the outer chamber of incubation pool, opening of inlet pipe to incubation chamber should be about 0.3 meter above the base of incubation chamber so that eggs are received into a water cushion to prevent injury to the eggs. Water holding surface of pools and cisterns should be smooth with neat cement finishing and painting. Provision has to be proivided for showers over the breeding as well as the hatching pools.

Brood fish raising/raring tank can be of various sizes *viz.* 0.2 to 1 ha preferable size being 0.3 to 0.5 ha. Depth can be 2 to 2.5 meter. Nursery pond may be of 0.2 to 0.06 ha size and of 1 meter to 1.5 meter depth. For both stocking as well as nursery tank, drainable system is preferable. Bottom may be sloping towards the outlet.

In places where there is water scarcity, facilities for reuse of water after filtration can be set up. Such recirculatory system has various types of filters such as drum screen filter, biofilter, foam fractionators, sand filters etc. Such systems are not yet popular in India due to the high cost involved in installation and operation.

Chapter 3

Seed Production Technology of Freshwater Cat Fish

INTRODUCTION

Fishes belonging to the order Siluriformes are commonly known as cat fishes. About 2211 species belonging to 400 genera and 31 families are reported under this group so far. These fishes are characterised by a scaleless body. But in some cat fishes body may be covered with bony plates. They possess barbels which arise from the upper jaw and also from lower jaw in some cases. Further, in most of the cat fishes an adipose dorsal fin is present (except in Claridae, Heteropneustedae and Siluridae). Some of the cat fishes can inflict severe injury to man and other animals by means of pectoral fin spine and can inject poison produced from poison gland cells modified from the epidermal cells covering the spine. Sting of a cat fish, *Plotossus lineatus* can lead to death. Most of them are fresh water fishes except those of the family Plotossidae and Aridae. Apart from their food value, they are also popular as a sport fish and aquarium fish. *Silurus glanis* found in Europe is the largest of all cat fishes that can reach 5 meter long and 330 kg weight. (Nelson, J. S. 1984). Seed production technology of three cat fishes such as Channel cat fish *(Ictalurus punctatus)*, Magur *(Clarias batrachus)* and stinging cat fish *(Heteropneustes fossilis)* are described in this chapter.

CHANNEL CAT FISH *(Ictalurus punctatus)*

The channel cat fish, *Ictalurus punctatus*, of family Ictaluridae, order siluriformes (Fig. 3.1) is a commercially important fish cultured extensively in U. S. A. (Fig. 3.2) and also in other countries such as Cuba, Guatemala, Paraguay, Russia etc. Out of the total 270760 mt of channel cat produced through farming in the year 1999 in the world, 270629 mt was contributed by USA. It is a hardy fish that withstands crowded conditions, accepts artificial feed easily and can be easily induced to spawn facilitating fry and fingerling production under controlled conditions. Other species of the genus Ictalurus that are of culture value are *I. furcatus* (blue cat fish), *I. catus* (white cat fish), *I. nebulosus* (brown bull head), *I. natalis* (yellow bull head) and *I. melas* (black bull head).

Fig. 3.1: Channel Cat Fish, *Ictalurus punctatus*.

Fig. 3.2: Channel Cat Fish Farm (Cage Culture) at Louisberg, N.Carolina, USA

Reproductive Biology

Sexual Dimorphism

Sexes are separate. Male and female can be distinguished externally during breeding season. In case of male, the genital papilla which looks like a fleshy nipple, is located just behind the anus. There is a single aperture for ejection of urine and milt, called urino-genital aperture located on the papilla. Further the male has a muscular head wider than the body and there is dark pigmentation on the ventral side of jaw and abdomen. In female, there are separate genital and urinary apertures and the same are located in a groove and the genital papilla is not conspicuous. Further, the head which has no muscular pads is narrower than the body and the pigmentation on the ventral side of jaw and abdomen is not dark. In a ready to spawn female, abdominal wall will be soft and distended and genital area will be swollen and red.

Age at First Sexual Maturity

The fish becomes sexually mature on attaining 2 years age; weighing even 340gm. But for breeding, fish weighing at least 1.4 kg and 3 years old are preferred. Males grow faster than female. Fish weighing more than 4 kg and above 5 years are not useful as brood fish.

Parental Care

The channel cat fish exhibits parental care. The male usually clears a place of debris or soft mud usually near or under submerged log, stump of trees or roots. After mating and egg laying, the male chases the female away and guards the eggs and also fry after hatching

Spawning

The female cat fish spawns only once in an year where as males exhibit viable sperms in almost all months of the year. At Mississipi state, USA it it is reported to start breeding from mid April to mid May when the water temperature exceeds 21°C for several days and the breeding period extends up to mid August. The optimum temperature for spawning is 21-29°C, 30°C and above are lethal for fish. Larger fishes breeds earlier than younger ones.

Courtship and Mating

During mating, the male and female remain side by side in opposite directions *i.e.* tail end of the male fish will be near the head end of female and *vice versa.* Then both the fishes wrap the caudal fin around the head of the other and then the male exhibits shuddering reflex for several seconds at the end of which it ejects out the milt. Instantly the female also exhibits similar shuddering reflex resulting in extrusion of a few eggs. Thus spermiation immediately precedes ovulation. The eggs settle to the bottom and soon after coming in contact with water it developes stickiness. This mating usually takes place at 4-10 minutes interval but each time the female may not ovulate. According to Clemens and Sneed (1957) and Nelson, B.A. (1960) the female releases eggs 5-10 times in one hour and it takes 4-12 hours for a pair to complete the production of an egg mass. A female weighing 0.45 to 1.8 kg can produce about 8800 eggs per kg body weight but larger fishes may produce only 6600 eggs per kg body weight.

Hatchery Technology

Brood Stock Raising

Brood Stock from Wild Sources

In case of channel cat fish, brood fish collected from wild are not preferred as they do not adjust themselves to crowded conditions and develop diseases easily. Their spawning performance in captivity is also poor.

Brood Stock from Perennial Ponds

Usually perennial, non drainable ponds contain fish of larger sizes as each harvesting done earlier shall be only partial. Sometimes the slow growing ones that evades the nest are reported to be left out in the pond and hence it may not have the desirable genetic traits to be used as brood fish. Hence male and female fish collected from different grow out ponds are to be used for breeding so that inbreeding is avoided. Brood fish are to be collected far in advance of the spawning season so that it can be maintained on a balanced diet and in desirable stocking density.

Raising Brood Fish from Fingerlings Obtained from Hatchery

Potential brood stock can be selected and raised from fingerlings obtained from hatchery. Meyer *et. al* (1973) suggested to choose 5 to 10% of the largest size ones from a population of fingerlings. But according to Brooks *et. al* (1982) 10% of the largest fingerlings of any population may be only males as males are faster growing than females. Norton *et. al* (1976) suggested to determine the sex of the fish by microscopic examination of genital area so that all male population can be avoided and desirable female to male ratio (1:1 to 3:2) can be obtained.

Brood fish are to be stocked at low density and fed on balanced diet to get the desired results. Fish weighing 1 to 1.5 kg can be stocked @ 500 to 1000 kg/ha Feeding is to be done during morning hours @ 3% of the body weight on alternate days. Pelletted feed containing 25 to 36% protein can be broadcast on water surface. Some workers prefer to supplement artificial feed with natural food like cut pieces of fresh/ frozen fish. During colder times when water temperature is less than 10°c feeding is recommended only once in 7-10 days @ 0.75 to 1% of the body weight. If temperature is 10°c to 12.8oc or more feed may be given once in 3 to 4 days. Floating type of artificial feed is found better than sinking type.

Several channel cat fish culturists resort to aeration to increase the dissolved oxygen levels either by spraying water over the water surface (Fig. 3.2) or splashing the water in to the air by paddle wheel aerators.

Egg Collection

Egg Collection by Providing Artificial Spawning Receptacle in Ponds/pens

Brood fish are transferred from rearing tanks to spawning tanks which are provided with spawning receptacle or spawning containers. Males to female ratio in a spawning tank can be 1:2 or 2:3. Spawning receptacle may be earthern crocks, tiles, wooden boxes or milk cans. Different containers are kept 1 to 10 meter apart at 0.3 to 0.75 meter depth to facilitate easy checking of containers for eggs. Each container can be marked with a float or a stake for easy identification when it is kept at the bottom. The fish lays eggs in the spawning receptacle, the containers can be checked once in 3 to 4 days for eggs. Usually, the male will be found guarding the eggs and hence, it has to be removed carefully before the eggs are collected. This egg mass can be carried to a hatchery for incubation or allowed to hatch in the spawning containers itself. While transporting to the hatchery for incubation, eggs should not be exposed to direct sunlight and the temperature of the water in which eggs are carried should not exceed 30^0c. The dissolved oxygen should not be below 5ppm and period of transport should not exceed 30 minutes. If it exceeds 30 minutes water in which eggs are kept should be aerated.

Egg Collection by Stripping

Injection of hormone

The female fish is usually injected with either pituitary extract/H. C. G. or LHRH-a either intra peritoneally or by intramuscular route. Most females require 3 injections of 4.4mg pituitary per kg of body weight at 24 hr intervals. Most of the fishes shall spawn 16 to 24 hours after last injection. Sneed and Clemens (1959) observed spawning of channel cat fish administered H. C. G @ 1760 I. U. per kg either in single dose or multiple dose Dupree *et. al* (1965) used H. C. G 1100 I. U. per kg body weight and later repeating the same on every alternate days for several days (up to 10 days) for inducing ovulation. LHRH-a @ 0.1 mg per kg body weight or in 2 doses with initial (primer) dose of 0.01 mg/kg body weight followed after 12 hours by 0.09 mg/kg body weight induced ovulation in channel cat fish

Stripping the female

Injected females are released in to a tank along with male When the fish exhibits pre spawning behaviour the female is collected, wiped with a dry cloth and stripped and the eggs are collected in to a tray taking care not to contaminate the eggs with urine or faecal matter from the fish.

Preparation of Sperm Suspension

In channel cat fish, milt cannot be obtained by stripping the male. Hence the male is sacrificed and the testes are dissected out and grinded in 0.6 % saline using mortar and pestle. This sperm suspension is used for artificial fertilization of the stripped eggs.

Fertilization

The sperm suspension is added to stripped eggs and mixed. Sperms present in the sperm suspension fertilize the eggs The fertilized eggs are washed with fresh water and treated with malachite green as a prophylactic measure.

Incubation of Fertilized Eggs

Incubation Devices

Incubation of fertilized eggs is done in specially designed hatchery troughs. The troughs can be made of wood/zinc free metals/fibre glass and it may measure 2.4 to 3.6meter length, 0.5 to 0.6 meters width and 0.2 to 0.3 meter depth. Each trough will have provision for water inlet and outlet (drainage pipe). From the sides of the trough, egg baskets made of 6 mm galvanised or rubber coated hard ware cloth is suspended by hooks in such a way that it is completely submerged in the water in the hatching trough. Each basket may measure 30cm wide, 56cm long and 10cm deep. A rod of 2.5cm diameter is fixed on bearings on the top along the length of trough in the center. This rod functions as a revolving shaft. Paddles are attached to the shaft in such a way that one paddle will be on each side of the egg basket. This rod is connected to a ¼ H. P motor (1750 r. p. m) through a series of pulleys in such a way that finally the shaft will rotate at 30 r. p. m. The paddles are long enough to extend beneath the basket bottom so that their rotation would bring about water movements on the egg similar to those made by male fish when eggs are hatched naturally. In a single trough of size given above, 2 lakhs eggs can be incubated at a time. Usually, 2 fry troughs are prepared for each egg hatching trough so that after hatching the fry can be shifted to the fry trough. Some workers raise the fry also in the same hatching trough after removing the egg basket and paddles.

Incubation period ranges from 5-8 days depending on temperature. The freshly taken egg mass is yellowish in colour but as they mature to hatch, it changes to pink or brown colour.

Care of Eggs During Incubation

Water quality

Maintenance of water quality is of prime importance in a hatchery. Both surface water and ground water can be used for incubation of channel cat fish eggs. while using surface water, care must be taken to get rid of pathogenic organisms and also to address fluctuating temperatures, turbidity and pollution problems. Compared to surface water, ground water can be of consistent quality. Ground water may also need pretreatment incase temperature is too high or too low, or it is dificient in DO or calcium or it has excess amount of CO_2, Fe or H_2S. Usually, pretreatment involves aeration which will increase DO_2 and pH and reduce CO_2 and H_2S concentrations. If temperature is too low, it can be heated to raise the temperature to the desired level. If the water temperature is too high, water from different sources can be mixed to bring down the temperature. Sedimentation and filtration to remove suspended solids including precipitated ferric iron is also practiced.

Desirable water quality parameters for incubation are : DO 5ppm, CO_2 :0-15ppm, total alkalinity 50-40 mg/l as $CaC0_3$, pH 6.5 to 9, total hardness 50-400 mg/l as $CaCO_3$, Calcium 10-160 mg/l, total iron 0-0.5 mg/l and H_2S should be nil.

Water exchange

A water flow of 7.5 liter/minute to each trough is recommended for maintaining desired level of DO and to ensure removal of metabolites.

Temperature

Optimum temperature for incubation has been shown to be 25-28°C. At lower temperature incubation period increases and also eggs are prone to more fungus attack. Higher temperature leads to formation of deformed embryos.

Hygeine

If proper temperature, waterflow and aeration are not maintained and crowding of eggs mass is not avoided eggs get affected with bacteria or fungus. Saprolegnia is the common fungus that affect the eggs hence dead eggs or affected eggs are to be removed soon after the same is noticed. The common bacteria that affect the eggs are Aeromonas, Flavobacterium and Acinetobacter. As a matter of fact, any aquatic bacteria can colonise the exterior surface of egg, if it is a damaged or dead. In order to facilitate proper aeration and waste removal egg mass has to be split and turned daily. Slightly shaking the egg mass may also remove any debris settled on it. 10-15 minutes bath in formalin (100-250 mg/litrs) is recommended as a therapeutic measure. Egg shell debris should be removed from the trough daily. Trough should be scrubbed, disinfected and rinsed before every operation.

Larval Care

The hatchlings are initially golden colured and it falls through the mesh of the egg basket into the trough. By 3rd or 4th day after hatching, the yolk sac gets absorbed and the young starts swimming up. At this stage they look black in colour and they start feeding.

Usually, yolk sac larvae are kept in the fry trough till the yolk sac is fully absorbed and swim up fry stage is reached when it starts feeding. If yolk sac larvae are directly released into the nursery tank heavy mortality may occur.

Swim up fry, after it starts feeding can be released into the nursery tank. Stocking density affects the growth of the fry. During 120-150 days rearing, fry may grow to fingerlings of 18-25cm size if the stocking density is only 24700 fry/ha. If stocking density is incresed to 494000 frys/ha growth was only up to 5-8cm long.

A nursery tank should be prepared throughly to eradicate predatory fishes, and weed fishes from the pond. Further, the pond must be fertilized to ensure growth of fish food organsims. Fry has to be fed frequently. Feeding rate and feeding frequency are related to temperature-at lower temperature feeding rate and feeding frequency can be less as it consumes less food at lower temperature (Tables 3.1 & 3.2).

Table 3.1: Suggested Feeding Rate (Percentage of Body Weight) and Feeding Frequency at Different Water Temperatures.

Water Temperature	*Feeding Frequency 3 to 4th Day*	*Feeding Rate (% of Body Weight*
10° -13°C	Alternate days	2
14 - 19°C	1 time/day	2
20-25°C	2 times/day	3
26-30°C	4 times/day	6
31°C and above	2 times/a day	2

Feed is formulated with ingredients such as fish meal, soybean meal, peanut meal, Ground corn and vitamin and mineral mix. According to Dupree and Huner, feeding rate for fry can be 11-28 kg feed/ha per day. Which may be devided between 2-3 feedings the optimum size of feed particle in relation to size of the fish as recommended by Dupree and Huner (1984) is given in Table. 3.2

Table 3.2: Optimum Size of Feed Particle Recommended for Feeding Different Stages of Fish (Dupree & Huner, 1984).

Length of fry (Cm)	*Size of feed particle (mm)*
< 1.25	0.42 – 0.59
1. 25 - 2.5	0.595 – 0.841
2. 5 - 3.75	0.841 – 1.19
3. 75 -6. 25	1. 19 – 1.68
6. 25 – 10	1. 68 – 2.38
10.5 - 15.0	2. 38 – 3.36

MAGUR (*Clarias batrachus*)

Introduction

There are three cultivable species of *Clarias* such as *Clarias batrachus*, *C. gariepinus C. macrocephalus. C. batrachus,* commonly known in Inida as magur (Fig. 3.3)and *C. macrocephalus* are smaller whereas *C. gariepinus* (Fig. 3.4), commonly called African catfish grows to a giant size of 1.5 m in length. The 3 species are easily distinguishable observing the shape of occipital process of skull. In *C. gariepinus* there is a distinct conical occipital process arising from the middle part of the occipital region of the skull and at its base on either side a deep depression is also found where as in batrachus central occipital process is not very distinct and depressions on either side are also not distinct. In macrocephalus occipital process is bow shaped. Batrachus is widely distributed in India, Ceylon, Bangaldesh, Pakistan, Burma, Malaya, Thailand, Indochina, Philippines, Hongkong and South China. They are found in derelict water bodies, having low DO high carbondioxide, methane and ammonia. They are having accessory respiratory organs by means of which they can also breath air.

Reproductive Biology

Age at first sexual maturity

It grows to 200-250 g in 6-8 months. Fishes in the age groups 1^+ year weighing 150 g attain sexual maturity

Sexual Dimorphism

Males are distinguishable from female externally. Genital papilla in male is long and pointed whereas in female it is round or oval button shaped. Further, the vent in the mature female is reddish, round and bulging but in male it is slender and whitish. The belly of a mature female is also bulging where as in male the abdomen is not bulging during the breeding season. If the belly of a mature female is pressed, ova

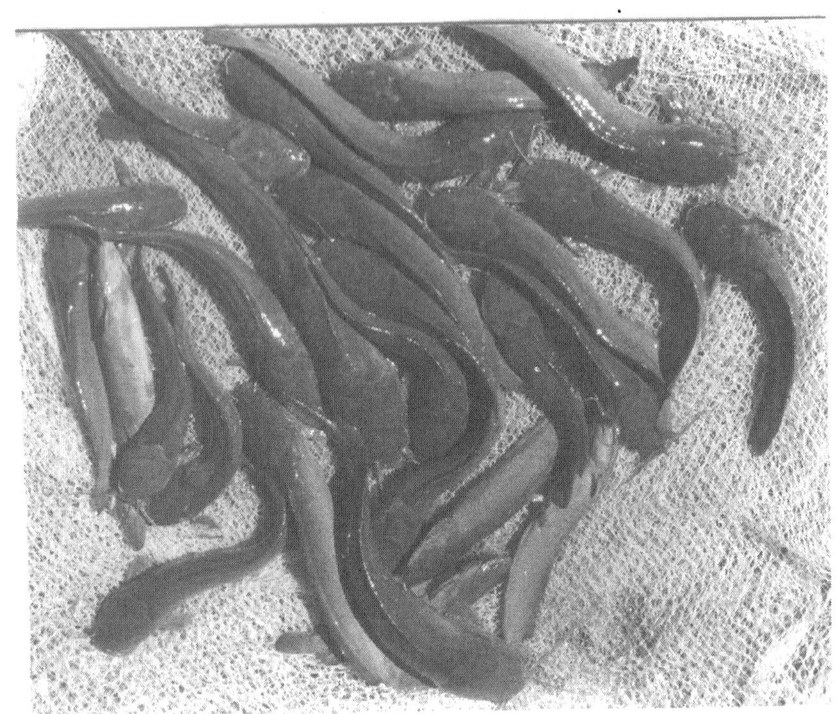

Fig. 3.3: Adult *Clarias batrachus*
(Courtesy, CIFA, Kausalyaganga)

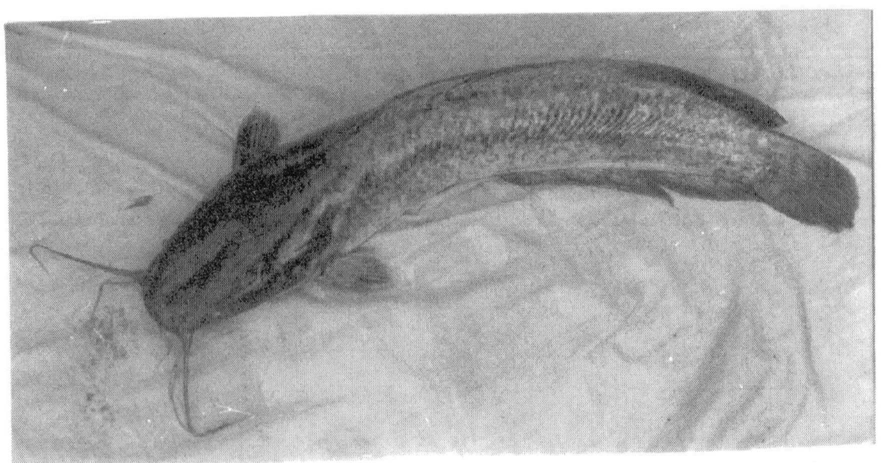

Fig. 3.4: The African Cat Fish, *Clarias gariepinus*

oozes out through the genital opening but in case of male no milt oozes out when belly is pressed.

Fecuntity

The fecundity of this fish is very low as it is in the range of 15000 to 20000.

Breeding Season

It breeds only once in an year *i.e.* from June to August, peak period being July.

Parental Care

It exhibits parental care. The females lay eggs in pits made along the margin of water bodies. The female moves out soon after spawning and the male guard the egg and young ones for a few days. Soon after yolk sac absorption, fry start moving around feeding zooplankton.

Hatchery Technology

Maintenance of Brood Stock

Fishes weighing above 150 g belonging to the age group 1 + year are suitable as brood stock. 2-3 months prior to breeding season, they are to be stocked in stocking tanks (Cement cistern) and fed on a feed containing trash fish: rice bran (9:1). Feeding may be done @ 10% of the body weight. A gentle flow of water 2 litre/minute is advisable and once in every fortnight entire water may be replaced and the tank is to be cleaned and disinfected and fish should be dipped in prophylactic agents

During the breeding season, breeders are selected from the brood stock. Females with soft bulging abdomen are selected and a sample of eggs can be collected using a catheter. On microscopic examination if the eggs are found to be 1-1. 2 mm in diameter uniformly and if irregular shaped eggs are not found, such females can be selected for breeding. Males are selected observing the genital papilla which is elongated and pointed. Usually, selected breeders are kept starved in plastic tubs at least for one day prior to injecting hormone. This will help in keeping the alimentary canal empty at the time of stripping.

Hormone Administration

Crude pituitary gland extract (CPGE)

A single dose of carp pituitary @ 30mg/kg is found effective in inducing Clarias to spawn. Procedure for the preparation of pituitary gland extract is same as described under carp breeding.

Ovaprim

The synthetic hormone preparation from Syndel laboratories Canada @ 0.4 ml/kg has been found very effective in this fish.

HCG @ 4000 IU/kg has been also used in Clarias successfully.

The hormone is administered intramuscularly below the dorsal fin and above lateral line in the posterior region of the body. It can be also administered intra peritoneally, at the inner side of the base of pectoral fin of the fish.

Preparation of Sperm Suspension

In case of Clarias, as milt dose not ooze out through stripping, collection of milt is done by grinding the testes in 0.9% saline after sacrificing the mature male. The procedure is as follows:

(a) Sacrifice a male fish by decapitation and remove the testis.

(b) The testis is taken into a mortar containing little saline (0.9%)

(c) Cut the testis into pieces by a fine scissor and crush it with pestle.

(d) This suspension is used for fertilizing the eggs.

This suspension can be examined under a microscope after adding a drop of water into the suspension taken on a slide. On the addition of water the sperms start moving fast which can be seen under the microscope. If the sperm are not moving after the addition of a drop of water, the suspension should be discarded.

Egg Collection by Stripping the Female

Stripping the female is done about 14 to 15 hours after the the injection of CPGE/ ovaprim at 27^{0}c to 30^{0}c. The females are taken and cleaned by means of a towel. It is held above a clean dry enamel basin. Gently press the belly, passing the fingers from front to backwards. Mature eggs shall fall in to the tray in a jet. Care should be taken to avoid contamination of eggs with blood, urine or water from the body of the fish. Weight of the stripped eggs is to be recorded. Total number of eggs can be calculated by counting the number of eggs in a given weight (1 g.) of eggs and multiplying the same with total weight of eggs. Fully mature eggs look dark brown or brownish green in colour. The stripped fish is bathed for 2 to 3 minutes in potassium permanganate solution after which it is released in to the stocking tank.

Fertilization

Milt suspension is added to the eggs and mixed thoroughly by means of a feather or by gently shaking the tray. Add a little fresh water and shake thoroughly. When the freshwater is added, sperms get activated and becomes motile. The sperms fertilize the eggs. Then add freshwater and wash after 1-2 minutes. Remove pieces of testis or any tissue from among the eggs. The fertilized eggs are now transferred to trays for incubation.

Incubation

Before incubation of fertilized eggs the rate of fertilization is to be calculated taking a sample of eggs. Unfertilized eggs can be easily distinguished from the fertilized ones as unfertilized eggs look paler and opaque. The fertilized eggs are demersal, adhesive and spherical. From a given sample, the total rate of fertilization can be calculated.

The eggs are also treated with malachite green as a prophylactic measure before placing the same in a flow through system for incubation.

Flow-through system for incubation consist of a stand on which a row of small plastic tubs are placed. Each tub is provided with an outlet at a height of about 4 cm and the same is connected to a common pipe for draining out the water. Each tub is provided with a control tap connected to a pipe receiving water from an overhead tank. Control tap is opened in such a way that a gentle flow of water is maintained over the egg kept for incubation in the tub.

Development and Hatching

The fertilized eggs undergo development and young hatches out within 25-26 hours at 27° –30°C. During development, the egg reaches 4 cell stage within one hour, morula stage within 4 hours, and head and tail ends become identifiable within 10-11 hours, Twitching movement of the embryo is seen within 23 hours.

The newly hatched larva measures 4-5 mm and weighs 2.8 to 3.2 mg. This larva retains a large oval yolk sac (2 mm × 1.5 mm) which is absorbed within 4 days. Newly hatched larva usually rests at the bottom and as the yolk sac gets absorved it moves in the water and starts feeding.

Larval Rearing

The spawn can be reared in a shallow container for 10 to 12 days. The larvae should be stocked at a density of 2000 to 4000 per square meter. The water should be continuously aerated and about 50% of the water should be siphoned out twice a day and replenished with good water to eliminate accumulated metabolites feed residue and faecal matter.

Aeration ensures adequate dissolved oxygen in the rearing medium and water replenishment ensures removal of feed residue and accumulated metabolites from the water. Aerial respiration starts after 10-12 days of hatching after which the fry makes vertical trips to surface of water to take in atmospheric air. In view of this behaviour, water column has to be kept low (8-10 cm). If the depth is more, the fry will have to sustain heavy loss of energy to reach the water surface to breath atmospheric air as they will have to travel a long distance from the bottom where they remain mostly. If they do not move to the surface to gulp in air they may die.

The water quality to be maintained in the rearing medium is as follows:

DO- 4.0 ppm to saturation, Water temperature – 26 –31 °C, PH – 7-8. 5, CO_2 - < 15 ppm, NH_3 - < 0.05ppm, NO_2 - < 0.25 ppm.

Feeding

Feeding has to be started on the 4th day. To start with the organism or food particles ranging in size from 20-30 µ are ideal which can be increased to 50 to 60 µ for one weak old fry. Artemia nauplii is to be fed *adlibitum* for one week to ten days. At least 50 nauplii per fry per day is required and they are to be fed 3-4 times a day, each time a fry may eat 10 to 15 nauplii.

In addition to the Artemia nauplii, zooplankters like cladocerans, moina, are also preferred by clarias fry. They also feed on Tubifex worms and the same is given when fry becomes 7 to 10 days old. Artificial food such as egg custard also has been recommended for the fry. Egg custard is prepared as follows:

Mix contents of a chicken/ duck egg and equal quantity of condensed milk powder and 0.5 g vitamin mineral pre mix. Make a homogenous paste and steam bake. Allow it to cool and store in a refrigerator.

The feed should be filtered through the required seive and given to fry.

Advanced Fry Rearing

10-12 day old fry can be shifted to a larger containers and stocked @ 1000 fry/sq. m. Artemia nauplii can be reduced at this stage and zooplankton, tubifex, molluscan meat etc can be given. Continuous aeration is not required at this stage. A feeble water flow may be maintained. It can be reared for 10-12 days more by the end of which the fry reaches 2 to 2.5 cm and weighs 0.8 to 1.0 g.

If rearing is continued in the above containers, their growth is found slowing down. The fry from this stage onwards needs the natural habitat *i.e.* pond with clayey bottom and weed. Hence the fry at this stage (20-24 days old) are released into nursery tank.

Rearing in Nursery Tanks

Rearing tanks shall be of small size (50 sq. m.) having sides cemented or stone pitched. Water level may be 25-30 cm having an embankment of 1 meter high. To prevent the escape of fingerlings a fence (made of closely knit split bamboo) of 60-100 cm height should be fixed in side the pond along the margin.

Apply lime to maintain pH at 7 to 8.5, fertilize with cowdung @ 10, 000 kg/ha and provide patches of Eichhornia or Lemna plants along the margin. Those plants should occupy only ¼ of the pond area. The plants control algal blooms and also provide shelter to insects which is a choice feed for advanced fry

The fry can be stocked @ 100-200 per sq. meter and reared for another 15 days. Feeding can be done with finely minced trash fishes or molluscan meat &rice bran (1:1) @ 5-10% of biomass daily in the evening hours. The fingerlings grow to a size of 4-5 cm weighing 1 to 2 g during this period. They can be taken to grow out ponds.

STINGING CAT FISH *(Heteropneustes fossilis)*

Introduction

The stinging cat fish, *Heteropneustes fossilis* (Fig. 3.5) is another air breathing cat fish of commercial value distributed in several Asian countries such as India, Nepal, Pakistan, Bangaldesh, Sri Lanka, Myanmar, Indonesia, Singapore, Philippines and Thailand. This fish is commonly known as singhi in several parts of India. It can be easily distinguished from *Clarias batrachus* by its smaller dorsal fin whereas in Clarias, the dorsal fin is found extending from anterior region of trunk to the caudal end. This fish lives in derelict water bodies such as ditches, ponds, swampy and marshy areas. It can live even in semidried conditions during summer. It fetches high price in the market due to the medicinal value of its flesh. It can inflict painful injuries to man with its pectoral fin spine by injecting a poison secreted by poison gland cells which are modified epidermal cells covering pectoral fin spine.

Fig. 3.5: Adult *Heteropnustes fossilis.*
(Courtesy, CIFA, Kausalyaganga)

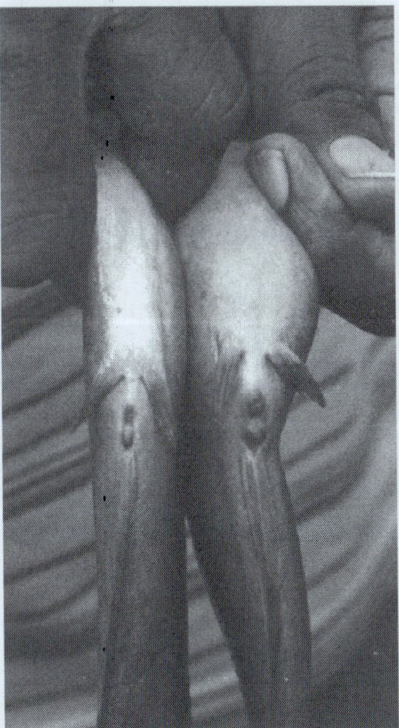

Fig. 3.6: Sexual Dimorphism in *H. fossilis.*
Note the Elongated Genital Papilla of Male (Left) and Oval Genital Papilla, Swollen Belly and Prominent Reddish Vent of Female (Right).

Reproductive Biology

Age and Size at First Sexual Maturity

It attains first sexual maturity when one year old reaching 8-12 cm length.

Sexual Dimorphism

Male and female can be distinguished from each other only during breeding season. The abdomen of the female is swollen whereas in male the body is streamlined. Further, on pressing the belly of such females the eggs ooze out but in male milt does not ooze out. The genital papilla of female is more prominent, round and blunt with a slit like opening in the middle, but in male, genital papilla is pointed. (Fig. 3.6).

Breeding Season

It is a seasonal breeder. It breeds in confiend waters during monsoon months between June and August, peak being July. There are reports of breeding extending up to October in West Bengal and Orissa (Nayak, P. K. *et al*, 2000).

Fecundity

Its fecundity has been estimated at 1500-2000 eggs per gram ovary (Nayak *et al*, 2000).

SEED PRODUCTION

Brood Stock Maintenance

Adult brood fishes weighing 50-100 g can be procured from natural waters and reared in cement cisterns. Stocking density can be about 10-12 fish per cubic meter Feeding can be done with molluscan meat or formulated feed containing rice bran, fish meal, soybean meal and ground nut oil cake.

Injection of Hormones

A female with the soft and distended abdomen and having oozing ova on applying gentle pressure on the belley is selected for stripping. In case of male, development of genital papilla, is the only criterion for selection as the milt does not ooze out on applying pressure on the belly

A female is administered either pituitary gland extract @ 15-20 mg per kg body weight or salmon gonadotropin releasing hormone analoge (D-lys^{6}sGnRH-A) @25 – 500 mg/kg body weight or ovaprim @ 0.6 to 0.9 ml per kg body weight. Intra muscular injection beneath the dorsal fin and above the lateral line is recommended. The injected fishes are left in the plastic tub in running water. The fish to be injected are starved for one day just prior to the day of hormone administration. This will help in preventing the contamination of eggs with faecal matter while undertaking stripping. According to Alok, D. *et al.* (1993), this did not respond to D-Lys^{6}s GnRH-A at temperatures below 15^{0}C. Optimum temperature was reported at 25^0 to30^{0}C. Further these workers also did not observe any potentiating role of dopamine antagonist in GnRH induced ovulation in this fish.

Stripping the Female

The female is stripped 14 to 18 hours after hormone injection. The female is wiped with the wet clothes and the readiness of the female is tested by gently pressing the belly. In a fish which is ready to ovulate eggs shall pass out through the genital pore like a jet. The eggs are collected into a clean dry container such as galvainised tray. The eggs are then counted either by gravimetric or volumetric method.

Preparation of Sperm Suspension

Just prior to stipping the female, the male brood fish is cut upon and testes are dissected out into a mortar and cut into pieces in 0.6 percent saline and crushed using a pestle. After this the pieces of testicular tissue are removed from the mortar by means of a fine forceps. Alok D. *et al.* (1993) used Holtfreters solution (0.67 mm KCl, 60.2 mm NaCl, 1.3 mm $CaCl_2$, 2.3 mm $NaHCO_3$) for preparing sperm suspension.

Fertilization

The milt suspension, prepared as noted above is added to the eggs stripped in to the tray and mixed my means of feather for 2-3 minutes. The fertilized edggs are washed repeatedly in fresh water. Usually the fertilized eggs are greenish blue in colour and they settle to the bottom whereas the unfertilized eggs are somewhat white and are found floating above the fertilized ones. The fertilized eggs are moderately adhesive, demersal and spherical in shape. They measure 1.4 to 1.6 mm in diameter having a narrow perivitelline space of 0.1 to 0.2 mm width.

Incubation of Fertilized Eggs

The fertilized eggs are given prophylactic treatment by washing them in malachite green (5 ppm for 10 minutes) & oxytetracycline (5 ppm for 10 minutes). Then the rate of fertilization is calculated by counting the number of unfertilized eggs in a known volume or weight. Then the fertilized eggs are spread on a nylon net kept in a tray provided with the running water in a flow through system. The incubation period varied from 16 to 19 hours at 28-30^{0}C. The hatchlings wriggle out and fall to the bottom of container through the mesh of the nylon net leaving the remains of the egg cells and dead eggs over the net. Periodically, in the course of incubation, dead eggs are to be removed from the tray as dead eggs are attacked by micorbial organisms such as bacteria and fungi and later these organisms shall affect the developing eggs also.

The newly hatched larvae measures 2.72 mm in length, having a round yolk sac which forms almost 42% of the total length. It is transperant having a laterally compressed body. 16-28 myomeres are visible. They do not feed for 3-4 days i.e. till the yolk sac is fully absorbed. They remain at the bottom resting on their sides.

Rearing of Fry

By fifth day after hatching, yolk sac is fully absorbed. Pectoral fin get differentiated behind operculum. Caudal fin rays start appearing. By tenth day it measures 7.5 mm in length. Then the dorsal fin differentiates. -the spine and fin rays are developed in the pectoral fin. The pelvic fin is not yet differentiated. Anal fin is continuous with the larval fin fold. At this stage they start vertical movement to gulp in air. They swim actively and feed vigorously. Within 19 to 21 days larva becomes a juvenile. The fins become distinct and get completely separated from caudal fin.

Care is taken to rear the spawn in shallow containers at a density of 3000 to 5000 per sq. mt. As the fry moves from the bottom to the surface to gulp in atmospheric air, in case the depth is more it is disadvantageous to the fry as it has to travel a longer distance for aerial respiration. Consequently, the energy required for movement to gulp in air shall also be more This wastage of energy can be checked if rearing is done in shallow waters. Maintenance of water flow is ideal even at this stage. Aeration and prophylactic treatments have been shown to increase the survival rate Air breathing habit is started when they are 10 to 12 days. As they reach 10 to 12 mm size they are taken to bigger containers.

Feeding

Soon after the larvae start feeding they are given adequate quantity of live feed. Live rotifers and ciliates form the choise food of the larvae for three to twelve days. The feeding is done 4 to 6 times a day, *adlibitum*. In addition, Artemia nauplii also can be given to the fry for 7 to 10 days. Cut pieces of Tubifex worms are also found good to the larvae at this stage. In addition, egg custard and molluscan meat also have been tried and found satisfactory for the fry.

Preparation of Egg Custard

Collect the egg content (white along with yellow) and mix equal quantity of milk powder. Add 0.5 g vitamine-mineral premix and make a homogenate. Stem bake the same, cool and store in a refrigerator, the feed is filtered in a sieve and given to the fry.

Rearing of Advanced Fry

10 to 12 days old fry are transferred to large cement cisterns or plastic pool containing 15 cm deep water. Stocking rate can be 3000-5000 fry per square meter. Before releasing the fry, water is aerated. A feeble water flow also can be maintained. As the fry develops air breathing habit, aeration can be stopped. In this container, fry can be grown for another 10 to 12 days. By 20th or 22nd day it becomes juvenile and can be stocked in earthern nursery ponds of 0.01 to 0.04 ha size. @ 300 – 500 fry per square meter for another 15-30 days. During these period feeding can be done with finely minced trash fish or molluscan meat and rice bran (1:1) @ 5-10% of biomass daily during evening hours. These ponds shall be drainable as it facilitates collection of fingerling at the end of rearing. The fingerling can be marketed for rearing in grow out ponds.

Chapter 4
Tilapia

INTRODUCTION

Among the farmed species of fishes in inland waters, contribution of tilapia, a cichild fish, is perhaps next only to that of carps and other cyprinids. About 80 species of tilapia have been described out of which about 10 species are reported to be used for culture (Macintosh and Little 1995). During recent years, tilapias have been classified into three groups (genera) such as (1) *Oreochromis* (maternal mouthbreeders) (2). *Sarotherodon* (biparental-male/female – mouthbreeders) and (3) *Tilapia* which are substrate spawners (Pullin, R. S. V 1988, Trewas, 1982). Though originally a native of Africa, it is distributed all over the world at present because of certain distinct characteristics of this fish which are favourable for farming. It is an easy growing fish having remarkable capacity to adapt to different climatic conditions, salinities, food and ecosystems. As a matter of fact, it undergoes precocious sexual maturation under certain adverse conditions (Pullin, 1982). It is highly resistant to diseases and reproduces prolifically under natural conditions. Because of its easy breeding habits, tilapia have been extensively used for hybridization, hormonal sex reversal, induction of polyploidy etc. It exhibits good growth rate and feed conversion efficiency. The common species of Oreochromis cultured are *O. niloticus, O. mossambicus, O. aureus, O. spilurus niger, O. urolepis hornorum* and *O. macrochir.* Species of cultural importance of Sarotherodon are *S. galileus* & *S. melanotheron* and of Tilapia are *T. rendalli* and *T.* zilli. As certain details of reproductive biology, growth rate etc. vary in different species, the same are described separately for each species below. In all species, males exhibit faster growth rate than females.

Oreochromis niloticus (Nile tilapia)

Of all species of tilapia *O. niloticus* or Nile tilapia (Fig. 4.1) is the most favoured species for culture. Out of the total farmed tilapia production of 1099268 metric tones in the year 1999, *O. niloticus* alone contributed 888368 metric tons *i.e.* about 80.8% of the total (FAO statistics). It originally belonged to Nile valley areas up to Central and Western Africa and south to Ethiopean lakes and lake Turkana. (Philippart and Ruwet, 1982). About seven sub species of this species have been recognised in these

Fig. 4.1: ***Oreochromis niloticus.***
A Sample of GIFT (Genetically Improved Farmed Tilapia)
(*Courtesy, CIFA, Kausalyaganga*)

areas. Normal colour of this fish is sliver gray with markings of dark gray or black. Recently, a red variety of *O. niloticus* has been reported and the same has gained much popularity due to their consumer acceptability. The term red variety is used for various colour types such as gold, pink, crimson, orange and several other intermediate colours (Galman *et al*, 1988). These red forms are either a pure strain of this fish from Egyptian stock as reported by McAndrew *et al.* (1988) or progeny of a cross of red mutant variety of *O. mossambicus* with *O. niloticus* and progency of this hybrid crossed with either *O. aureus* or *O. urolepis* (El Gamal *et al*, 1988, Kuo, 1988).

It grows more than 600 gs within 6 months and attains upto 3 kg weight and 50 cm length in the wild. The males exhibit a growth rate 2.5 times more than female. It is an omnivorous fish feeding mainly on phytoplankton such as green algae, blue green algae and diatoms and also on detritus. It also feeds on benthic fauna. It can tolerate salinity upto 29 ppt. Temperature below 12^0C is found lethal. Because of their adaptability to wide range of culture conditions such as fresh water and brackish water, high growth rate and wide consumer acceptability this fish is commonly called as "aquatic chicken"

Reproductive biology

Sexual Dimorphism

The sexes are separate. The male has two apertures on the ventral side of the belly, one is anus and the other is urinogenital aperture. The female has three apertures, anus, urinary aperture and genital aperture. The urino-genital aperture of male is smaller when compared to the prominent genital aperture of female.

Age at First Sexual Maturity

It attains first sexual maturity on reaching 10 to 17 cm length *i.e.* 4 to 5 months old in farm conditions. In the wild, it matures on attaining 20 to 39 cm.

Court ship and Mating

The males engage themselves in active display in a breeding territory which is referred to as "lek" (Macintosh and Little, 1995) where nests are prepared by males by scooping out depressions at pond bottom. The mature female visits a "lek" when there shall be immediate courtship and mating resulting in spawning, spermiation and fertilization.

Spawning and Fecundity

A mature fish spawn 6-12 times in an year. According to Macintosh, (1985), it produces 6 batches of eggs in an year but according to Philippart and Ruwet (1982) it produces 10-12 batches of eggs annually. The number of eggs in one batch is reported to vary from a few hundreds to 2000.Larger females produce more eggs per batch and the eggs are also larger in size. They also incubate their eggs better. But the smaller females produce less no. of eggs and eggs are also smaller in size. Though the number of eggs in a batch is less, the smaller fishes (less aged) spawn more frequently as a result, smaller brood fish yields more eggs/fry collectively.

Parental Care

Soon after the eggs are laid and fertilized the same are picked up by the female in its mouth for incubation. Thus this fish is a maternal mouth brooder. Soon after taking the eggs in its mouth the female leaves the breeding territory, "lek" or nesting area. The incubation period varies from 4 to 10 days depending on temperature. It is 6 days at 20^{0}C, 4 days at 28^{0}C and only 3 days at 30^{0}C After hatching, the yolksac larvae takes 4 to 6 days for yolksac absorption. After yolksac absorption the larvae leave the mouth of the mother. At this stage they are called swimup fry and they start feeding. The mother continues to shelter the larvae for another 1 to 4 days. Larvae takes 10 to 12 days to become fully independent. During the period of incubation the mother does not feed and due to starvation during this period the female loses weight. Soon after the incubation period is over the female starts feeding vigorosusly.

Characteristics of Eggs and Larvae

The eggs are large yolky and nonadhesive. The volume of the eggs range from 2.25 to 11.15 mm^3. As stated above size of the eggs and fry vary depending on the age of the parent fish.

Reproductive Cycle

A female may take about one month to complete one reproductive cycle. This includes 4 to 10 days for oral incubation of eggs, 1-10 days for nursing the young fry, 2-4 weeks of intensive feeding and recovery phase of ovary and the period of courtship and mating.

Oreochromis mossambicus (Java Tilapia)

Next to Nile tilapia this species is widely cultured. In the year 1999, production of mossambicus through farming was 45909 tones (FAO statistics)

It is a monogamous, maternal mouth breeder attaining sexual maturity in ponds within 2-3 months (6-10 cm long). But it takes 6-9 months to mature. They breed in shallow waters having 30-90 cm water column. The males prepare the nest by scooping out saucer shaped depressions of 35-cm diameter at the pond bottom. The female lays the egg and the eggs are fertilized by the male. The eggs varying from 75-100 are taken into the mouth of the female. The eggs hatch out within 2-5 days and the larvae are retained in the mouth till the yolk sac in fully absorbed (5-8 days). The offsprings tend to remain with the parents for 2-3 weeks. When threatened, they return back to the mouth of the female for 10-15 days. During incubation of eggs and till the larvae swim out of the mouth the female seldom eats. They spawn 6-11 times in an year *i.e.* once in every 22-40 days. It breeds when the temperature is 21-23^0 C and when the temperature lowers to 11-13^0 C it stops breeding.

It tolerates salinity up to 13 to 20 ppt and feeds on phytoplankton and epiphytic algae. It grows to 14 cm in six months and grows up to 2.5 kg (40 cm). The male grows 1.4 terms more than females.

Oreochromis aureus

It is an omnivorous, euryhaline species growing well in salinities up to 19 ppt.

Temperatures less than 13^0 c is lethal for the fish and suitable temperature is $31\text{-}37^0$c. Males are faster growing, it attain a maximum size of 31. 5 cm.

They attain sexual maturity at 13.5 to 16 cm. (during second year) In ponds it matures on reaching 7.6 cm. They lay eggs in batches once in every 4-9 weeks 300 to 2000 eggs are orally incubated by the mother at a time. Incubation period is 7-8 days and young are cared for 8-10 days.

Oreochromis spilurus niger

It is also an omnivorous fish, feeding mainly on epiphytic, filamentous algae and benthic organisms. This euryhaline fish tolerates salinity up to 33 ppt. Temperature below 8^0c is lethal. Males grow faster and the maximum size recorded is 1 Kg (38 cm).

They start breeding on attaining 4 months (11-16 cm) and spawns at 3 monthly intervals.

Orechromis urolepis hornorum (Zangibar Tilapia)

A native of Tanganian river system, it tolerates salinity up to 40 ppt. Females and non breeding males are silvery or steel gray coloured with 2-4 mid lateral blotches. During breeding season, male turns black with red margined dorsal fin. It breeds at temperature between $25\text{-}35^0$c. Other breeding habits are similar to that of *O. mossambicus.*

Oreochromis macrochir (Green headed bream)

A phytoplankto phagous fish, feeds on planktonic and epiphytic algae. Tolerates salinity range of 13 to 20 ppt and grows well at $23\text{-}24^0$c. Less than $11\text{-}13^0$c is lethal. Spawns above $21\text{-}23^0$c. Males grow 1.4 times faster than than female reaching 14 cm in 6 months and attains maximum size of 2.5 Kg (40 cm).

In wild they mature reaching 10-20 cm long (8-12 months). 1000 to 1500 eggs are orally incubated at a time by mother. Spawning at 4-6 weeks intervals, 6-11 times per year. Incubation period is 5 days and hatchlings are cared for 2-3 weeks.

Sarotherodon galileus (Galilee cichild)

A planktophagous fish, tolerating salinity up to 29 ppt feed and grows between $11\text{-}16^0$c. Male grows faster attains maximum size of 0.8 Kg (40 cm).

Attain first sexual maturity at 80 g (12 cm) size. Nest making and brooding is shared by both sexes. They incubate 150-1100 eggs at a time and hatchlings are cared by parents for 10-15 days.

Sarotherodon melanotheron

They are mainly brackish water in habitat. Eggs and larvae are brooded by one or both the parents.

Tilapia rendalli (Red breasted Tilapia)

They feed on higher plants, do not tolerate salinity above 13-19 ppt, and temperature below 11-13^{0}c is lethal. Ideal temperature for growth is 31-39^{0}c but spawns at 21-23^{0}c. Grows to 2.5 Kg (40 cm) in the wild. Male crowfoots than female.

A monogamous substrate spawner, reaches maturity on reaching 12 cm (7 months). It spawns 4-8 times per year at temperature above 21^{0}c. Female deposits the egg in nest prepared by both sexes. Nest is in the form of 5-10 holes, each of 10 cm diameter made on the pond bottom. The female lays 1000-6000 eggs in a hole and the eggs are removed from hole to hole during incubation period which extends to 15-20 days. The larvae are cared for 2-3 weeks.

Tilapia zilli

It is a phytophagous fish feeding on roots and stem of rotted aquatic vegetation found in both fresh water and saline water (up to 45ppt). In saline water it reproduces when salinity is between 11-29 ppt. Grows well at temperature 20-31^{0}c. Males exhibit 35% more fast growth than female, reaches 10cm. in 3 months 26-29 cm (300-450grm.) in one year. Maximum size recorded is 3Kg.

It attains sexual maturity on completing one year (25-27cm) in wild but in pond on reaching 25-50 g within 3-5 months. They make saucer shaped nets of 20-25 cm diameter at pond bottom and lay about 7000 eggs. It spawns 6 times a year. The female guards the egg and larvae.

ENVIRONMENTAL FACTORS AFFECTING REPRODUCTION

Several workers have reported that temperature, light and salinity are the environmental factors that affect the fecundity, hatching and development of these fishes. Optimum temperature requirement of different species have been already mentioned. To generalise, it can be stated that Tilapias breed at temperature above 20-23^{0}C. Hence in tropical regions, they spawn throughout the year but in sub tropical of higher altitudes of tropics, reproduction does not take place during cold months. In case of *O. niloticus* it is seen that prolonged exposure to cold months (below 20^{0}C) causes gonadal resorptions (Mishrigi & Kubo, 1978). Some workers were successful in inducing spawning in *O. niloticus* by lowering temperature from 25^{0}C to 18^{0}C for 2 weeks and then raising back again (McAndrew as cited in Balarin and Haller, 1982).

Some scientists (Chimits, 1955; Ray, 1978; Legner, 1978) have reported that bright sun light stimulates breeding of Tilapia in ponds. But Cridland (1962) found that when *T. zilli* was exposed to strong illumination, its sexual maturity was delayed. According to him prolonged photoperiod stimulates breeding.

Salinity above 20 ppt has been shown to decrease fecundity. For optimum fecundity, isotonic concentration is reported to be ideal. For example, for *O. aureus* 4.6 ppt and for *O. mossambicus* 10-13 ppt are ideal. (Uchida & King, 1962).

Fry production is said to decrease after three successive spawning periods.

PRODUCTION OF FRY FOR STOCKING

As mentioned above, all species of tilapia are easy growing and easy breeding in different ecosystems such as fresh water, brackish water marine and in open water bodies as well as confined water. Most of the species mature very early and spawns several times in a year. In spite of all these favourable traits adequate quantity of seed for stocking is not available and this remained as a major constraint in developing tilapia farming on a commercial scale. As a matter of fact, the ability of the fish for uncontrolled reproduction became a hindrance in the culture practice because uncontrolled reproduction causes over population of the fish in aquatic ecosystem and consequent stunted growth and uneconomic returns from a given farm. Hence researches in this field were aimed at the following objectives.

(a) Standardise a seed production system from where tilapia seed can be procured and used for farming.

(b) Generate all male population of tilapia seed (monosex seed) as males are faster growing. It will also check over populating the ecosystem due to uncontrolled reproduction by female.

(c) Produce inter generic as well as inter specific hybrids that may have hybrid vigour in terms of higher growth rate and other desirable traits.

(d) Generate polyploidy in offsprings so that faster growing triploids or tetraploids are made available for culture.

Seed Production in Tilapia

Brood Stock

Though tilapia breeds prolifically in confined waters, and fry are released by the mother into the pond continuously, collection of the naturally produced fry was not sustainable because of the following reasons.

(a) There was cannibalism among the fry. Older and larger fry shall consume the smaller and weaker stage fry.

(b) Asynchronous spawning habit, Even brood fish of same age group do not spawn at the same time or synchronously. As a result of this adequate numbers of fry may not be available in the pond ecosystem at a given time.

(c) Over crowding, stunted growth and decline in the reproductive efficiency. As these fishes continue to reproduce prolifically, there shall be overcowding and stunted growth in the population. This results in the decline of reproductive efficiency of the brood fishes which results in decline in frequency of spawning and fecundity of individuals.

Because of the above mentioned limitations, for large scale production of fry under controlled conditions, advanced techniques for brood stock management became necessary. These techniques included (i) conditioning brood fish before spawning and (ii) replacement of spawned females with conditioned brood fish.

Brood Fish Conditioning

Stocking the brood fish at optimum stocking density and feeding the fish on good quality feed to meet the actual nutritional requirement at different stages of reproductive cycle are the essential aspects brood fish conditioning. As mentioned earlier, the mother fish during incubation of eggs do not feed. As a matter of fact, in the course of reproductive cycle of one month it is reported to feed only for a short duration between two incubation periods. So in order to regain the lost weight during incubation and to effectively bring about the rapid ovarian growth, the female feeds voraciously during this period. *O. mossambicus* is reported to consume feed upto 40% of their body weight 48 hours immediately after releasing a batch of fry.

Several studies have been conducted to find out the optimum protein level required for best reproductive performance. Wee and Tuan (1988) observed a crude protein level of 35% for optimum growth and spawning in *O. niloticus*. When the protein level was increased to 42-50%, earlier maturation, large eggs and higher hatching rate were observed but spawning frequency and fecundity was reduced.

A few workers have suggested to condition the brood fish of tilapia in "green water ponds" (*i. e.* ponds containing adequate quantity of algae). This provides for the natural food base of the pond as *O. niloticus* is phytoplanktophagous. This ensures healthy growth of ovary. In view of this in certain countries like Philippines, well fertilized ponds are used for conditioning the brood fish as such ponds shall be rich in green algae, blue green algae, diatoms etc. In addition, feeding on pelletted feed with 30% protein, 3 times daily till satiation, has given good results in Thailand (Little, 1989). Floating feed has been shown better for tilapia. Diet with 25-30% protein are sufficient for green water systems where as in clear water feed with 30-40% protein are found necessary.

Natural Fry Production

Traditionally followed method for obtaining fry for stocking was by netting out the swim up fry daily from a pond stocked with tilapia brood fish. Swim up fry usually shoal at the pond edge. It has been observed that fry production by this method is not sustainable as there is decline in the availability of fry in course of time. This decline is attributed to various factors such as cannibalism by older recruits, asynchronous spawning of brood fish and reduction in spawning frequency due to over crowded condition.

Collecting the Eggs/Larvae from the Mouth of the Mouth Brooding Female

Another method for obtaining the seed (eggs/larvae) of this fish was by collecting mouth brooding female and robbing the seed from its mouth. To collect the female, water level in the brood fish holding tank is lowered by draining out the water to a depth of 20-30 cm. This facilitates easy collection of females for examination of mouth. After netting out the female carefully, it is forced to spit out the eggs from the mouth and the same is collected into a basin. The eggs are taken for artificial incubation (see below). The spawned females were replaced with the ready to spawn (mature) females once in every 5-10 days. It was found that if fresh females are introduced once in every ten days, seed production was 274 no. per kg per day and when replacement was done once in every five days seed production increased to 322 no. per kg per day (Macintosh and Little, 1995).

Stripping the Brood Fish

The female fish can be stripped for obtaining eggs and milt was also collected from males by stripping. (Rana *et al.*, 1990). The fertilization is done artificially by mixing the milt with the eggs.

Artificial Incubation of Eggs

The eggs collected from the female either from the mouth while mouth brooding or by striping the female and fertilised artificially can be incubated in an incubation unit.

Incubation Unit

A recirculatory system designed for egg incubation is shown in Fig. 4.2. This consists of the following

Incubation device

Round bottom incubation container are proved good as this avoids damage to egg chorion and subsequent loss due to attack by micro organisms. 20 lit. incubation vessel is provided with an inlet pipe at the centre and an outlet pipe at the side and it can hold about 80000 eggs for incubation. (Fig. 4.2 and 4.4a).

Fry trays

The out let water from incubation vessel is allowed to flow into the fry tray which enables the collection of any hatchling that are carried out with the over flowing water.

Filtration unit

The water from the fry tray is allowed to flow to a filtration unit by gravity flow. The filtration unit may consists of a prefilter and a slow sand filter. A U.V. sterilizer also may be used after sand filtration.

Sump or reservoir

The water from the sand filter is allowed to flow into a sump.

Over head tank and pump

The water from the sump is pumped into an over head tank from where the outlet pipe is connected to the incubation vessel.

Incubation period varies depending on temperature. (6 days at 20^0C, 2-3 days at 34.5^0C, optimum being 28-29^0C at which temperature incubation period is only 4 days.

Larval Care

The newly hatched larvae or hatchlings are very delicate and it is reared in trays in a flow through system upto 20 days after hatching (Fig. 4.3).

A tray made of aluminium or plastic of the size 40 cm × 25 cm × 8 cm or 40 cm × 30 cm × 10 cm having holes drilled at the two lateral sides are used (Fig. 4.4b). Fine meshed plastic or nylon net is fixed on the inside of the tray. The water shall flow out through the sides of the tray, but the hatchlings are prevented from escaping because of the fine meshed net. The holes are drilled in such a way that 3 cm water depth is maintained in the tray. Studies conducted by Rab (1989) has indicated that the time taken for yolk sac absorption is related to flow rate. If the water flow was maintained at a higher rate it took lesser time for yolk sac absorption. On an average the time taken for yolk sac absorption was 4 to 5.5 days. It was also found that growth rate and

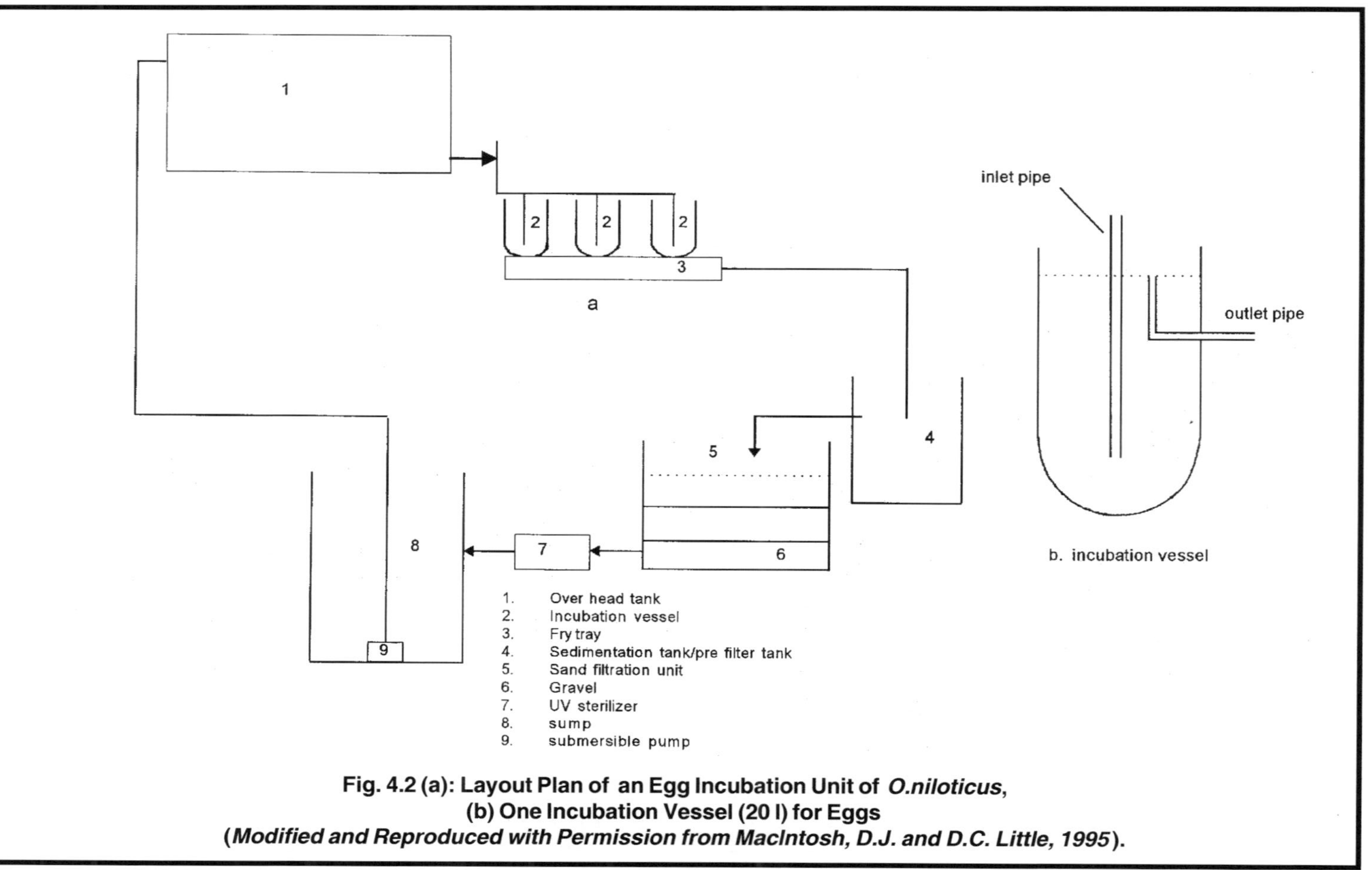

Fig. 4.2 (a): Layout Plan of an Egg Incubation Unit of *O.niloticus*, (b) One Incubation Vessel (20 l) for Eggs
(*Modified and Reproduced with Permission from MacIntosh, D.J. and D.C. Little, 1995*).

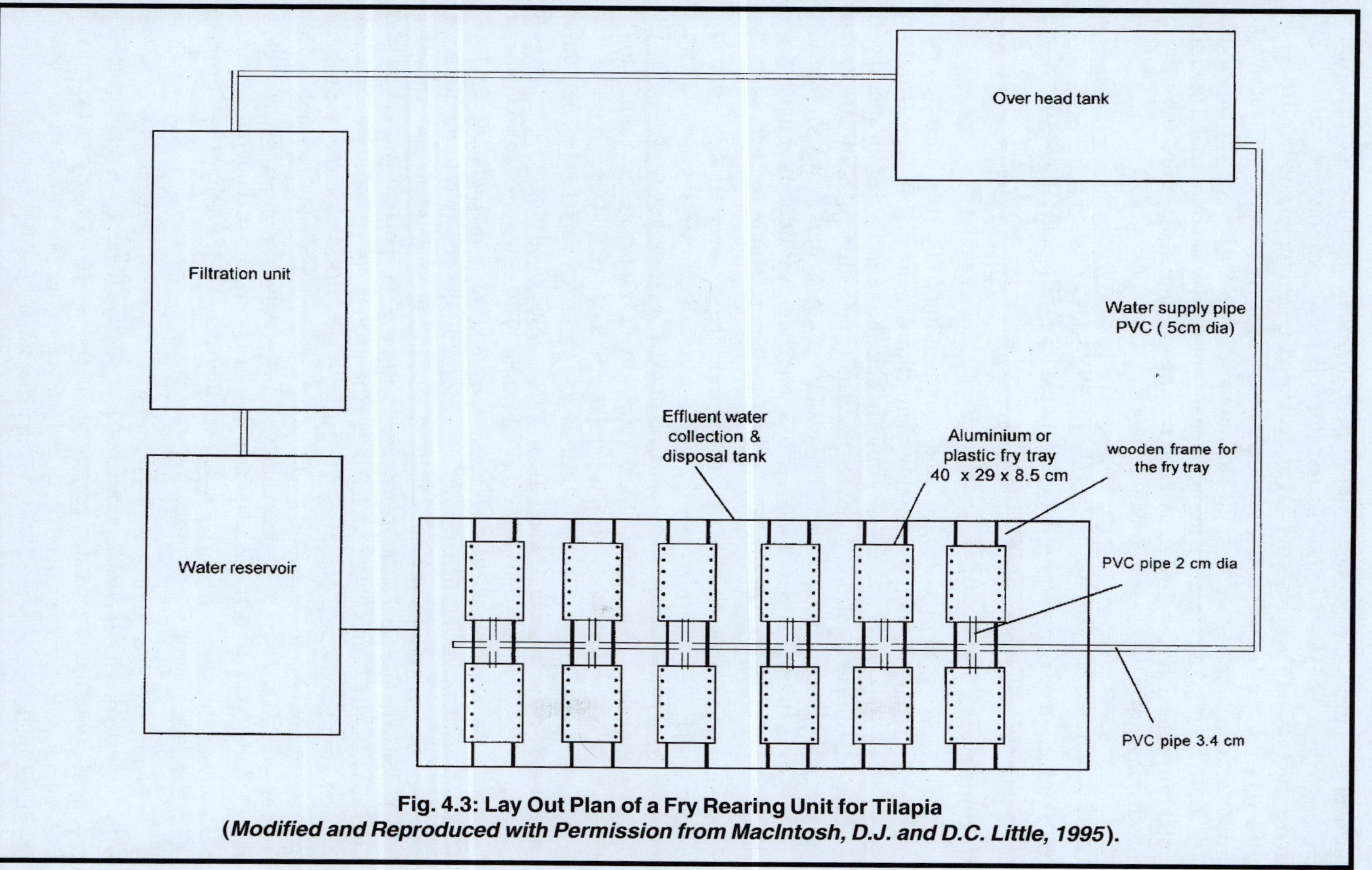

Fig. 4.3: Lay Out Plan of a Fry Rearing Unit for Tilapia
(*Modified and Reproduced with Permission from MacIntosh, D.J. and D.C. Little, 1995*).

survival of fry produced through artificial incubation was better than that of naturally produced once. The swim up fries *i.e.* fry after yolk sac absorption shall start feeding. Hence they are fed 4 to 8 times per day and retained in the tray for another 10 days on artificial feed. If a larval rearing unit contain 64 such fry rearing trays, about 0.5 million fry can be produced with the stocking rate of 10, 000 fry per tray. (assuming 80% survival rate). Thus through the maintenance of good water quality and optimum water flow rate in the incubation unit and rearing unit and also feeding the swimup fry adequate no. of tilapia seed can be produced.

Production of All Male (Monosex) Seed

Production of all male population of tilapia fry is achieved either by hormonal sex reversal of females or by hybridization.

Hormonal Sex Reversal

Sex steroids such as androgens and estrogens are utilized to bring about sex reversal *i.e.* to change genotypic female into functional males or genotypic males into functional females respectively. The hormone can be administered orally mixed with feed. Oral administration along with feed is usually done during the initial stages of larval rearing *i.e.* between 10th and 20th day of hatching (in mossambicus) when the gonad is in an undifferentiated stage. If hormone therapy is done after the gonad is differentiated in the male or female direction it will not result in sex reversal.

Results obtained by different workers in producing sex reversal using 17α-methyl testosterone [17αMT] are not consistent with regard to the dose of hormone, duration of treatment or percentage of sex reversal obtained. For example, Clemens (1968) fed 17αMT at the rate of 10mg/gm diet and 30mg/gm diet for 69 days *i.e.* from 7th to 75th day, feeding @ 6% of the body weight per day and obtained 100% sex reversal. The same author in another experiment obtained only 95% sex reversal by feeding 20mg/gm diet and 81% reversal by feeding 40mg/gm diet, though all other parameters such as duration, feeding rate, etc. were kept constant. Nakamura (1975) and Guerrero and Abella (1976) obtained 100% sex reversal by feeding 17 αMT @ 50mg/gm for 19 days *i.e.* from 7th to 36th day of hatching. Thus the minimum effective dose required for 100% sex reversal is not yet standardised. According to Pandian and Varadaraj (1988) if the fish *(O. mossambicus)* consumes 1.5 mg 17αMT/gm body weight per day, 100% masculanisation is obtained, if it is less than 1.5 mg/gm/day, some females or inter sex are also obtained. If the uptake of hormone by the fish is more *i.e.* 8 mg/gm fish day though resultant sex reversal is 100%, rate of mortality is found higher. Further, stunted growth and abnormalities in liver were noticed in sex reversed tilapia subjected to higher dose. If effective lower dose is administered, it not only results in 100% masculanisation but also faster growth than control. 17α-ethynyl testosterone (17 ET) is also used for sex reversal.

Feminization of genetic males through estrogen administration is practiced with a view to produce all male population by breeding them. If a functional female produced through sex reversal of a genotypic male is crossed with another gonotypic male all male offsprings shall be produced provided the males are homogemetic, if the males are heterogametic 25% of the offsprings shall be females.

Monosex Production through Hybridization

Hickling (1960) was the first to produce all male offsprings by crossing *O. mossambicus* with *O. urolepis hornorum*. Ever since this discovery, cross breeding works have been done between several species to produce either all male fry or all female fry. (Table 4.1).

Table 4.1: Hybridization of Tilapias* for Monosex Production (Pandian Varadaraj, 1987)

Species used in hybridization		*Sex of F_1 offspring*		*Author*
Male parent	*Female parent*	*Male*	*Female*	
O. aureus	*O. niloticus*	100%	-	Fishelson (1962)
O. u. hornorum	*O. niloticus*	100%	-	Pruginin (1967)
O. u. hornorum	*O. mossambicus*	100%	-	Hickling (1960).
O. macrochir	*O. niloticus*	100%	-	Lessent (1968)
O. variabilis	*O. niloticus*	100%	-	Welcomme (1962/63/64)
O. aureus hornorum	*O. niloticus*	100%	-	Wohlfarth (1983)
O. mossambicus	*O. spilurus niger*	-	100%	White head (1960)
O. u. hornorum	*O. aureus*	-	100%	Lee (1979)
O. niloticus	*O. leucostictus*	-	100%	Pruginin (1967)
O. niloticus	*O. spilurus niger*	-	100%	Pruginin (1967)
O. niloticus	*O. mosambicus*	-	100%	Chen (1976)

+ hybrid between *O. aureus* and *O. hornorum*

Hickling suggested the existance of at least 2 types of sex determining mechanisms in tilapia. *. In some species males are heterogametic (XY) while females are homogametic (XX) where as in some others males are homogametic (ZZ) and females are heterogametic (WZ). According to Chen, T.P. (1969) in a hybridisation, if both parents are homogametic, all male offsprings shall be obtained but all results of hybridization experiment could not be explained based on chromosome theory. According to Kornfield (1984) strong chromosomal sex determination mechanism does not exist in tilapia. According to some scientist (Avtalion and Hammerman, 1978, Tave, D. 1988, Wohlfarth and Hulata, 1983) autosomal genes play a major role in sex determination.

In most of the cases, hybrids have been shown to possess hyrbid vigour in terms of better growth rate, greater body depth and width so as to increase the fillet size. Another important point to be considered is regarding the purity of parental stock used for hybrid production, which is a necessity for generating single sex offsprings. As this is a remote possibility in an ecosystem, hybridization to produce all male progeny has not become successful always. Further poor spawning success, and low fertility due to incompatibility of breeders are also reported in these works.

As stated earlier, one variety that is used extensively for culture is called as red tilapia. According to Kuo, H (1987) red tilapia was originally discovered and genetically improved through hybridization at Taiwan fisheries research institute from the year 1968. They have produced red tilapias by crossing a mutant variety of *O. mossambicus* (having red body colour and red eyes) with *O. niloticus*. Red strain developed by them exhibited very fast growth rate *i.e.* 500-600 g within 5 months and upto 1200gm in 18 months. Further they obtained tilapias of golden yellow, brown, and white colours and some of these hybrids grew to 2-3 kg. They also got all male offsprings from a cross of female hybrid red tilapia with male *O. urolepis hornorum*. Another variety of red tilapia has been developed at Florida by crossing the red varient *O. mossambicus* with *O. urolepis hornorum* (Perschbacher, P.W. and R.B. McGeachin, 1988). According to some workers red tilapias are of unknown parentage (Payne, A.I. *et al*, 1988).

INDUCTION OF POLYPLOIDY

Another approach to generate genetically improved seed for stocking was by induction of polyploidy. This is achieved by either suppressing the extrusion of second polar body or by inhibiting second meiotic division. Chourrout and Itskovich (1983) induced triploidy in *O. niloticus* by exposing four minute old fertilized eggs to 40-41. 5^{0}C for 2.7 minutes. Pandian and Varadragan (1987) induced triploidy in *O. mossambicus* by subjecting the fertilized egg (2.5 minutes after fertilization) to 42^{0}C for 3 minutes. Valenti (1975) obtained tetraploidy in *O. aureus* by exposing freshly fertilized eggs to 4^{0}C for 15 minutes, to 11^{0}C for 60 minutes or 38^{0}C for 60 minutes. In case of *O. mossambicus* first cleavage is completed within 65 minutes; *O. aureus* requires 30 minutes for the same at 32^{0}C. Tetraploidy is the result of supression of 1st cleavage. Triploidy is also produced by mating a tetraploid with diploid.

Hepher and Pruginin (1981) observed that various species of tilapia do not mate with each other easily in containers and hence they have suggested for stripping the female and male and mixing the gametes for fertilization followed by incubation for hybrid production. This is practiced in Israel.

* *the term 'tilapia' is used in a broad sense to include all species categorised under Oreochromis, Sarotherodon and Tilapia*

Fig. 4.4(a): Close Up View of 20 Litre Incubation Vessel Forming Part of Recirculatory System for Incubation of Tilapia Eggs Shown Diagrammitically in Fig. 4.2

Fig. 4.4(b): Close Up View of Fry Rearing Tray for Tilapia in Operation (Part of the System Shown Diagrammitically in Fig. 4.3).

(*Reproduced with Permission from McIntosh, D.J. and Little, D.C. 1989*).

Chapter 5

Seed Production Technology of Brackish Water/Marine Fishes

GREY MULLETS *(Mugil cephalus)*

All species of fishes of the family Mugilidae are referred to as grey mullets. Of the 13 species of genus Mugil (Jhingran and Gopalkrishanan, 1974), *Mugil cephalus,* commonly known as striped mullets (Fig. 5.1) is the most widely distributed and of maximum commercial importance. It is because of its faster growth rate, large size attained reaching adult hood and also consumer preference in several regions of the world. Other species of mugil of cultural value are *M. capito, M auratus, M. saliens, M. chelo, M. parsia, M. tade* and *M. macrolepis*. Another species of the genus Rhinomugil *(R. corsula)* is also of cultural importance. Being a euryhaline fish, it is cultured in brakish water farms either alone or in combination with other fishes such as tilapia/ milkfish/pearl spot/Chinese carps.

Reproductive Biology

Sexual Dimorphism

The sexes are separate. The male and female are distinguishable only during the breeding season. Males are generally smaller and slender whereas, in females belly shall be distended and swollen during spawning season. On gently pressing the belly, the eggs ooze out through the genital aperture in case of female, but in case of male, the milt oozes out. The left gonad is smaller than the right one, the right one being 12.8 to 15% larger than the left. The gonad of a gravid female may weigh 1/4 of the total weight of the fish. In certain countries like West Africa, Algeria, Tunisia, Adriatic region, south of USSR, Eastern Mediterranian sea etc. ripe roes of grey mullets are of high market value in the dried and salted form.

Age at First Sexual Maturity

The age at first sexual maturity of *M. cephalus* vary depending on the ecological factors and food supply. Thong (1969) provided evidence for the first sexual maturity

Fig. 5.1: Grey Mullet *(Mugil cephalus).*

in the beginning of 4th year at Brittain. In Israel, Abraham (1963) observed first sexual maturity in 3 year old fish in Mediterranean coast whereas in lake Tiberias, the same author found sexual maturity before the end of second year. In the Tamiahua lagoon of Mexico it takes 3 years (Marquez, 1974). Males mature earlier than female.

Spawning Season and Spawning Habits

They migrate from the lakes and estuarine waters to the sea for spawning. *Mugil cephalus* breeds in India during the post monsoon period (Kowtal, 1967). In Taiwan it spawns from October to January. In Honkong it is reported to have 2 spawnings one at the end of November and another by the end of December (Bromhall, 1954). In North Carolina, USA it spawns during autumn or winter when temperatures are falling (Anderson, 1958). Thus spawning period varies in different geographical areas. Spawning usually takes place at night and it lays egg on the water surface where depth of water ranges from 15 to 900 fathoms. Usually in the spawning act one female is surrounded by several males. During spawning act, males stay at the side of female, close to the tail so that eggs are fertilized soon after it is laid.

Fecundity

Absolute fecundity (number of ripening ova in a female before spawning) and relative fecundity (number of eggs per unit weight or length of fish) have been studied. Absolute fecundity of *M. cephalus* has been seen to vary in different countries –Australia 1572000 to 2781000 (Grant and Spain, 1975) India 1320000 (Jacob and Krishnamoorti, 1948), Hawaii, 1000000 effectively released – (Nash e*t al.*, 1974). Relative fecundity of *M. cephalus* in fish caught from Hawaii has been shown as 849 ± 62 eggs per g body weight (Nash *et al.*, 1974).

Eggs and Incubation Period

Eggs of mullets are spherical and transparent of 0.93 mm diameter covered by an egg shell. In the cytoplasm one oil globule of 0.33 mm dia is present. Presence of oil globule make the egg buoyant (pelagic). The optimum temperature for incubation and hatching is between 22°C and 24°C and optimum salinity is 32ppt. Incubation period is 36-44 hours at the temperature of 22-23°C. At lower temperature, the incubation period also increases, for example at 21°C incubation period is 60-65 hours.

The hatchlings measure 2 to 3.4 mm in length. The newly hatched larvae do not feed for the first four days of larval life. The newly hatched larva remains suspended in the water column in an inclined position with the ventral side directed upwards. When 2.5 days old, they sink due to rapid absorption of yolk sac. From fifth day onwards they start feeding on zooplankton. They attain 20 mm size within 3-4 weeks.

The fry move from the sea to bays, gulfs, estuaries, lakes and lagoons where they feed and grow. Juveniles may even migrate to the rivers, canals and lakes where they feed and grow. They feed on detritus and benthic diatoms which are abundant in these areas.

Artificial propagation

Brood Stock Development

The brood stock is collected either from the natural spawning ground or are reared in farms. The brood stock has to be reared in full saline conditions (32-35 ppt). Selected breeders can be stocked in indoor tanks. The breeders are carried in a blue or black plastic bag (60 × 90 cm) provided with 30 litre of sea water and inflated with oxygen and released into stocking tanks of size 5 × 7 × 1.5 m. Fresh sea water is pumped into the tanks with the water exchange speed @ 50 litres/minute.

Induced Breeding

Mature, females of 4 to 6 year old are injected with mullet pituitary gland extract (2-5 pituitaries) along with 10-60 rabbit units of synahorin, twice at 24 hour interval. The males do not need any injection of hormone at the initial stages of breeding season. The treated fish is stripped and fertilized by the milt taken from male by stripping. The eggs are incubated in running water.

Kuo *et al.* (1975) have induced spawning in *M. cephalus* by the administration of partially purified salmon gonadotropin. HCG also has been used to induce spawning. Female needs aproxmately 60 IU HCG per gram body weight in 2 instalments first @ 20 IU per gram bodyweight followed after 24 hours with 40 IU per gram body weight (Kuo *et al.,* 1973) Sebastian and Nair (1975)successfully induced *M. macrolepis* to spawn by the pituitary of same species. 3-4 glands per female of 40 to 130 g weight through I.M. injection at 6 hr intervals induced spawning, when the females were kpet in hapas together with males (uninjected) in salt water of 29-31 ppt. salinity. But poor fertilization was reported. By stripping and artificial fertilization better percentage of fertilization was noticed. The hatching took place in a day at 26-29°C. Carp pituitary extracts have been also used to induce spawning in *M. cephalus* raised in freshwater ponds. Three injections were given to the fish first @ ½ carp pituitary per kg, second after seven hours 1 pituitary per kg and third injection after 14 hours at the rate of 2 pituitaries per kg. At the time of last injection, males are also injected at the rate of ½ pituitary per kg. The females are stripped 16-24 hours after 3rd injection and eggs are fertilized with milt stripped from male. Prior to hormone administration, the breeders were acclimatised to sea water for at least 2 weeks, as oocytes donot undergo vitellogenesis in fresh water (Abraham *et al.,* 1966, 1967). After rinsing in sea water, the fertilized eggs were transferred to incubators for hatching.

During recent years, scientists of CIBA, Madras have induced *M. cephalus* to spawn with the aministration of ovaprim at the rate of 0.5 ml/kg after priming the fish with a dose of HCG @ 1000 IU/kg of fish.

Incubation and Hatching

Incubation and hatching can be done in round plastic tanks of 0.5 to 1.0 ton capacity. In the flowing water type, 2-3 conical hatching nets were hung in the water. Their tail end was connected to inlet pipes though which fresh sea water is let in. The eggs are released into the conical nets. The upward flow of water inside the conical bag cause the continuous rolling movement of eggs which helps in ensuring adequate oxygen supply to the developing embryo and removal of metabolic wastes from the hatching unit. If incubation is done in stagnant water, water has to be changed when water quality becomes poor. Continuous aeration is also done in the hatching tank. The incubation period is 16-30 hours at 20-24°C. The larvae are small in size measuring 2.5-3. 5 mm. They avoid light.

Larval Rearing

Larval rearing is done in plastic tanks of 0.5 ton capacity kept indoors. Larger out door tanks can also be used for the purpose but it should be covered with plastic sheets to prevent sudden rise and fall in temperature.

The newly hatched larvae do not feed for 2 days and the feeding begin on the third day as the yolk sac get absorbed and distinct mouth is developed. There are 2 critical periods in the developmental stages, first during 3rd and 4th day of rearing (which may extend even up to 7th or 8th days) and another on 11 th and 12th day. Heavy mortality is reported during these periods. 45 days old young ones of 32.8 mm length are found suitale for stocking as by this time the larva develop high environmental resistance and survives upto 4.15 ppt. Initially, salinity of the rearing medium is maintained at 32-33 ppt whih is gradually reduced in the course of 45 days rearing to 4.15 ppt. During the course of rearing the larvae not only increase in size but also there is marked changes in colouration, habit and phototaxicity.

The larvae exhibit phototaxis on 4th day and they congregate at water surface during night. From 3rd day, eggs and larvae of oyster, blue green algae and ciliates are provided as food. By 6th day onwards, it makes up and down movement in water and rotifers and copepods are incorporated in the feed. By 13th day, the larva measures 6.6 mm length and develops black spots scattered on the body and silvery green appearance on the belly. 20th day, it grows to 7.65 mm in length when the body colour is changed to brown or silvery green and it exhibits phototaxis during day time, but floats on the water surface during night. Artemia nauplii are given from 20th to 45th day. By 24th day, silvery scales start appearing and they remain scattered on the surface but get congregated under light. 28 day ol larva measures 15 mm having all scales and fin rays well developed and silvery green colour also is developed. By 29th and 32nd day, they swim in the middle or lower level of water in daytime but floats on the surface at night and measures 18. 6 mm in length. By 34th day, it changes to green colour and swims at the bottom during day time but floats on the surface during night. By 45th day, when they attain 32. 8 mm length they are ready for stocking.

MILK FISH *(Chanos chanos)*

Introduction

Milk fish, *Chanos chanos* (Fig. 5.2) is cultivated extensively in many South East Asian Countries like Indonesia. Philippines, Taiwan etc. It is a euryhaline fish that can tolerate not only wide range of salinity varying from hypersaline to freshwater but also low oxygen levels in water and wide ranges in temperature varying from 15°C to 40°C. Though they grow well in freshwater and brackishwater, adults move to the sea for spawning. They spawn offshore near coral reefs or small islands. The eggs and larvae are pelagic. 2-3 week old larvae move inshore both by passive advection and active migration. After passing through shore waters and surf zones, they settle in mangrove swamps and coral lagoons where they stay for a few months. Some juveniles may enter freshwater lakes also.

Out of the 400000 ha of milk fish farm that has been reported to exist in the world, 183000 ha. are in Indonesia, 176000 ha. are in Philippines and 15600 ha. are in Taiwan. Brackishwater ponds and freshwater pens in South East Asia produce about 346000 metric tones of milkfish annually. During recent years, some attempts have been made on a small scale for culture of this species on a small scale in the coastal regions of India and SriLanka. To meet the seed requirement of the milk fish farms, fry collection from natural resources was followed in all countries earlier. But by this time, techniques have been developed to induce breeding and obtain seed of desirable quality of this commercially important fish.

It is a long living fish, having a life span of more than 18 years. A pelagic fish exhibiting *schooling* behaviour it grows to attain a length of about 1.5 meters and a weight of 15kg.

Reproductive Biology

Age at Sexual Maturity

In nature and in large floating cages, milk fish reach sexual maturity in 3-5 years but in ponds and concrete tanks it may take 8-10 years (Bagarinovo, T. 1994) According

Fig. 5.2: Milk Fish (*Chanos chanos*).

to Liao and Chen (1984) females take 6 years for ovarian maturation whereas males mature at 5 years but according to Lin (1984) successful spontaneous spawning occurs only at the age of 10 years. Farm reared fishes exhibit mature gonad on attaining a weight of 3 kg and reaching above six years of age. Hence, for seed production brood fish of 4 kg weight and above are recommended.

Sex differentiation

Males can be distinguished from female by examining the urinogenital region. Males have two pores externally where as females have three pores (Choudhuri *et al.*, 1976).

Breeding Season

Breeding season varies in different georgaphical areas. In Indonesia there are two breeding seasons a year, one from March to May and another from September to December peak being October to November. In Philippines Taiwan and India it extends from March/April to August-peak in Philippines being May-June and in Taiwan and India being April-May.

Brood Stock

Brood Stock Raising

The fish that are to be used for seed production can be either raised in the farm or can be caught from the natural habitat. Collection of wild brood stock in the live condition without injury is uncertain and hence farm rearing of juveniles or sub adults to maturity is desirable. But the long period of 5 years of rearing required to develop a breeder is a major set back in the expansion of milkfish culture on a commercial scale. In spite of this, polyculture of milkfish along with other species of fish such as mullet, groupers and Tilapia is practiced in many countries

Brood Stock Feed

As milk fish is an omnivorous fish, it feeds on a wide variety of food items. They feed mainly on algae lab lab (bacteria, diatoms, green algae-chlorophyceae), animal components like protozoans, flat worms, larvae and adults of mulluscs, polychaete worms. copepods, large decapods, and insects. Hence the pond is prepared to ensure growth of best natural food of the fish (lab lab) in the pond. Usually, before stocking the fingerlings the ponds are dried and all predatory fauna are completely eradicated. Chicken manure @ 2 tones per ha. is spread on the pond bottom and water is let in so that it forms a layer covering the bottom. Then inorganic fertilizer (N:P:K = 16:20:0) @ 150 kg/ha or N. P. K. (18:46:0) @ 75 kg/ha is added after 2–3 days. Then urea @ 25 kg. /ha. is applied to the pond. This helps in speeding up the breakdown of chicken manure within one week. After this treatment lab- lab growth starts in the pond. At this point water level is gradually raised by instalment raised to 25-30 cm within a period of 1 to 1.5 months, each time raising the water level by 3-4 cm. If the water level is suddently increased to 30 cm, lab lab may detach from the bottom. Other organisms that feed on or disturb the lab lab are also to be eradicated. If salinity is lowered with the addition of freshwater, growth of filamentous algae dominated by chaetomorpha, cladophora and enteromorpha occur. This is called lumut in Philippines. As chaetomorpha is too coarse and fibrous for fry and fingerlings it is not recommended for these stages. In addition to natural food, formulated feed consisting of rice bran, wheat meal, soybean meal, and formulated eel meal was given as food for breeders (Lin, 1985). Another formulated feed consisting of 70% formulated eel feed, 14.75% wheat germ meal, 14.75% soybean and 0.5 % vitamin E was also used as feed for spawners.

Brood Stock Conditioning

About 2 months prior to the breeding season, breeders are transferred to the spawning pond. If spawning pond is small and if stocking density is higher, the fish may jump out of the pond and get injured. To prevent this, a fence of 1 meter height made of nylon net is fixed along the dyke. In order to transfer the fish from one pond to another a plastic bag with water is used for carrying the brood stock (Su-Lean Chang *et al.*, 1993).

Spawning

The brood fish can be either induced to spawn through the administration of pituitary extract/ other ovulation inducing agents or they can be allowed to have spontaneous spawning in the pond.

Inducing Spawning Through Hormone Therapy

Milk fish has been induced to spawn by the injection of partially purified salmon gonadotropin (Vanstone *et al.*, 1976, Nash and Kuo, 1976), mullet pituitary homogenate along with HCG (Liao and Chen, 1979), carp pituitary homogenate with HCG (Kuo *et al.*, 1979), Salmon pituitary homogenate with HCG (Liao *et al.*, 1979) or HCG alone (Lin, 1984). Marte *et al.* (1987) induced the milk fish to spawn using either mammalian or salmon gonadotropin releasing hormone analogue (m GnRH-A/s GnRH-A). Either the hormone pellets (100 µg GnRH-A/fish) were implanted intra-peritoneally or it was administered by intramuscular injection (10 µg/kg m GnRH-A). Treated females along with oozing males were released into floating net cages lined with plastic to retain spawned eggs. When the spawning was over, the eggs were collected with a plankton net and incubated. Several other workers have collected the eggs from female by stripping and then fertilized it with the milt collected from male. (Liao, I. C. *et al.*, 1979; Vanstone *et al.*, 1977., Kuo, C. M. *et al.*, 1979)

Spontaneous Spawning Without Hormone Therapy

Mature milkfish also spawns spontaneoulsy in ponds either in first quarter (1-7th) last quarter (24-30) and mid month (12-18) of a lunar cycle. Usually the fish spawns after mid night though the chasing behaviour of males start before sunset. During courtship, very often the dorsal fin and caudal fin are found projecting out of water when male chases the female. Optimum temperature for spontaneous spawning was 29°C-33°C and optimum salinity was 26 ppt. The first spawning in Taiwan was reported in April, peak being May to July, from August to October the spawning declined.

Incubation of Eggs

The fertilized eggs of milk fish are spherical, transluscent and floating measuring 1.5 to 2.5 cm in diameter. It has a narrow perivitelline space and there is no oil gobule in the ooplasm.

From a spawning tank/ floating net eages where breeders are released, the eggs are collected before dawn by a scoop net. Some workers prefer to collect the eggs by installing small paddle wheel aerators in the four corners of the pond in such a way that the movement of the aerators create circular water current. An egg collecting net is fixed at the water surface in such a way that that the eggs that are carried along with the water current enters in to the net along with water current. The eggs that are cleared of the detritus are measured and tranferred to well aerated tanks of 0.5 ton capacity for incubation @ 1-3 kg of eggs per tank. One kilogram of fertilized eggs shall contain about 800000 eggs. The incubation period is 20-35 hour at temperatures 26-32°C and salinity 29-34 ppt.

Development and Hatching

Development of egg and larvae of milk fish has been studied by Liao, I. C. *et al.* (1979)at 27-32°c and 34 ppt salinity. According to these authors the fertilized eggs undergo segmentation and reach blastula stage about five and half hours after fertilization, gastrula stage with germ ring formation six hours after and late gastrula stage after 8 hours C-shaped embryo with 22 myomeres and optic vesicles in seen after 14.35 hrs. twitching movement after 21. 30 hrs and hatching of fully formed embyo after 25. 45 hrs.

Larval Development

The newly hatched yolk sac larva measure 3.4 mm long with a large yolk sac of 2.2 mm length and 0.28 mm width. Larva is slightly curled. Mouth and anus are not opened, eyes are unpigmented 2 hrs after hatching dorsal fin fold get broadened, body length also increases. It takes three days for the yolk sac to be completely absorbed. Feeding behaviour is conspicuous from this stage onwards. Critical period in the larval development is between 4th and 6th day when 80% of the mortality is observed.

For the development of hatchling into the fry stage, about 21 days are taken. During this period, changes in the body size, food habit and behaviour can be noticed. Growth of the differentiation of body parts also takes place.

Changes in Body Size

On the day of hatching, the hatchling measures 3.2-5. 3 mm in length but 21 says after hatching it measures 13.5 –16. 5 mm in length.

Larval Behaviour

From the day of hatching upto one day old, the larvae remain suspended in water with head down near the water surface then sinks slowly in an obleque position with the belly up, then make quick upward turn by 360° and then swim rapidly to water surface. On day 2, the larvae make 360^{0} horizontal turn and rest in mid water near the surface or near the bottom.

From day 3 to day 13, the larvae exhibit phototaxis and rheotaxis (swimming against the current) during day time but during night they drift with the current.

On day 14 and onwards they swim swiftly and circularly in daytime and swim slowly to counteract the current during night. From 18th to 20th day, larvae are not sensitive to direct exposure to sunlight

Growth and Differentiation of Body

On 1st day after hatching, caudal part of finfold is more constricted. Rudimentary pectoral fins appear. Eyes are not pigmented. Anus is closed. Eyes and caudal fin fold (posterior part) are unpigmented. On 2nd day, pigmentation of eye is started, rudimentary lower jaw appears. Mouth and anus are well demarkated. On 3rd day, pigmentation of eye is conspicuous yolk is completely absorbed, air bladder is formed.

On day 4& 5, further differentiation in fin fold is not seen, heart gets developed, optic lens is also well formed. 6th & 7th day, gill filaments start appearing, pectoral fin are well developed, caudal fin start getting bifurcated. 8th & 9th day operculum start developing. On 10th day, dorsal and anal fin fold get differentiated. Body becomes transparent and stomach contents are visible. On the 11th day, dorsal and anal fin get separated from caudal fin. On 12th and 13 th days caudal fin folds get completely separated from the dorsal and anal fins and finrays also start developing. By 21 st day, pelvic fins also start developing.

Larval Rearing Tanks

Each out door larval rearing tank is of 200-300 square meter size haivng 1.1 to 1.2 meter depth. Bottom of the tank may be made of 20 cm thick clay which is meant to prevent water leakage. Over this, 30 cm thick sand is put which keeps the water clear and makes the harvesting of fry easy. The water is aerated by means of air stones kept at every 2.5 meter interval. The tank is covered by means of plastic shade to prevent strong sunlight into the tank before the larva is 10 days old. 2-3 larvae per litre is the suitable density of stocking. Light intensity for the larvae range for 1000-3000 lux.

Larvae reared in out door tanks have been shown to growth fastes than those cultured in indoor system. Faster growth rate is observed from 8th day onwards

Larval Feed

Algae (chlorella) rotifers, eggs and and larvae of oysters, fish meal, micropellet feed and cultured micro-organisms are used as feed for the larve. There is no evidence of fish larva feeding directly on chlorella but chlorella may form the food of rotifers on which the fish fry shall feed. On the first day after hatching only pond water rich in chlorella (green water) is added to the pond. Density of chlorella can be between 50 and 350×10^4 cells/litre. From 3rd day to 8th day, fertilized eggs and larvae of oyster are given as feed. Fish meal/micropllet feeds can be given from 15th day onwards. At the end of 20day it grows to 1.5 cm.

Through milk fish larvae are not prone to diseases, they are found to suffer from a disease called bubble disease. Affected larvae float on the water surface and die soon after. Bubble disease is checked by controlling light intensity, algal species and algal density. Diatom bloom is considered harmful as it causes the above disease.

SEA BASS (*Lates calcarifer*)

Introduction

Sea bass (Fig. 5.3) is a euryhaline fish living in brackish water estuaries and also in freshwater. Sea bass found in the mediterranian and Atlantic are called European Sea-bass *(Dicentrarchus labrax)* where as the Asian Sea-bass *(Lates calcarifer)* is found in the littoral waters of Indo-West Pacific region extending from Iran to Australia. Culture of sea-bass is popular in countries like Thailand, Hongkong, Singapore and Malaysia. It fetches high market price as it is in high demand. It is also called giant perch (family centropomidae). Culture of this fish has received much attention only during recent years in countries like, India and Philippines. For spawning they migrate to sea and the juveniles migrate back to the brackish water for feeding and growth.

Reproductive biology

Sexuality

L. calcarifer is a protandrous hermaphrodite, being a male initially changing to a female later. All the males may not become female and some are also primary females. Hence, in the smaller size groups *i.e.* 1.5-2.5 kg body weight, males predominate. In the 3-4 year old population both sexes can be found. Male has a slender body and its abdomen does not bulge out. It has a slightly curved snout and milt oozes out on slight pressure on abdomen. The belly of female is soft and it it bulges out. A red pink papilla also extends out at the urinogenital aperture of female

Spawning Habits

Natural spawning ground is the sea near the mouth of big rivers and lakes where the water is 5-10m deep at a salinity range of 28 ppt to 30 ppt and water temperature of 27-34°C. The spawning usually takes place 3-7 days after full moon and new moon between 6pm and 11 pm during low tide. *L. calcarifer* is an year round spawner with the peak season being from April to August. In Thailand, natural spawning of sea bass in tanks and in floating net cages has been reported during late June to October with the peak breeding in June and July (Parazo *et al.*, 1998). Sea bass

Fig. 5.3: Sea Bass (*Lates calcarifer*).

is a protracted spawner *i.e.* eggs are released in batches by the same fish. Hence the same female has been induced to spawn three times in a season at an interval of fifteen days.

Brood Stock Maintenance

Brood fish can be collected from fish farm or from natural waters and reared in separate earthen tanks/ cages/ concrete tanks. 2-6 year old fish having an average weight of 3-5 kg shall be useful for brood stock developments. Wild brood stock are transported to hatchery or holding tanks (cages). The fish are treated with antibiotic @20mg/kg of fish by injection or in 2 ppm medium for dipping for 24 hours. They are fed on trash fish. In Indonesia, floating cages are used for brood stock maintenance. Polytheylene net (mesh size 4 to 8cm) attached to G. I. pipe or wooden frame form the cage and the same is kept afloat by styrofoam drum and anchored in calm Bay area. Cage may be of the size 5x2x2m or 10x10x2m. Fish is stocked at the rate of 1 fish per m^3 of water. Concrete tanks when used for holding the brood stock can be of size 5x10x2m or 10x10x2m stocking density in a concrete tank may be 1 fish per $2m^3$ of water. Water replenishment @30% of the water daily is recommended for the concrete tanks. They are fed on fresh trash fish @5% of the body weight by 4pm. Active, healthy, injury free and parasite free males and females of sea bass weighing 3-5 kg body weight are selected as spawners and are released the pre-spawning tank. The ratio of male and female shall be 1:1. The feeding on trash fish is reduced to 1% of the body weight and reared for 2 months before undertaking spawning operation.

Assessing the Stage of Ovarian Maturation

A catheter of 1.2 mm diameter is introduced into the oviduct through the urinogenital aperture. The eggs are sucked into the cannula orally. Then the cannula is withdrawn. The eggs are taken on a slide and diameter of the eggs is examined. If it is 0.4 to 0.5mm, the females are ready for injection.

Spawning, Fertilization and Incubation

There are 2 methods of seed production such as:

(a) Stripping and artificial fertilization, and

(b) Induced spawning.

Stripping and Artificial Fertilization

Egg collection from female and milt collection from male is done by stripping the fully mature fishes. Dry method is usually followed as the egg is collected into a container having no water and then milt is added into it and mixed by means of a quill feather for a few minutes (5 minutes). Filtered sea water is added into this mixture and the same is allowed to stand undisturbed for 5 minutes. The eggs are counted, and rate of fertilization is also calculated. Fertilized eggs are transferred to incubators for hatching.

Induced Spawning

Two methods are followed for inducing sea bass to spawn in captivity such as:

(a) Hormonal therapy, and

(b) Environmental manipulation.

Hormone Therapy

Hormones used for inducing spawning in sea bass are HCG with carp pituitary, Puberogen and LHRHa with dopamine inhibitors (ovaprim)

HCG with carp pituitary : The spawners are usually given 2 intraperitonial injections at the base of pectoral fin. The first injection is of 50 IU HCG and 0.5-1 pituitary gland/kg body weight. The second injection is given after 12 hours with 100-260 I. U. HCG and 1.5-2 pituitary gland/kg body weight. Within 10-12 hours after second injection spawning occurs. (Pillai, T. V. R. 1998) The fish repeatedly spawn over a period of 3-5 days and a total of 2-17 million eggs can be obtained from a single spawner, depending on the size of fish. Some authors have recommended carp pituitary 2-3 mg/kg of fish with HCG at 250-1000 I. U. /kg of fish

Puberogen

It contains 63% FSH (Follicle Stimulating Hormone) and 34% LH (Luteinizing Hormone). The spawners are given two intra muscular injections 50-200 I. U. /kg body weight of the female and 20-25 I. U. /kg fish for the male. The fish usually spawn within 36 hours after injection. If no spawning occurs, a second injection may be administered 48 hours after the first injection. The dosage of second injection shall be double that of first injection for female but for male it may be 20-50 IU/ kg/fish. The female spawns 12-15 hours after the second injection.

LHRH- A (analogues)

During recent years LHRH analogues have been used to advance gonadal maturity and also for inducing spawning in sea bass.

Inducing Precocious Gonadal Maturation

Parazo *et al.* (1998) reported advancement of gonadal maturation at least 2 months earlier than the known peak period in the wild by implantation of LHRH a pellet at a dose of 0.1 mg/kg body weight in fish maintained in floating net cages at South east Asian Fisheries Development Centre, Phillippines.

Inducing Spawning

Through LHRH-a Pellet Implantation

Nacario, J. F. *et al.* (1984) reported that implantation of LHRH analogue in pellet (50-100 mg/fish) in 'ready to spawn fish' induced spawning 2-4 days after the implantation. AT SEAFDEC, Phillippines, Parazo *et al.* reported that LHRHa pellets containing hormone 5-75mg/kg body weight if implanted into ripe male and female and released into spawning cage (ratio of male:female 2:1) the fish spawns 2 nights after implantation. Mature female implanted with 5-10 mg LHRHa/kg spawned once, 20-40 mg/kg group spawned 2-3 consecutive nights and 40-70 mg/kg group up to four consecutive nights.

Through LHRH-a Injection

Nacario J. F. *et. al* (1984) reported spawning in sea bass 2-4 days after injections of LHRHa @10m g/kg body weight 24 hours apart. Parazo *et al. (1998)* recommended injection of 20 to 100 mg of LHRH-a per kg, body weight to female and 40 mg. /kg. body weight to male. Fish administered above 20 mg. /kg spawned in three consecutive nights. According to scientists of Central institute of Brackish water Aquaculture LHRH-a is administered intramuscularly in a single dose @60 to 70 mg/ kg. body weight for females and 30 to 35 m g/kg. body weight for males and released in to spawning tanks of 15 to 20 tons capacity. Care is taken for maintenance of good water quality and providing aeration. The fish spawns after 35 to 36 hours of hormone therapy.

Induced Spawning through Environmental Manipulation

This is done by simulating the changing environmental condition during natural spawning activity. When the fish migrates from estuaries/brackish waters to the sea there is an increase in salinity, decrease in water temperature after rains and also a change in water level during rising tide. Hence in the prespawning tank, salinity is increased from 20-25 ppt (before stocking) to 30-32 ppt within 2 weeks after stocking. This will simulate the increasing salinity from growing ground to spawning ground. The fish behaviour is monitored closely. When the fish exhibits prespawning behaviour, segregation of sexes may be done, and feeding is stopped one week prior to spawning. As the natural spawning takes place during full moon or new moon male is released with the female, 2-3 days before the same in the spawning tank. Male and female swim together near the surface as spawning time aproaches. At the beginning of new moon/full moon, the water temperature in the spawning tank is manipulated by reducing the water level in the tank to 30cm deep at noon time and exposing to the sun for 2-3 hours. This results in increase in water temperature to 31-32°C. Then filtered sea water is added to the tank and rise in water level simulates the rising tide in the sea. This also reduces the temperature drastically to 27°C to 28°C. This induces spawning in the succeeding night. In case there is no spawning entire process is repeated untill the spawning is achieved.

Incubation and Hatching

The fertilized eggs of sea bass measure 0.78 to 0.8 mm in diameter and are found floating and are transparent. The unfertilized eggs are opaque and sink to the bottom. In a single spawning 0.75 to 1.5 million eggs can be obtained. The eggs are collected in the morning by means of a soft and fine meshed seine net (mesh size 0.4 to 0.5mm) and transferred to the sea water in a fibre glass tank or hatching jars. Before the eggs are kept for incubation dip them in 5 ppm acriflavine followed by repeated rinsing by salt water. Density of eggs in hatching jar can be 1200 nos/litre or less. Gentle aeration is done in the jar. Hatching occurs 17- 18 hours after fertilization at 26°C-28°C and 32-33 ppt. Best hatching is found between 20 and 30 ppt.

Larval Rearing

Stocking

The newly hatched larva is 1.5 mm in length. They are transferred to larval rearing tank and stocked at a density of 30 nos/litre or even up to 90 individuals per littre. When larva is 10 days old density may be reduced to 15 nos/litre and further to 6 numbers when 20 days old.

Larval Rearing Tanks

Rectangular concrete tanks having facilities for water inlet and the outlet located usually opposite the inlet and aeration are used for larval rearing. Circular tanks having a conical bottom and drainage pipes located centrally are also used for the purpose. The level of drainage pipe is adjusted to maintain the water depth.

Feeding

The newly hatched larva do not feed up to 50 hours or for three daysafter hatching. Then they start feeding. If no feed is given within the next 10 hours, mass mortality may occur. Hence approximately 36 hours after the hatching rotifers (Brachionus) is added in the rearing tank (at the rate of 15-20 rotifers/ml) each day until the 12th day. Addition of 60 litres of algae (Chlorella) is also done to the rearing medium to attain a density of $1\text{-}3 \times 10^5$ cells/ml. This serves as food for rotifers. From day 12 to 15th day, rotifer density is reduced so that by 15th day no rotifer is found. Instead from day 12, add newly hatched Artemia nauplii in increasing densities of 0.5 to 2.0 nauplii per ml per day. From day 15 to 17 increase artemia nauplii density to 5-10 individuals/ml/day. From day 18 to 23, enriched artemia nauplii at a density of 5-10 individuals/ml/day is provided. From day 23 onwards they are fed subadult/adult Artemia biomass at the rate of 1 individual per ml or more. From 17th to 20th day, Moina also can be fed @1 individual/ml. Reduce the salinity to 10 ppt during this period and feed the larvae at least 4 times daily. Brackish water cladoceran; *Diaphanosoma celebensis* can also be given from 15th day @2 nos/ml at 32-35 ppt.

In addition to the live feed mentioned above, frozen Artemia, frozen Moina, minsed trash fish and artificial diets are also tried during the larval rearing.

Water Quality

From 4th to 14th day, 15 to 50% of the water is replaced daily. From the 15th day onwards, 50-75% of the water is replenished. If artificial feed or any other non live feed is given 100% of the water is to be replaced. Every day, from the 3rd day onwards, faecal matter, feed residues etc. are to be siphoned out from the bottom of the tank.

Size Grading of Larvae

Sea bass larvae are highly cannibalistic. This reduces the percentage of survival Some among a population of fry grows faster and these are called "shooters". These shooters usually consume the smaller size fry. So in order to reduce cannibalism, the frys are segregated size wise into 3 categories - large, medium and small using a 'fry sorter box'. Inner most box made of net with biggest mesh size is placed in a middle box of intermediate mesh size and this is kept in the outer most box of fine mesh net. Using a wide hose pipe of 5cm diameter, siphon out the fry into the inner most box. Smaller fry shall pass through the large mesh while the bigger ones are retained inside. In the same way, the middle box shall retain the medium sized one and the outer most shall contain the smallest one. The three size groups are to be released into 3 separate rearing tanks. As cannibalism is possible only when another fry is less than 2/3 of the length of a given fry, size grading reduces the chances of mortality due to it.

First grading is done a few days after the Artemia feeding is started as shooters usually develop at this time.

21 day old fry reaches an average of 1 cm in length and the same are harvested for nursery rearing.

GROUPER (*Epinephelus*)

The grouper (fam. *serranidae*) is a marine fish of economic value as it is highly relished as a food fish in many parts of South-east Asia and middle east. There are different species of groupers such as greasy grouper or estuarine grouper (*Epinephelus tauvina*), brown marbled grouper (*E. fuscoguttatus*), tiger grouper or small teethed grouper (*E. microdon*), red grouper (*E. akaar*), Malabar grouper (*E. malabaricus*), red spotted grouper (*E. suillus*), coral trout (*Plectropomus leopardus*) and *Promicrops lanceolatus*. Of these different species, the estuarine or greasy grouper and brown marbled grouper are the important species cultured in countries like Singapore, Hongkong etc. These fishes are reported to grow to a large size. For instance, *E. tauvina* of 300kg weight and 2.36 meter long, *E. malabaricus* weighing 197kg and 2.3 meter long, and *Promicrops lanceolatus* of weight 250 kg and of length 2.41 meter have been recorded from Indian coasts such as Visakhapatnam, Goa and Kanyakumari respectively. In view of its high demand as a food fish and in view of the fast depleting natural resources due to over fishing, attempts have been made to culture this fish in floating net cages in sea in countries like Singapore, Taiwan, Hongkong, Malaysia and Kuwait. The net cages are stocked with fry collected from wild by fishermen. Some of these countries have made some progress in the artificial propagation of these species through induced spawning and larval rearing under controlled conditions.

Reproductive Biology

Groupers are protogynous hermaphrodites, that is to say, a fish matures as a female but later as it grows to 74cm long and 11kg weight, changes into a male. Fish of size range between 66 to 72cm are usually intersex. The sexual dimorphism is not distinct except that the mature females have more distended and bulging abdomen than the male. *E. tauvina* takes more than five years to become a functional male. The female grouper grows faster than male. It has been observed that all females of a given population do not undergo sex reversal, some times sex reversal may be limitted to only 2% of the female population of age group 4-7 years. Hence sex ratio of these fishes is biassed towards female.

Groupers are year round spawners, though a given fish may not spawn in some months of an year such as May and August or June to September as reported from Taiwan in case of brown marbled grouper. They also exhibit lunar spawning rhythm as spawning occurred once in a month *i.e.* from the last quarter of full moon to just before the beginning of new moon. Spawning usually takes place in the night and each spawning run lasts for 2-5 days. It may lay about 3.7 million eggs per spawning run as reported from the experimental fish farm located in the east Johore straits of Taiwan. Size of the egg varies in different species - 0.7 to 0.83 mm dia in red spotted grouper, 0.77 to 0.9 mm in greasy grouper and 0.86 mm in black spotted grouper. The fertilized eggs are buoyant where as dead eggs sink to the bottom.

As there is scarcity in the availability of males in a given population, attempts have been made to produce male brood stock by administering the male hormone, methyl testosterone to females either orally (1mg/kg body weight through feed given three times a week for 2 months) or by implantation (Chao and Chow, 1990, Cheng *et al.*, 1977)

Natural Spawning in Captivity

Mature brood stock of brown marbled grouper (females weighing 10.7kg and 78. 9cm length and males weighing 9.9kg and 71cm long) were collected from the wild and stocked in floating net cages of 5 × 5 × 3 meter size at a stocking density of 0.8 kg fish per m^2. A spawning net of 0.5 mm mesh size was set inside the net cage so that eggs will not be lost into the surrounding water. The fish is fed on trash fish @2. 5% of the body weight daily. The fishes spawned naturally in the floating net cages as noted above. Similar natural spawning of *E. tauvina* in captivity was reported by Hussain, N. A. and Masaki Huguchi (1980) at Kuwait institute of scientific research. These authors have stocked the brood stock of this species in spawning tanks of 90m^3 size provided with continuous flow of sea water.

Induced Spawning

Ripe females of size 6.6 to 9.41 kg and males weighing 3.7 to 5.7 kg were injected with H. C. G. (53 to 76 I. U. per kg for males and 88 to 135 I. U. per kg for females) and stocked in circular spawning tanks of 18 m^3 size (2. 4 meter dia and 2.5 meter height) on shore. The fish started spawning in the second night after injection and spawning lasted for 2 days. The egg production was to the tune of 0.66 million per spawner. (Chao, T. M. et al., 1993). Kohno *et al.* (1990) obtained 4 million eggs from a female of 5-7 kg size in a 2 day spawning run after induction of spawning.

LHRH analogue has been used by some workers to induce spawning in groupers stocked in floating net cages (Chao, T. M. *et. al*, 1993). The spawning net cages of 0.5 mm mesh size net was provided inside each cage net. According to them administration of LHRH-a @0.15mg/kg resulted in spawning for 8 days. An average egg production of 1.41 million eggs per spawner was recorded. A note worthy observation of the authors is that, in the first 3 days of spawning eggs were of poor quality *i.e.* buoyant eggs were 0-25. 2% and fertilization rate was 0-4% only. According to these workers, poor quality eggs were due to the effect of LHRH-a where as in the next few days, better quality eggs were induced by the pheromones disseminated into the water due to the first 3 days spawning. This has to be confirmed by further studies.

Incubation Hatching and Larval Development

Rectangular fibre glass tanks of 2 m^3 capacity was used for incubating grouper eggs by some workers. Provision for continuous flow of filtered sea water and moderate aeration (10 litre/minute) are given in the tank. The eggs may be stocked @250 buoyant eggs per litre.

Dead eggs that sink to the bottom are to be siphoned out once in every 2-3 hours. Incubation period for brown marbled grouper is 18-19 hours at 28-30°C, 19 hours for malabar grouper at 29-30°C, 23-25 hours for red spotted groupers at 25-27°C, 35 hours at 28°C for brown spotted grouper. The size of the newly hatched larva of brown marbled grouper is 1.8-1. 9 mm. It takes 3 days for the asorption of yolk sac. The duration of larval rearing is 35-40 days, when it will be ready for stocking in grow out ponds.

The larval development of *E. tauvina* has been studied by Hussain, N. A. and M. Higuchi (1980). The newly hatched larvae of *E. tauvina* varies in size from 2 to 2.4 mm (average 2.25mm). It has a large yolk sac, unpigmented eye, optic vesicle, heart and narrow gut. At the posterior end of yolk sac is an oil globule. As the yolk resorption goes on, oil globule moves anteriorly, when it is 1½ days old, pectoral fin and jaw buds appear and gut increases in thickness.

By 10th day, the larva develops long and projecting dorsal and ventral spine (0.65 and 0.75 mm respectively). This is characteristic of grouper larva. (Fig. 5.4) Initially the external nostrils start appearing as a single opening. By 25th day, the larva possess adult dorsal and anal fins and caudal fin rays also develop. The dorsal and ventral spine reach a length of 3.9 mm and 3.1 mm respectively. These spine are serrated with well demarkated spines on their anterior and posterior margin. The head is enlarged, and external nostrils get constricted. The body colouration is not yet developed fully, though melanophores appear on the top of the head, maxilla and operculum 31 day old larva possess adult pectoral and ventral fins. The length of dorsal and ventral spines are shorter when compared to the length of the body. 4 short spines appear on the operculum. Melanophores get well developed in the head, dorsal edge of the body and median line of caudal peduncle. By 50th day, metamorphosis is completed. All fins develop adult shape and structure.

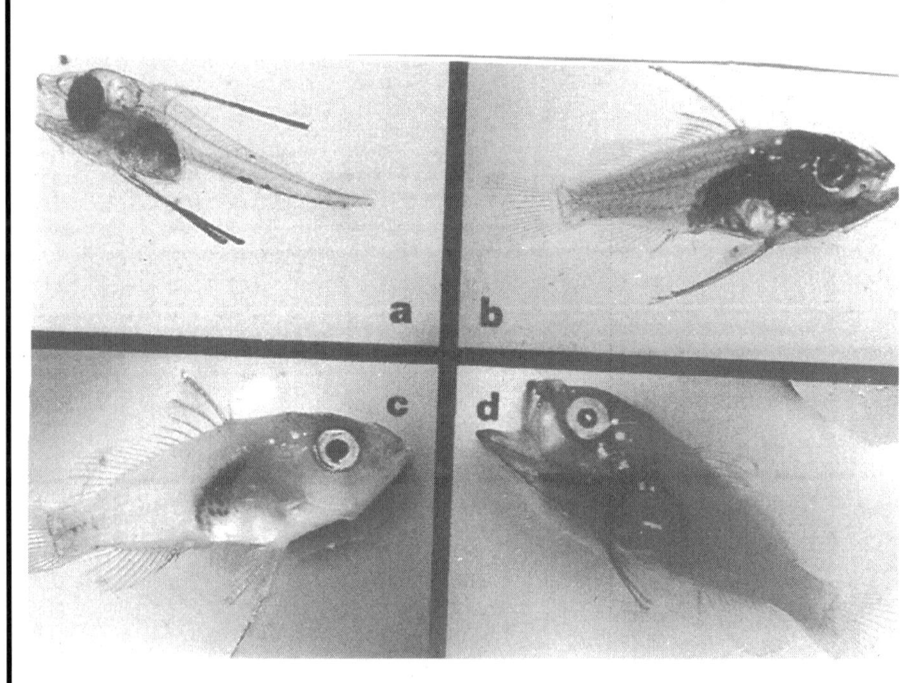

Fig. 5.4: Grouper Larvae, Note the Long and Projecting Dorsal and Ventral Spines.

(Reproduced with permission from I.C. Liao, 1993)

Till the larva reach 20 mm size, they swim in the middle of water, but by the time it reaches 30 mm size it settles to the bottom and become demersal. When they reach 50 mm size, they take shelter under objects found in water and they go out only for food.

Larval Rearing

Larval rearing of groupers can be done in concrete tanks of $10m^3$ capacity at a stocking density of 15000 to 387000 larvae per m^3. (Hussain, N.A. and M. Higuchi, 1980) Filtered sea water of salinity ranging from 27 to 31 ppt was used for rearing, Aeration: was provided @0.1-0.2 litre per minute upto 3rd day, Which may be increased to 0.6-0.8 litre per minute from 4th day. Daily water exchange. It was @20% from 4th to 5th day, 30% from day 6-8, 50% from day 9-11. From 12th day onwards, continuous flow of sea water is maintained. Water temperature is to be maintained stable, especially during first 3 days of larval rearing. In case of sudden drop in temperature, especially during night hours, there may be mass mortality. To prevent this, larval rearing tanks can be covered with canvas cloth both day and night during first 3 days and later only during night. Removal of debris: Accumulated debris from pond bottom can be removed by siphoning, twice in a week from day 7-24 and on alternate days from day 25. Feeding: From 3rd to 12th day, the larvae are fed on rotifers @5-10 number per ml of water. The brine shrimp nauplii are supplied from 9th day to 40th day feeding rate being 0.5-1 per ml from day 9-10, 1-2 per ml from 11-15, 2-3 per ml from 16-24, 3-5 per ml from 24-40. Green water containing Nannochloropsis is added daily to the tank at a rate of 0.2 to 0.3 million cells per ml. This serves as a food for rotifers and also as a shade for larvae, which are sensitive to sunlight. The larva grows to fingerlings within 35-40 days.

Chapter 6

Seed Production of Cold Water Fishes

Introduction

Salmomid fishes *viz.* Pacific salmon *(Oncorhynchus* sp.) Atlantic salmon *(Salmo* sp.) and trouts are grown in cold areas of world. Both Pacific salmon and Atlantic salmon are anadromous fishes and they need saline environment for growth. But trouts can be grown in inland waters. Contribution of salmonids to global aquaculture production of fishes is next to that of carps. During the year 1999 production of salmonids from culture was 1391615 tonnes (FAO statistics). Hundreds of hatcheries operate in several parts of cold areas of the world which are meant not only for providing stocking materials for aquafarms but also for stock enhancement in natural waters through ranching. Hatcheries of *Oncorhynchus* and *Salmo* sp. are engaged in the production of smolt for farming and ranching. Trout hatcheries are established in several cold regions of the world including India. In India Trouts were introduced from other countries. Mahseers found in the cold streams and lakes are endogenous fishes in India

TROUTS

Trouts are grown in cold waters. There are 3 genera of cultivable trouts such as rainbow trout *(Salmo gairdnerii* or *Oncorhynchus mykiss*), Brown trout *(salmo trutta fario)* and Brook trout *(Salvelinus fontinalis*) (Fig. 6.1). In India, trouts are exotic fishes, introduced into the cold waters mainly of Jammu-Kashmir, Himachal Pradesh and Penninsular India (Nilgiri of TN and Munnar High range of Kerala). Eyed eggs of brown trout was brought to India (Kashmir) from Scotland in the year 1900 by F. J. Mitchell. In 1912, Mitchell succeeded in hatching rainbow trout eggs obtained from England. Brook trout was also introduced from Canada later.

Till early fiftees trout culture was done to meet the requirement of sport fisheries. With the standardisation of artificial propagation technique commercial scale culture became popular in the cold areas of India. During 1905-1906, a brown trout hatchery was constructed at Harwan near Itinagar, later in the year 1909 another hatchery

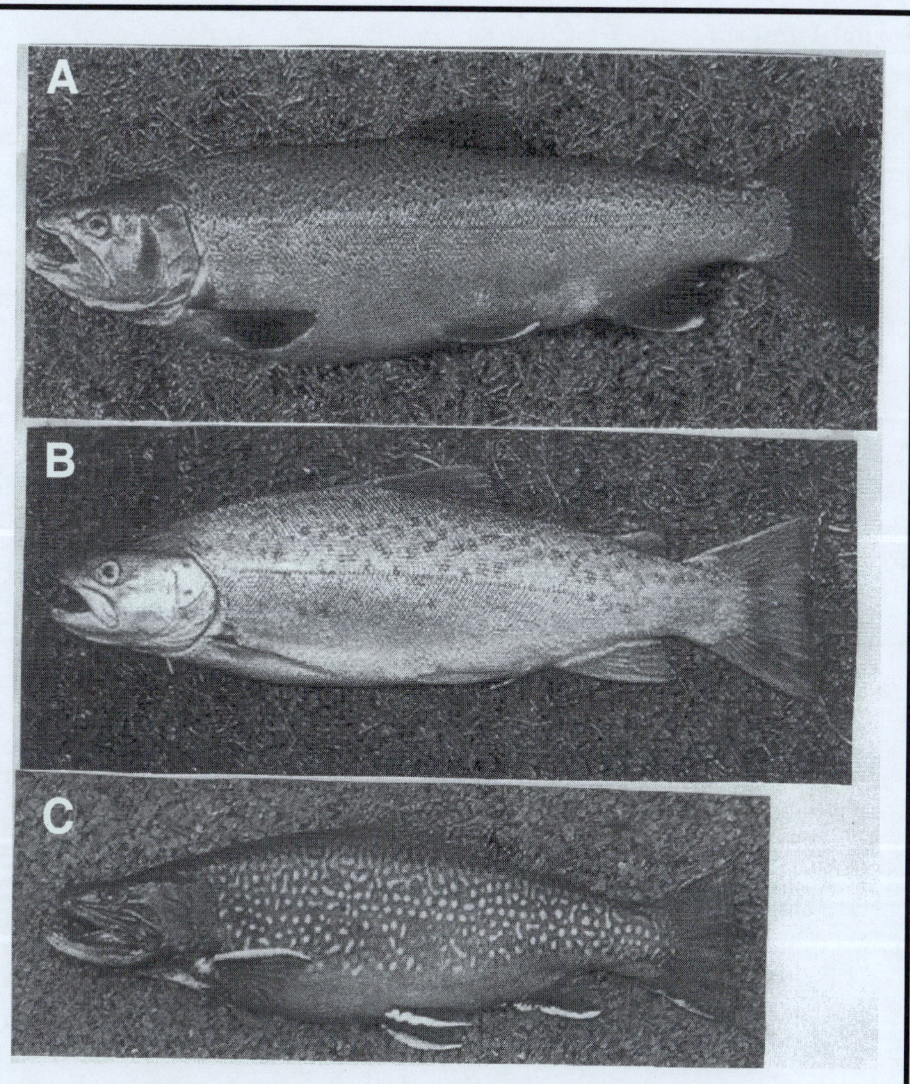

Fig. 6.1: Trouts Commonly Culvitated in Cold Waters.
(a) Rainbow trout (*Oncorhynchus mykiss*), (b) Brown trout (*Salmo trutta*), (c) Brook trout (*Salvelinus fontinalis*).

came up at Katrain, Himachal Pradesh (Mahili hatchery) and during 1909-1910 at Avalanche in Nilgiris a trout hatchery was established.

Reproductive biology

Sexual Dimorphism

The sexes can be distinguished during the breeding season. In all the 3 genera of trouts, the lower jaw of the males is turned up into a kind of hooked beak shape where as in female it is not found. Further a genital papilla is present in the fully ripe female but the same is absent in male. Again, there is change in body colouration in male - colour becoming brighter during breeding season - in rainbow trout males have lateral iridisent bands and vivid stripes on lateral side which increase in number, in brown trout belly turns blackish and white stripes at the outer margin of anal fin and in brook trout it becomes bright red colour. If the belly of the ripe female is pressed the eggs shall flow out where as gentle pressure on the abdomen of male causes the milt flow. If the milt is watery it is indicative of poor quality. If it is white and creamy it is of good quality.

Age at First Sexual Maturity

Trouts attain maturity at the end of second year. In case of male, milt is best in the age group between 2 and 4 years. Female are used for breeding from 3+ years onwards upto sixth year. As the age increases, yield of eggs also increase but part of their descendants may be sterile (Huet, 1986)

Spawning Season

Rainbow trout spawns in spring *i.e.* from January to May where as brown trout and brook trout spawn from October to January *i.e.* Fall/winter in USA/Europe. In India Brown trout and brook trout breeds during December-January, where as rainbow trout breeds during February and March except in Avalanche, T. N. where it breeds from September to February.

Fecundity

The spawning fecundity of trouts is very low *i.e.* 1500- 2000 eggs per kg body weight.

A trout hatchery should have the following units:

(a) Brood stock unit

(b) Egg taking unit

(c) Incubation unit

(d) Larval rearing unit

Brood Stock Unit

Adequate care is taken to rear the brood fish so that healthy, viable eggs are obtained from it.

The brood fish can be collected either from natural, open waters or they can be raised in ponds. They are collected from open water while they are swimming upstream prior to spawning. Pond raised breeders are reared in race ways of circular, rectangular or hexagonal shape or in growout ponds.

For raising the brood stock, a trout farm should have the following:

(a) Optimum climatic conditions,
(b) A source of adequate water supply for maintaining the required water flow in the pond, and
(c) Good water of quality.

Optimum Climatic Conditions

Temperature

As trouts are cold water fishes, tolerable limits of temperature for adult is 12° to 14°C for brook trout, 10°C to 20°C for rainbow trout and Brown trout. So there has to be moderate and uniform temperature.

Rainfall

Rainfall has to be moderate since heavy rain fall areas are deficient in mineral contents.

Gradient

Mountainous terrain with steep gradient is also deficient in mineral contents.

Adequate lime stone and other mineral deposits in the area are desirable to ensure mineral rich water as water containing calcium is richer in natural food and is the most productive. Water from granite sources, though pure is not rich in natural food.

Source of Water

Water from rheocrene and limnocrene spring is considered as ideal for rearing of trout:

There are 3 types of spring.

Rheocrene spring

In this category, on emerging the water runs down towards the valley through a slope.

Limnocrene spring

The spring is located in a depression as a result the spring water fills the natural depression.

Helocrene spring

The water seeps through a thick layer of earth and transforms the ground into a marsh. The water will be rich in humic acid and organic matter. This is not suitable for trout farm.

While spring water shall be ideal for hatchery, stream water can be used for grow out ponds.

Water Quality

Water should be pure, free from pollutants, transparent and free from dirt and suspended particles.

Turbidity

Turbidity @100 ppm and above choke the gills of fry and fingerling. Even 25 ppm hampers normal development of trout eggs, delays hatching and causes deformities in the hatchlings.

Dissolved oxygen

The DO level has to be 9 ppm *i.e.* almost double that is required for cyprinids. If it is below 5 ppm fish develop respiratory difficulties. Normally, in spring water it may be 8.5-9. 4 ppm where as in stream it may be 7.4 to 10.2 ppm.

pH

If the pH is about or above 9, it causes mortality among eggs and alevins. Usually, spring water of Himalaya has pH of 7.9 to 8.4.

Quantitative Requirement of Water

As the running water system is recommended for trout culture water flow has to be maintained at 1 litre of water per minute per age month for every 1000 fry/fingerlings (Huet, 1970)

Some other German workers have suggested the following:

For extensive trout farming water flow shall be 100 lit/sec/ha.

For semintensive farming it shall be - 200-300 lit. /sec/ha

For intensive farming it shall be - 300-500 lit. /sec/ha

Feeding

Feeding the brood stock with adequate feed of good quality is of prime importance in raising good quality brood stock. According to Huet (1986) if trouts are fed exclusively on artificial food, their sexual products shall not be of high value, where as fish fed on their natural food give eggs and milt of good quality. As it will be impossible to obtain and feed only natural food to fish, good quality artificial food is usually recommended. Feeding the brood fish @1-2% of body weight with 1:1 mixture of meat balls of snow trout and mirror carp has been found effective. For example, Brown trout fed on the above diet has shown 14.5% increase in egg number per kg body weight where as in Brown trout there was 12% increase in egg number.

Number of breeders needed for producing the required quality of fry have to be worked out earlier as the fecundity is very low in trouts *i.e.* 1500 to 2000 eggs/kg body weight.

Sex Ratio for the Brood Stock

The number of males can be half the number or even one third of the number of females. This is because of the fact that males give milt several times *i.e.* three to eight times in a season provided that successive spawnings are separated by 2 weeks. Usually milt from one male is sufficient to fertilize eggs from 2 females. Males are usually segregated from female at least for 3 months during the breeding season as they fight between each other if left with females.

Egg Taking and Milt Collection

Wet Method vs Dry Method

The eggs from female and milt from male are collected by stripping *i.e.* by applying gentle pressure on both sides of the belly moving the fingers from front to back. In the wet method, eggs are collected in a pan containing water and then the milt is added to it for fertilizing eggs. In the dry method, the eggs are collected into a dry receptacle into which the milt is added. After mixing milt with eggs *i.e.* after fertilisation is over, the mixture is poured into a pan half filled with water. Dry method is preferable for the following reasons. (1) If the milt is not mixed in water spermatozoa of trout remain viable for a longer duration (even up to 2-10 days at 4°C to 8°C). If the milt is mixed in water it is viable only for maximum 90 seconds but greater part of motility caeses after 30 seconds. (2) If the eggs come into contact with water, eggs absorb water and swell as a result the micropyle - the route by which the sperm enters into the egg - closes. As a result of this the sperm penetration and subsequent fertilization become impossible.

Egg Taking Procedure

There are several methods of egg taking of trouts such as (1) Hand stripping without incision (one man method and two men method (2)Hand stripping after incision (3) Australian method (4) Swedish method.

Hand Stripping Without Incision

One man method

The female fish in the ripe condition is wiped gently so that water from the body will not drop into the container kept for collecting eggs. After wiping, the female is held with the ventral side directed down above the pan by one hand of the operator whereas by the other hand he exerts a pressure on the belly which causes the egg to flow through the vent which is taken in to the spawning pan.

Two men method

In two men method one person holds the fish and the other does the stripping operation. The fish is held in a steeply inclined position with the head held up, back towards the operator and underside over the pan. Pressing with the thumb and index finger on the belly, hand is moved down on the abdomen and as it descends over the lower end of the body eggs starts flowing through the genital opening. The process can be repeated several times so that the female is completely emptied of the eggs. Care is taken not to exert excess pressure on the abdomen so that internal haemorrhages do not occur. If any vein is ruptured it causes the flow of blood and the same may contaminate the eggs and harm the mother fish.

Milt collection is also done in a similar way. The body of the male is wiped so that water does not drip from the body into the container. Under side is then turned downwards and sides are pressed. Milt flows out through the urino genital aperture and the same is taken in to a sterile tube.

Hand Stripping After Incision

Incision is made in the vent region to facilitate easy flow of sexual products. It is practiced in species of salmon which die after spawning *eg. Oncorhynchus*. Milt collection is also done in a similar way. The male is wiped so that water does not drip from the body into the container. Underside is then turned downwards and sides are pressed. Milt flows out through the urinogenital aperture and the same is taken into a sterilised tube

Australian Method

In the year 1950, an Australian farmer named Worton introduced this method. Hence it is called Worton's method.

An hypodermic injection needle is attached to a bicycle pump through a rubber tube. The needle is inserted into the body cavity near the pelvic fin. The needle is held firmly in position and the air is gently pumped into the body cavity through the needle which expels the egg through the vent. After all the eggs exude out, the air pumped inside the abdominal cavity of the fish is carefully siphoned out by the mouth of the operator and the fish is returned to the holding tank. In case of male, the needle is inserted into vasdeferens and the milt is sucked out and taken to a test tube.

In some of the trout farms of USA a modified method is used. In this method, instead of cycle pump, air compressor (1. 6 H. P. motor) is used which is connected to a rubber tube connected to the outer barrel of a hypodermic syringe which is inserted into the body cavity near pelvic girdle. When the compressor is turned on, compressed air enters the abdominal cavity at a pressure of 0.4 kg/cm^2. Then the fish in inverted and gently shaken to release the eggs. After the eggs flow out, the same pump is used to suck out the air from inside also. In this, the entire process is finished within 15 seconds. The eggs are allowed to fall on a net sieve through which waste materials pass out, clean eggs without faecal matter is transferred to a dry container.

Swedish Method

The female is placed in a double walled rubber sac having a holder to hold the fish closely. While keeping the fish, the tail part up to the vent is left protruding which is held by the operator. The space between the two walls is connected to a water tap. As the tap is opened, the water enters into the sac at an appropriate pressure by which the inner wall of the air sac presses against the whole fish. The fish readily releases the eggs which is taken into the container. Milt is also taken in a similar way.

Fertilization

One cubic centimeter of trout milt may have 10, 000 million spermatozoa. Hence only a few drops are required for a large quantity of eggs. Usually milt from 2 males is used to fertilize eggs from 2-4 females. The eggs and milt are mixed well with the aid of a feather and this is poured into a pan having water. Then they are again mixed 3-4 times with feather carefully. Fertilization takes place immediately. Only one sperm is allowed to enter through the micropyle and form the zygote nucleus. This is set aside for an hour for the eggs to harden. Fertilized eggs have a greenish tinge and hence it is called "green eggs".

Structure of the Trout Egg

Fresh Egg

Fresh trout egg is 3.5-5 mm in diameter and it is covered by an outer shell which is not firm but soft, porous and adhesive. The yolk mass is enveloped by a yolk membrane. The germinal disc with nucleus is located at the animal pole of the yolk mass. There is a space between the outershell and yolk membrane which is called perivitelline space. There is an opening called micropyle located at one point of outershell which is meant for the passage of sperm.

Water Hardened Egg [Fig. 6.2(a)]

As the eggs come in contact with water, water diffuses through the porous shell into the perivitelline space as a result the volume of the egg increases by 20% and swelling of eggs make the shell firm and turgid. This is the water hardened egg. As the eggs swell, the micropyle gets closed.

Determination of Egg Number and Calculation of Fertilization Rate

The total eggs collected for hatchery operation can be calculated by volumetric, gravimetric or Burrow's displacement method.

Volumetric Method

The eggs are measured in a cup. The number of eggs in a cup is counted and the same is multiplied with the total number of cups from which the total number of eggs can be obtained.

Gravimetric Method

The total weight is taken in a monopan balance. Then the number of eggs in a given weight of egg is counted and multiplying the same with the total weight of the egg, the total number of eggs can be calculated.

Displacement Method

This is based on finding out the total volume of water displaced by egg mass and calculating the number of eggs displacing one ml of water. That is, the total number of eggs = number of eggs displacing I ml of water × total volume of water displaced in ml.

Unfertilized eggs turn white. This process is very slow and by the time it is discernible fungus may attack the eggs. Hence, in order to quicken the whitening of the unfertilized eggs shocking or addling of trout eggs is done. This is done by stirring the eggs with a bare hand. Dead eggs are removed by a glass pipette having an inner diameter of 5-8mm. The number of unfertilized eggs in a given quantity of eggs is calculated and from this percentage of fertilization is calculated.

Incubation of Fertilized Eggs

After the trout eggs are water hardened and counted, they are transferred to a hatchery for incubation.

Incubation devises for trout eggs:

(a) Flat troughs and trays,
(b) Vertical flow incubators,
(c) Vibert box, and
(d) Jar incubators.

Flat Troughs and Trays [Fig. 6.2(b)]

The classical incubation devise of trout eggs consist of troughs in which one or many meshed trays are kept. The eggs are placed in the meshed trays. The trays are made of perforated stainlesssteel/Aluminium/ zinc/ plastic. The mesh size is 1.5 to 2.5 mm in diameter. Sometimes, the bottom of the tray is made of glass rods of 4-5 mm thickness and separated from each other by a space of 2 mm. The size of the tray should be in such a way that it will fit into the trough easily. Tray may rest on ridges fixed to the inside wall of the trough or tray may stand on legs resting on the bottom of the trough. The space between side of trough and tray must be as narrow as possible so that water flow shall be from bottom to top across the tray. The space can be filled with foam rubber to ensure this. A tray of 50 × 50cm can hold a layer of 10000 eggs. The trays are provided with wooden/metal/ eternite lids as hatching always takes place in dark. The troughs are either short (Californian type) or long shaped.

Californian Type [Fig. 6.3(c)]

These are short troughs made of zinc sheet or wood or plastic of size 0.5 × 0.2 × 0.15 m (20 × 8 × 6 inch) or 1 x. 5 × 0.25m (40 × 20 × 10 inch). The former can hold one tray where as the latter can hold 2 trays. These troughs can be arranged in tiers so that water can flow from the upper tier to the next lower tier. [6.3(d)]

Long Trough [Fig. 6.3(e)]

A long shaped trough may be 2 to 3 meters in length, 0.5 to 0.8 meter in width 0.2 to 0.35m in depth. Several hatching trays can be kept inside the trough as it is of large size. The trough may be made of concrete/cemented bricks/ zinc sheet/ wood or wood covered with zinc sheet/ aluminium or steel. The inside of the trough is usually painted with dark coloured water proof paint so that eggs will be protected from bright light.

This type of troughs are useful for both hatching and rearing of fry upto 3 months.

Vertical Incubators [Fig. 6.3(a)]

When incubation is done in flat troughs with trays inside, whether troughs are arranged in tiers or placed parallel to each other, incubation unit shall occupy a large part of the building. Hence in order to overcome this difficulty, vertical incubation models have been developed in countries like USA and Sweden. The most popular of this consist of one or two superimposed shelves, each one holding 8 incubation trays. Each tray is made up of an external basket which can hold water. The egg containing tray is kept inside this basket. This tray is covered with mesh lid. The whole devise is [Fig. 6.3 (b)] made in plastic. The shelf may be of the size 80cm height and 62 cm wide,

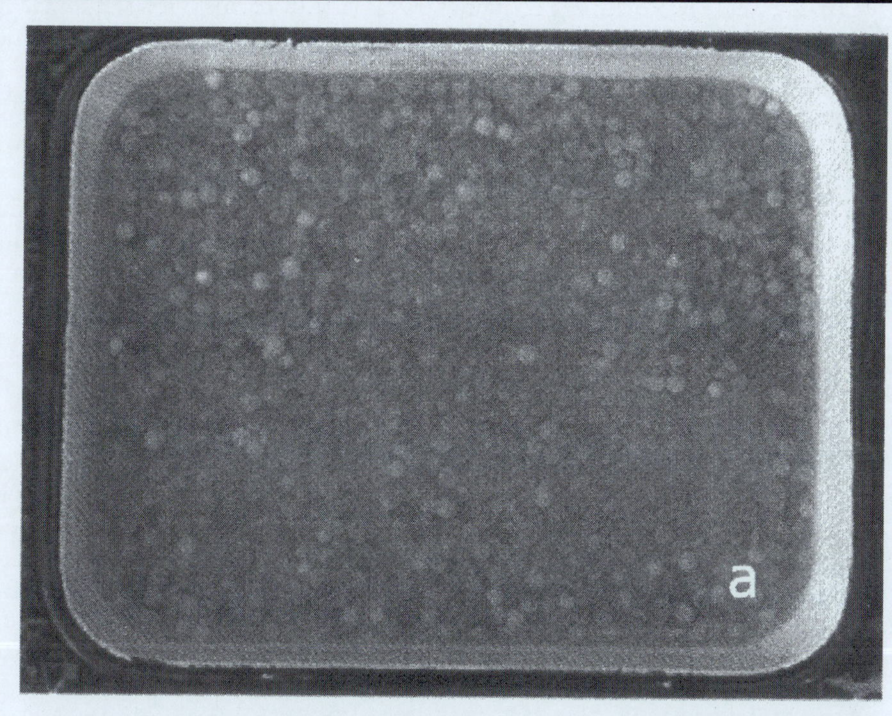

Fig. 6.2: Trout Eggs and its Incubation in a Trout Hatchery in Kashmir, India. (a) Fertilized Egg Mass of Trout, (b) Incubation of Eggs in Hatching Trays. (Source: *Basir Ahmed Chowan*).

the water tray is 53 × 62 × 9cm, egg container tray is 40 × 35 × 5cm. The water that falls into the top most water tray after filling the same shall flow into the tray below and ultimately passes out from the lower most tray.

A modified vertical flow incubator is being used in the Laribal atchery in Kashmir in India. It consists of a drum having a water inlet at the bottom and outlet at the top. Just above the water intake point, a filter made of metallic wire holding pebbles is placed. Another circular tray containing eggs is hung with nylon twine, below the outlet point. As the water is let into the drum from below, it passes through the filter which helps in removing suspended particles found in the water. As the filtered water swells up the drum, it bathes the eggs kept in the egg tray and then it passes out through the outlet. Thus the eggs are bathed in filtered, clean, clear water continuously.

Vibert System

In this system, eggs are kept in a cage called vibert box designed by Vibert (1975). Each box is of the size 70 × 63 × 94 mm and weighs about 45gm and is made of rigid, transparent plastic material. The holes are drilled in the walls diameter of which is smaller than the egg size but at the same time large enough for the fry to swim out and its holds about 1000 trout or 800 salmon eggs. The box is immersed in river bed where there is swift water current and clear water. Bottom of the stream should have pebbles of the size of hen's egg. A hole deeper than the box size is dug in the stream bed into which the vibert box with eggs is placed. Around the box, larger stones are kept to create empty space against the side of the box. The hole is then filled with smaller gravel, smaller gravel prevents predators from reaching the spot and also ensures complete darkness. Larger stone will have space in between which allows the alevin to escape in to the river.

Jar System

It was in Pennsylvania, USA, Buss and Fox (1961) adopted jar type hatcheries for salmonid eggs. Each jar is cylindro conical with the narrow end fixed to a water inlet. A layer of gravel is kept at the narrow part of the jar which prevents the swift movement of eggs thereby reducing the chances for injury to the eggs. In Italy, large jars of the size 35cm diameter (13 inch) 80cm height (31 inch) and 70 liters capacity are used. Supposing that 50 litres of used, with 8000 to 9000 eggs per liter, about 400000 eggs can be incubated at a time.

Development of Eggs

The fertilized egg undergoes development till the embryo is fully ready to hatch out of the egg. Unlike carps or catfishes, the trout has a prolonged incubation period extending to several days Duration of incubation depends on water temperature. The four stages of importance from the point of view of a pisciculturist are (1)Green egg (2)Eyed egg (3)Alevin or sac fry (4)swim up fry. The newly fertilized egg is called green egg because of the greenish tinge of colour of the egg. In course of development when the embryo develops eye which is visible through the egg envelope it is called eyed egg stage. The newly hatched out hatchling is called alevin or sac fry referring to the heavy yolk present on the lower side of belly. The hatchlings are 1.5 to 1.8 cm long, weighing 45 to 50 mg, 50 to 65% of the weight is due to heavy yolk sac. They remain

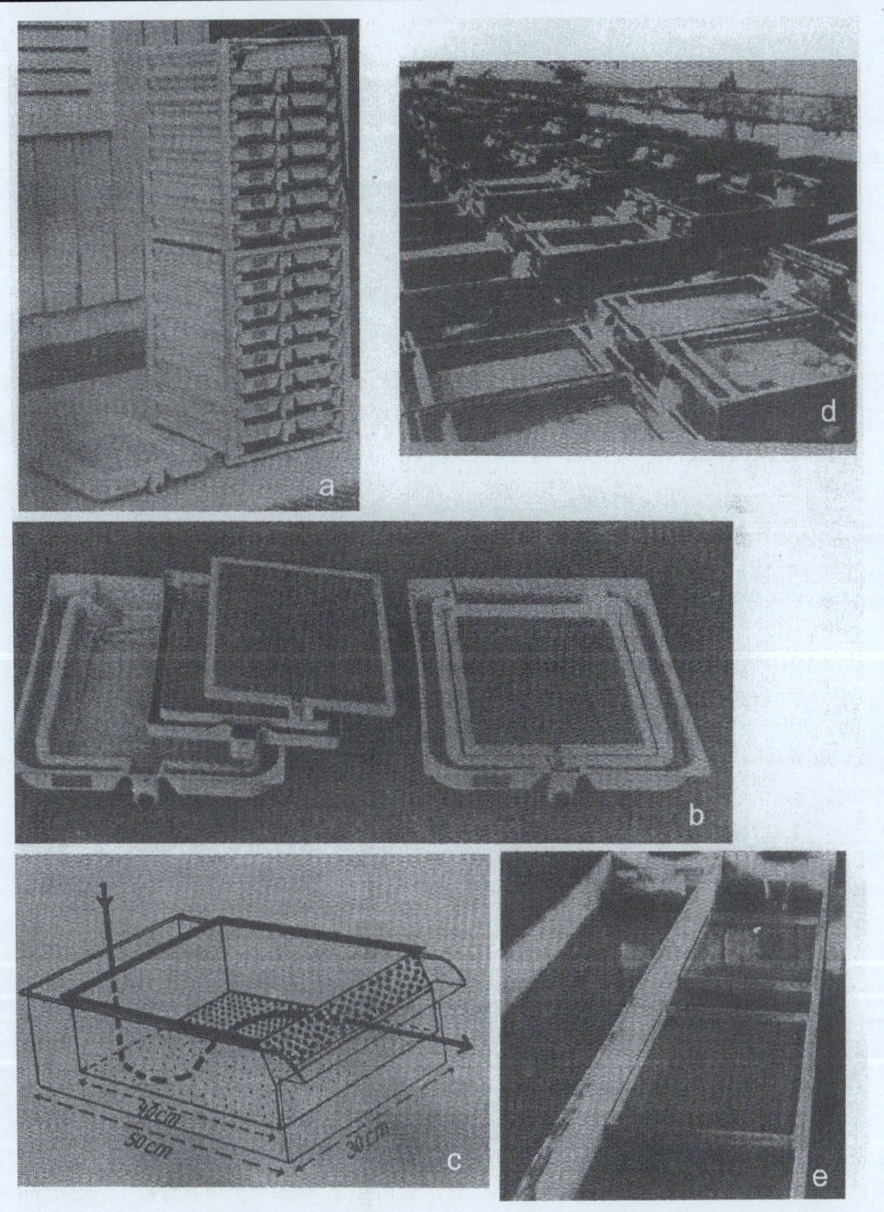

Fig. 6.3: Incubation Device for Trout Eggs (cont....)

(a) Vertical flow incubator, (b) Incubation tray used in vertical incubators. Each tray has an external basket which can hold water. Egg containing tray is kept inside this basket. The tray is covered with mesh lid, (c) Californian type trough and tray (diagrammatic), (d) Arrangements of troughs in tiers to allow continuous water flow by gravity, (e) Long troughs with egg trays inside.

(*from Huet, 1986*)

quiscent at the bottom of the incubation apparatus. It may take 15-18 days for the yolk sac to be absorbed in rainbow trout. When 2/3 of the yolk sac is absorbed, larva starts moving towards the water surface and this stage is called swim up fry. At this stage it starts active feeding.

Care of Eggs During Incubation

Continuous and Uniform Water Flow

The incubation unit has to be supplied with a continuous and uniform flow of water so that eggs are bathed in fresh water always. The rate of water flow can be 0.2 to 0.3 litre/minute/1000 eggs. This ensures not only the required dissolved oxygen level in the rearing medium but also the removal of metabolic waste from it. Water supply for the trout hatchery shall be from rheocrene or limnocrene spring. Water flow should not be too intense otherwise, water current may cause injury to the egg.

Water Temperature

Temperature of water should not be below 1.5°C and above 14°C. For Rainbow trout optimum temperature has to be 5°C to 9°C, for brown trout it can be 2°C to 3°C. The incubation period (period from fertilization to hatching) and period for transformation from alevin to swim up fry are 72 to 90 days and 32-40 days respectively for Brown trout at 3°C to 7°C, where as at 7-11°C, it takes only 36 to 53 days and 10-12 days respectively and at 11°C-13°C it is 26 to 35 days and 10-12 days respectively. In case of Brook trout the periods are 45 to 72 days and 30-32 days respectively at 3°C-7°C but in case of Rainbow trout, at 7-11°C temperature range, it is 41 to 56 days and 10-15 days respectively and 18 to 27 days and 10-12 days respectively at 11-13°C as reported by Seghal *et al.* (1978) at Laribal and Harwan hatcheries of Kashmir. The incubation period also depends on the quality of egg as eggs taken from different parents may have different periods of incubations at a given temperature. It may very as much as six days between eggs taken from different parents, even when the temperature is maintained constant

The incubation period has been also described in terms of temperature units and also as degree days. One temperature unit (T. U.) is defined as 1°F above freezing *i.e.* 32°F during 24 hours. For example of water temperature at the end of 1st day of incubation was 56°F, it will be equivalent to 24 temperature units (56-32=24). During second day of incubation if temperature was 52°F, it will amount to 20 T. U. thus giving a total of 44 T. U. The degree days refer to the product of incubation period expressed in days and average temperature of incubation. For Brown trout it is 400-460 degree days, at 4^{0} c for rainbow trout it is 320to 350 degree days at $6^{0\ \text{to}}$ 10^{0}c For absorption of the yolk sac it takes about 220 degree days for Brown trout and Brook trout and 180 degree days for rainbow trout.

Quality of Water

Water should be clear, free of silt particles as silt particles may settle on eggs and prevent the exchange of gases. There will be also higher risk of infection. Hence periodically, filters must be examined and cleaned of necessary. If silt has settled on eggs, it must be washed taking care to see that no mechanical injury occurs to the eggs. It is done by lifting and submerging the trays in water simultaneously spraying them with water from a sprinkler.

Oxygen Content of Water

It should be from 6-8 ppm.

Protection from Sunlight

Direct sunlight and electric light are harmful to developing enbryos. So the eggs must be kept in darkness.

Removal of Dead Eggs

Unless the dead eggs are removed, infection may set in. Common infection is of *Saprolegnia*. The disinfection is done at a time when the eggs are least sensitive to external stress. This safe period is either one day after fertilization or during the eyed egg stage.

Maintenance of General Hygiene

Both prophylactic as well as therapeutic measures are to be taken to avoid disease out break in an incubation units. Disinfectant used to prevent fungal infection are malachite green, acriflavin, gentian violet and Potassium permanganate. Prophylactic measure to prevent fungal attack is by treating the egg for about 30 minutes in a mixture of malachite green with water @5mg malachite green per litre of water. After this it is flushed and fresh water is pumped in taking care not to disturb the eggs. Before adding malachite green solution, utmost care is taken to ensure that every crystal is completely dissolved. The solution must be uniformily spread in the trough very carefully without disturbing the eggs. If it is not uniformly spread, there may be hyper concentration of the dye in localised areas in the trough which may prove toxic to the eggs. This treatment shall continue till the alevins emerge.

Transport of Trout Eggs

The trout eggs from the 'eyed egg stage' till about 5 days before hatching are transported from place to place. During this period, the eggs are found to withstand shocks and other mechanical stress during transport. It can also remain alive out of water provided the temperature is not too high, and humidity is sufficient and they get enough of oxygen.

Transportation can be done in boxes specially designed for the purpose. It consist of an insulated wooden cabinet of the dimension 51 × 25 × 41cm with a door that can be locked. At the top is a cup board with wire netting base. The cup board may be 11cm high. Cup board is meant for keeping moss with snow or ice. The eggs are kept is egg trays. Each egg tray is of the size 23 × 23 × 1.0 cm having a muslin cloth base and can hold about 1000 eyed eggs in a layer. The egg trays are kept in 2 vertical rows, each vertical row consisting of 21 trays. So in total there shall be 42 trays in a box. Above the egg trays and below the cup board, 2 trays having perforated zinc sheet base are kept with 'moss' to hold the water from the melting snow or ice kept in the cup board - Muslin cloth base of the egg tray can keep the eggs moist. In India, the transport box is named pahari box.

Rearing Unit for Raising Fingerlings

The swim up fry or alevin with 3/4 of the yolk sac absorbed are released into the rearing tanks directly or they are reared in the hatching trough after removing the hatching trays for 30-45 days and then released into the rearing tanks. The latter method is practiced in the hatcheries of USA which has some advantage over the former because after 30-45 days of rearing in a trough under controlled condition and intensive feeding on artificial food, the fry become stronger and healthier and hence are fit enough to tide over any adverse environmental conditions and predators in the rearing system. Further by this time they pass over the sensitive stage which is prone to the attack of whirling disease. They are also tuned to the artificial feeding and hence feeding becomes easy in the larger tanks as it accepts the artificial feed readily. In some hatcheries rearing of swim up fry for 30-45 days is done in special tanks either circular or rectangular, constructed along the same lines as hatching troughs. Stocking density in this system can be one per sq. cm or 10000 nos per square meter or even up to 30000 nos per square meter for rainbow trout. Care is taken to siphon out feed residues, dead fry and faecal matter from the tank every day. Feeding is done on natural food (zooplankton) and also on artificial feed. During this stage rearing is done under illumination as it grows better under illumination.

Raising of Trout fingerlings is done in nursery tanks. Different management practices exist for raising fingerlings. Rearing may be done in natural ponds which are not manured pond without artificial feeding, or manured ponds with artificial feeding or in ponds made of RCC walls or stone walled with only artificial intensive feeding and continuous water flow at the rate of 20-30 gallons per minute. In India, these tanks are of the size 3 × 1 × 0.75 to 10 × 1 × 0.75m in size. Stocking density may be 50-100 nos of swim up fry per sq. meter or 25-50 nos of 30-45 days old fry reared in hatchery trough. They grow to 8-12 cm (3-5 inch) in length at the end of 6 months of rearing.

MAHSEER

Introduction

Fishes commonly called mahseers belonging to 2 genera and seven species including 3 subspecies are found in both lentic and lotic water systems in countries such as India, Afghanistan, Bangladesh, China, Myanmar, Nepal and Pakistan. In India, this fish is mainly found in hill streams and upland rivers and is of high commercial value as a game fish and as a fish of consumer preference. The genus Tor has six species *viz. Tor putitora* (golden or putitor mahseer), *T. tor* (tor mahseer) (Fig. 6.4) , *T. mosal* (mosal mahseer), *T. progeneius* (Jungha mahseer), *T. Khudree* (Deccan or Khudree mahseer) and *T. mussullah* (hump back or mussillah mahseer). Another genus is *Acrossocheilus (Neolisschilus) hexagonolepis* commonly called copper or chocolate mahseer. The three subspecies reported so far are *Tor mosal mahanadicus, Tor Khudree malabarius* and *Tor Khudree longipines* but this is disputed by some workers. In their natural habitat they grow to large size of 1.5 meter long and 36.9 kg weight. In case of *T. putitora*, specimens of 2.75 meters length also has been recorded. They are mainly found in the upland waters of sub himalayan region extending from Kashmir

Fig. 6.4: Common Species of Mahseer.
(a) *Tor tor*, (b) *Tor putitora*.
(Courtesy: Bijoy Kumar Behera)

to Assam. *Tor khudree* is found in the river Mahanadi and riverine systems of penninsular India. *Tor tor* is also found is the rivers Narmada and Tapti. *N. hexagonolepis* (chocolate mahseer) is found in Assam, China, Myamar, Nepal. Mahseers have been lately identified as an endangered fish in India as its catch is fast declining in all riverine systems and lakes inspite of its remarkable capacity to adapt itself to different types of food and environment. In view of this, artificial seed production and culture has been recommended as a mitigating measure to conserve and propagate this endangered species which is known as the king of sport fishes or mighty mahseer.

Reproductive Biology

Size at First Sexual Maturity

It is reported by several workers that males mature on attaining 1. 5 to 2 years while the female matures at the age of 3 years. *Tor tor* attains first maturity on attaining 273 to 290 mm length (Karamchandani *et al*, 1967) where as Chaturvedi (1976) found that males of *T. tor* attain maturity on reaching 251 mm and females on reaching 320 mm.

Sexual Dimorphism

Pathani (1978) described the distinguishing features of male and female of *T. tor* and *T. putitora* as follows:

Size of the Body

Males are smaller and body narrower than females of same age. Head length-body width relation is higher in male than female during breeding season. The abdomen is bulkier in female. Ventral profile of abdomen is straight in male where as in female it is arched. On pressing the belly of mature male, milt oozes out.

Fins

The pectoral fin of male is longer than that of female. In male it reaches up to 6th lateral line scale but in female it reaches only up to 5th lateral line scale. But according to Chatruvedi (1976) in male it reaches up to 7th lateral line scale but in female up to 5 th or 6th lateral line scale. So also, base of anal fin projects out of the ventral profile in female but in male it does not project out of the ventral profile line.

Tubercle on Head

In males tubercle are present on either side of the head which is absent in female.

Body Colour

Male appears brighter in colour than female. Small black spots are seen on the lateral sides of mouth of male.

Breeding Season

Varying breeding seasons for different species have been reported by different workers. Even for the same species different authors have described different breeding seasons at different localities. For example, in case *T. putitora,* Thomas (1897) reported several breedings in a season, Dunsford (1911) observed the commencement of breeding with monsoon in Kumaon, Khan (1939) reported the occurrence of three breeding seasons in a year such as January and February, May and June and July to October, Quasim and Qayyum (1962) and Bhatnagar (1967) observed breeding several times a year, Joshi (1976) found in August to September in the streams of Jammu, Seghal (1978)during May-June and August-September in river Beas. Sunder and Joshi (1977) during August September in Yamuna river and Tandon (1981) 2 breeding season once in May-June and another in July-october in Himachal pradesh. For *T. khudree,* Kulkarni (1971) observed breeding from July to September in Lonovala. Ahmed (1953) found breeding of *T. mosal mahanadicus* in river Mahanadi from November to December and breeding of *A. hexagonolepis* from April to October.

Breeding Habits

It has been reported by several workers (Thomas, 1897, Khan, 1939, 1940) that mahseers migrate upstream into shallow running water streams of 13 to 14 cm water depth. They swim up the rivers during flood season and travel a long distance to reach the shallow streams. After laying eggs they usualy linger in these areas till the level of stream decrease when they move towards places of safety. As the eggs are demersal, eggs settle to the bottom gravel/sand instead of sand/gravel if there is loose mud the eggs shall perish.

Fecundity

The fecundity of maheer is very low. According to Karmachandani *et al.* (1967) fecundity of *Tor tor* varies from 6670 to 40780 for fish varying in length from 290 to 750mm. Desai (1973) found 42600 ova in a fish measuring 657 mm. Nantiyal and Lal (1985) reported the fecundity of *T. putitora* as varying from 26, 998 to 98583 in the fish of size range 3.5 to 23. 0 kg body weight. Joshi and Langer (1983) found that by stripping, eggs obtained from T. *putitora* was 3922 per kg body weight. In case of *T. khudree*, Kulkarni and Kulkarni (1971) recorded 20, 000 ova for a fish of 630 mm size.

Hatchery Technology

In India, at 2 places hatcheris are operating for Mahseer seed production. These area located at (1) National research centre on Cold Water Fisheries, Bhimtal, Nainital (dist) and (2) Tatapower company Ltd, Lonavla, Pune (dist), Maharastra.

Brood Stock Collection/Maintenance

Brood stock required for seed production are detained either from natural spawning ground during breeding season or are raised in farms. Female fish of 3 year old weighing more than 900 g and male fish 2+ years with oozing milt through genital aperture are used for breeding. Farming of mahaseers has been tried only by a few workers for raising the brood stock by rearing fry/fingerlings. Different systems such as concrete tanks earthern ponds and running water ponds or cages have been tried (Raina *etal*, 1999 for *T. putitora*, Langer *et al*, 1992 for *Tor tor).* Kulkarni and Ogale (1995) have grown it to 900 g in 3 years in concrete ponds at Lonovala on good proteinaceous artificial diet containing feed additives.

Induced Breeding/Stripping

Pond reared 3 year old female fish weighing 900 g or more and 2 year old male fish are administered injection of pituitary gland extract for induced spawning (Pathani, S. S and Das S. M, 1979 and Kulkarni and Ogale, 1986). For female 2 doses such as initial dose of 6mg/kg body weight and final dose of 12 mg/kg body weight are given. Ovaprim @ 0.6 to 0.8 ml per kg body weight for females and @0.2 ml to o. 3 ml per kg body wt. for maies. is also used instead of pituiitary gland (Kulkarni and Ogale, 1995). For male, only a single dose in necessary. For brood fish caught from spawning ground hypophysation in not required. They are released in to spawning tank and showers are provided on the top. Water level is raised from 0.2 to 0.7 meter. If the pair does not breed within 6-12 hours stripping is done. The female fish is stripped for eggs and eggs are collected into an enamel fray [Fig. 6.5 (a)]. The male fish is stripped and milt is collected [Fig. 6 5 (b)] The milt is mixed with the freshly taken eggs by means of a feather. This facilitates fertilization.

The fertiliized eggs are demersal, lemon yellow or brownish golden in colour [Fig. 6.5 (c)]. The fertilized eggs are washed 3-4 times. Trays containing eggs are filled with fresh water and allowed to stand for 15-20 minutes. After this trays are kept in

Fig. 6.5: Stripping and Artificial Fertilization in Golden Mahsser.

(a) Hand Stripping of Female Golden Mahseer. The Eggs are Collected into the Tray, (b) Stripping a Male. The Milt is Spread on Stripped Egg Mass, (c) Fertilized Egg Mass, (d) Circular Tank for Conditioning the Brood Fish Before Stripping.

(reproduced with permission from Fishing Chimes, vol. 19 ,no. 10,11)

incubation troughs for hatching. The increase in size due to water hardening is very little in mahseers. The eggs are measured volumetrically to calculate the number of eggs. The water hardened fertilized eggs measure 2.8 to 3.2 m in case of *T. khudree* and 2-3.6 mm for *T. putitora*.

Incubation

Usually, the fertilized eggs are kept in floating trays of 56x56x10cm size spread in a single layer at the bottom of the tray. The sides of the trays can be made of wood and bottom of synthetic netting cloth of 1 mm mesh size. These wooden trays are kept floating in water contained in cement tanks (2.5 x 1.2 x z0.75m) as is done in Lonovala fish farm or in tanks made of galvanised iron sheet or fibre glass (200x60x30cm) as is done at NRC, Bhimtal. Provision is provided for continuous sprinkling of water on the trays with eggs, either by fixing water sprinkler or a by perforating the inlet pipe kept on both sides of the tank. (Fig. 6.6) Provision is also made for the inlet pipe into the tank to come from the overhead tank and out let is also given from the cement tank. The water flow can be one liter per minute for incubation of 2000 eggs/hatching at 20-28^{0}c. Depending on the size of the tank and trays, number of trays that can be arranged in a tank varies from 5-8. Each tray can hold 5000 to 6000 eggs.

The incubation period varies from 76 to 96 hrs (average 82 hr) at a water temperature of 19°C to 24°C. After hatching the shells are pipetted out from the tray and the larvae are left undisturbed.

Larval Rearing

The newly hatched larva measures 8.5 to 9mm. Body is laterally compressed and some what transparent. The tail is thin and inconspicuous. Mouth is not differentiated. Pectoral fins are small without fin rays. It has very heavy yolk and hence remain at the bottom of the tray lying on the side. It vibrates tail when disturbed for 6-8 days. During this period larva utilizes the food reserved in yolk for nutrition. By 8th day the hatchling shall start moving. Yolk sac absorption is completed in 2 weeks or 10 to 12 days. The 2 week old larvae swim freely in the tray. If water temperature is lower than 20°C it may take upto 3 weeks. The 10 day old larva measures 10 mm to 12 mm in size. At this stage they are fed on minute zooplankton like Moina, Daphnia etc. Simultaneously artificial feed such as powdered ground nut cake and rice bran sieved through a small mesh is also given 3-5 times a day. At the time of feeding, sprinkling of water is stopped. On the 15th day, 12 mm yolk free swimming fry are shifted to cement nursery tank and stocked @ 240 numbers per sq. mt. and fed on artificial feed 5-6 times a day. Rice bran and oil cake at 1:1 proportion is soaked in water and made into balls. The ball is kept in enamel trays and hung by rope at six places, 6-9" below water level. They are acclimatised to artificial feed in 4 weeks. The feed residue, faecal matter and other debris are siphoned out twice a week, one third of the water is also replenished each time. They reach 25-35m size by this time and are ready to be released into earthern nursery tanks. They can be transported for stocking at the stage. Kulkarni and Ogale (1979) also recommended to transport the eggs packed in moist cotton wools to save expenses in transport in containers with water and oxygen.

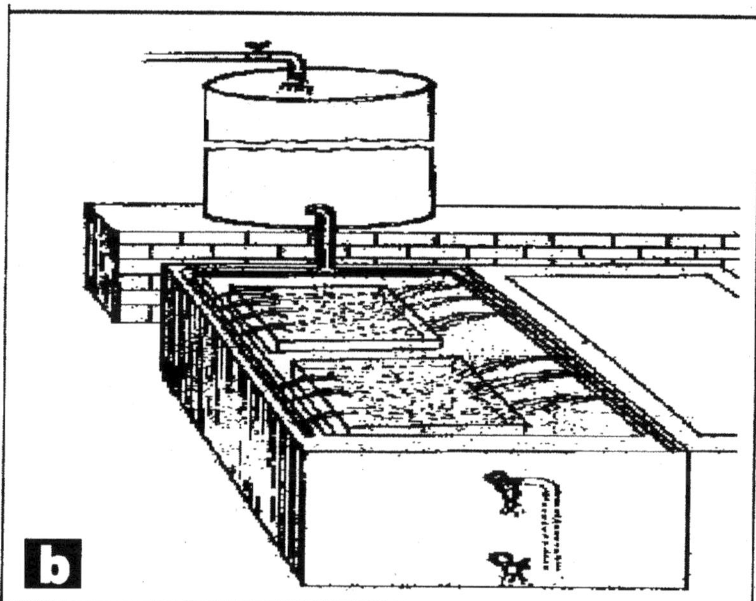

Fig. 6.6: Incubation Device for Mahseer Eggs (Diagrammatic).

(a) Mahseer Hatchery of Tata Electric Co. at Walwhan (Lonavla), (b) Diagrammatic Sketch of the Hatchery.

(reproduced with permission from Fishing Chimes, vol. 19 ,no. 10,11)

Chapter 7

Breeding of Ornamental Fishes

INTRODUCTION

The total international export trade in ornamental fishes during 1998 was worth 173. 87 million U. S. dollars (FAO fisheries statistics, 1998). Singapore, which ranked first among ornamental fish producing countries, contributed 24. 6% of the total export trade, as it was worth 43. 063 million U. S. dollars. India's share in the global ornamental fish market is negligible with only Rs. 1.58 crore worth export in the same year. USA was one of the major importers of ornamental fishes, as it imported worth US$ 67. 309 million in the year 1998.

The boom in global market for ornamental fishes is due to their increasing demand for keeping them in households as pets in decorative aquaria. The hobby of keeping fishes-mainly gold fish -as pets is said to have originated in China several hundred years ago but later on it got popularised in countries like U.K., USA, Japan, Belgium, Germany, Italy etc. It is reported that 8% of the estimated 90 million houses in USA, 15% of the 21 million houses in Britain, 4% of the Belgium and Italian houses, 3% of Japanese houses and 5% of German houses have ornamental fishes as pets (Abidi, S.A.H. and Thakur, N.K., 1997). It can also be a means of gainful self employment with the seed production, culture and export of ornamental fishes and also with the sale of accessories such as aquaria, aerators, decorative dolls, natural and artificial plants, lighting and heating equipments, filters and fish feed (natural and artificial). About 80% of the ornamental fishes are of fresh water origin where as the rest are of brackish water and marine origin. Hence the breeding of fresh water ornamental fishes is described below.

Devraj K.V. (1989) categorised the ornamental fishes into two broad categories based on their spawning habit such as (i) Oviparous fishes that lay eggs. (ii)Ovoviviparous/viviparous fishes (live bearers). In the fishes belonging to the latter category fertilization is internal and the development takes place in the body of the mother. After the development is completed young ones are released through the genital aperture of the female. Oviparous fishes can be further classified into three groups (a) Ornamental fishes that lay eggs and the eggs are allowed to hatch unattended (b) Ornamental fishes that lay eggs and the eggs are guarded by either of the parents or both but he developing eggs are not in direct contact with the body of the parent fishes (c) Ornamental fishes that lay eggs and the eggs remain in direct contact with the body of the parent fish (egg carriers)

Oviparous Fishes that Hatch Out Eggs Unattended

Four subgroups are identified under this category (Axelrod and Vordevwinkler, 1986) *viz.* egg scatterers, egg hangers, egg buryers and egg stowers.

Egg Scatterers

Fishes that scatter the eggs while spawning are of two types:

(a) Egg scatterers with adhesive eggs, and

(b) Egg scatterers with non-adhesive eggs.

Egg Scatters with Adhesive Eggs

Common examples of this category of fishes are Gold fishes, Barbs and Tetras.

Gold Fish *(Carassius auratus)*

It is one of the most wanted fishes for aquarium because of its graceful appearance. It is reported to grow upto 20-30cm. By this time, through selective breeding and hybridization several varieties as noted below have been developed with different and striking external morphological characteristics such as devided caudal and anal fins, thickening of skin on top of head (Wen), absence of dorsal fin, egg shaped body through shortened vertebrae, protruding eyes and various colour patterns. Ouf of about 126 breeds of fancy gold fish, common varieties are: (1) Shubunkin (2) Comet (3) Ryukin (Fringe tail) (4) Demekin (telescope moor) (5) Oranda (6) Ranchu (Lion head) (7) Celestial (8) Veil tail (Fig. 7.1).

Common gold fish is metallic orange red in colour, dorsal fin is having long base but caudal fin is only slightly forked and rounded. Shubunkin has nacreous scales with mixture of blue, purple, brown, yellow, red, orange and black. Caudal fin is large, lobed and errect. Comet is orange red or silvery variety with deep red cap on head. Caudal fin is attractive and deeply forked. Fantail has nacreous or metallic orange body, twin tail, rounded egg shaped body with top edge of caudal fin standing up without drooping below horizontal line. Veil tail a breed of fan tail has elaborate square cut caudal fin hanging gracefully in folds, flowing pectoral and pelvic fins having orange coloured body usually but metallic coloured with red cap are also found. Telescope moor has telesope eyes and velvety black body. Oranda has a hood over the entire head and red/white forms exists. Lion head are similar to oranda but does not possess a dorsal fin. Celestial has no dorsal fin but possess metallic orange body with upturned eyes and rudimentary anal/caudal fin.

Sexual Dimorphism

(a) During breeding season, tubercles appear on head, operculum and sometimes on pectoral fins and other fins of males. They appear as dots sometimes on scales also. Females do not exhibit breeding tubercles.

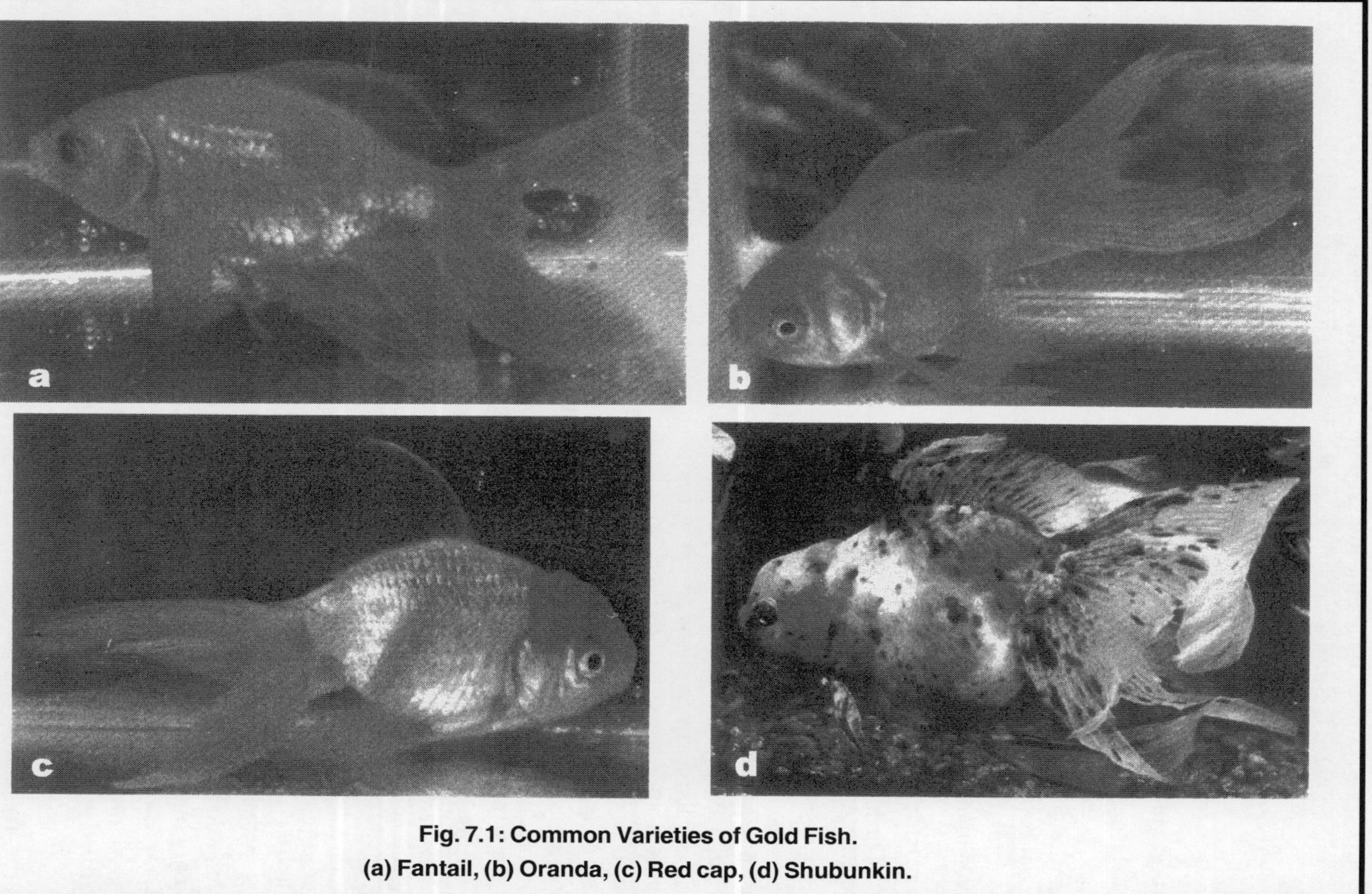

Fig. 7.1: Common Varieties of Gold Fish.
(a) Fantail, (b) Oranda, (c) Red cap, (d) Shubunkin.

(b) If belly line between the lowest point of belly to base of tail is observed from above, it is circular in outline in female but a slight curve in male. Belly of the female may be slightly bulging to one side.

(c) Genital opening is round and protruded in female while in male it is long and oval.

Brood Stock Development

Fishes of 8 to 15 months old ranging in size from 40 to 100 g can be used as brood stock. Feeding is done thrice daily-morning, noon and evening. During morning and evening live feed (preferably tubifex) and during noon formulated feed are recommended. Feeding may be done at the rate of 10% of the body weight per day. Of the total ration, 30% can be given in the morning and evening each and 40% can be given in the afternoon.

Courtship, Spawning and Larval Rearing

During breeding season, male and female move in pair, male taking a position below the posterior region of female with its snout nearer to the vent of the female and trunk below the caudal fin. Some times the courtship continues for several hours or even days. Then the female releases the egg which is fertilised by the milt released by the male simultaneously. As the eggs are adhesive, various types of water plants are utilized as egg collectors as they stick to these plants. Even polythene strips made into fibrillar form are also used as egg collectors. As the parent fish has the habit of eating its own eggs, it is advisable to separate the egg collectors containing eggs into a different container. Each female releases about 2500-3000 eggs. The fertilized eggs are transparent in the beginning but the unfertilized, dead eggs are cloudy. The dead eggs are to be taken out. As the development proceeds, transparency also decreases. Under ideal conditions, the incubation period is 72 hours, but it varies depending on water temperature. After all the eggs are hatched out, egg collectors are removed carefully so as to avoid mortality of newly hatched young. The larvae start feeding after the yolk sac is absorbed. For the first seven days, infusoria are given as feed, in the second week, it is supplemented with boiled egg yolk and microworms. From 3rd week onwards, Daphnia and Tufifex are provided till it attains 10-15 mm length. One month old fry can be transferred to grow out tanks.

Barbs

A variety of fishes of different colour patterns and size of the genus Puntius are available for aquarium keeping. They prefer water with pH 6.8, temperature 28°C, plenty of space and oxygen rich water. The common species are *Puntius tetrazona* (Tiger/Sumatran barb), *P. schuberti* (golden barb), *P. oligolepis* (chequer barb), *P. nigrofasciatus* (black ruby barb), *P. melamphyx* (Ember barb), *P. hexagona* (six zoned barb), *P. aurulius* (filament barb) and *P. conchonius* (rosy barb) etc.

P. tetrazona (Tiger/sumatran barb) are the most striking and colourful of all barbs. There are 4 vertical bars on an orange coloured body. In addition, pectoral and ventral fins are bright red, dorsal is also red above the black bar. Upper and lower lobes of the tail fin also has intense red streaks. All red markings are more pronounced in male than female. During breeding season, males have a red snout where as in female it is not seen. Males may also develop tiny, pin head size, prickle like projection on the operculum.

For breeding purpose, aquaria is provided with selected plants having feathery leaves like Meriophyllum/Ceratophyllum which serves as egg collectors. Plants are to be cleaned, washed in Pot. permanganate solution before introducing into breeding tanks. Ready to spawn female along with 2 males are introduced into the water during evening. Spawning usually take place in the next morning. Soon after spawning, parents are taken out as they have the habit of eating their own eggs. One female may lay about 150 to 200 eggs. Incubation period is 24-36 hours. Hatchlings hang on to the underside of leaves or wall of the aquarium till the yolk sac is absorbed which is usually 24 hours. Then they fall to the bottom and start darting here and there. After they start feeding, infusorians are given for 7 days after which brine shrimp nauplii can be given as feed followed by Daphnia and Cyclops.

Tetras

Common ornamental fishes of this group are Neon tetra *(Paracheirodon innesi)*, cardinal tetra *(P. axelrodi)*, emperor tetra *(Nematobrycon palmeri)*, Glowlight tetra *(Hyphessobrycon erythrozonus)*, flame tetra *(H. flammeus)*, lemon tetra *(H. pulchripinnis)*, bleeding heart tetra *(H. erythrostigma)*, rosy tetra *(H. rosaceus)*, gilded tetra *(Hemigrammus armstrongi)*, redtailed tetra *(H. caudo vittatus)*, beacon fish tetra *(H. ocellifer)*, red nosed tetra *(H. rhodostomus)*, feather fin tetra *(H. unilineatus)*, silver tetra *(H. nanus)*, spotted tetra *(Copeina guttata)*. These aquarium fishes are mostly from S. America where as some like rosy tetra, spotted tetra, neon tetra are from Brazil, lemon tetra, cardinal tetra are from amazon basin.

Breeding, egg collection and larval rearing are similar to that of barbs as they are egg scatterers having addhesive eggs.

Egg scatterers with Nonadhesive Eggs

Several species of Danio and Brachydanio belong to this group. Common examples of the genus Danio are giant Danio *(D. malabaricus)* of S. India and D. *devario* of N. India. The different species of Brachydanio are (1)Pearl Danio *(B. albolineatus)* of Sumatra, Burma and Thailand (2)Leopard Danio *(B. frankei)* of Thailand and Zebra Danio *(B. rerio)* of India.

Danios are prolific breeders. Because of their habit of eating their own eggs, it is advisable to have gravel or round pebbles at the bottom so that eggs that are scattered by the female can escape the notice of the parents as they fall in between the pebbles. The eggs are of the size of a pin head. Incubation period is 3-4 days. After egg laying the male and female are removed from the breeding tank. Hatchlings hang from the plant leaves or from sides of tank. After they start feeding infusoria is given as feed and later Daphnia and rotifers are given. They reach 1″ size in 8 weeks time.

Egg Hangers

The fishes of this group lay eggs singly and the same are attached to the objects by a fine sticky thread. The female does not release all the eggs at a time but lay 10-12 eggs in one day and repeat the same on the next day and so on till they lay 150-200 eggs. Incubation period is usually long *i.e.* 12-16 days. Incubation is usually done in dark condition.

These fishes prefer darkness for breeding and hence breeding tank has to be kept avoiding bright light and providing sufficient aquatic plants having fine leaves. Some breeders prefer to provide artificial surfaces such as "spawning mop" which is prepared out of nylon yarn cut into pieces of 6" length and tying together in the middle. A small cork attached to this keeps it in floating condition. After spawning is started allow it to continue for a week so that eggs will get accumulated on the spawning mop and then it is removed to a hatching container. After the fry hatches out, it starts feeding almost immediately on zoo- plankton such as Daphnia/Artemia nauplii.

Common examples of this group are given below:

Aphyosemion australe or Lyre Tail

It is a beautiful fish, the body of the male is chocolate brown with red coloured dots while the belly is yellow. Pectoral and ventral fins are bright orange. Anal fin is orange at the base with purple edge and white tip. Female has light brown body and round fins.

Other species are *A. bivittatum* (red lyre tail) of W. Africa, *A. cognatum* of Zaire and *A. calliurum* of Liberia and Africa.

Other fishes of such breeding habits are *Aplocheilichthys flavopinnis* or yellow finned panchax, a native of Nigeria, - *A. macrophthalmus* (lamp eye) of Nigeria, *A. myersi* of Zaire, *Aplochilus blocki* (dwarf panchax) of India and Srilanka, *A. lineatus* of India and Srilanka, *A. panchax* (blue panchax) of India, Burma, Malaysia and Thailand, *Epiplatys dageti* (fire mouth panchax) of Africa, *E. fasciolatus* (Banded panchax) of Nigeria, *Fundulus chrysotus* (golden ear) of S. Carolina to Texas, *F. dispar* of USA, *Pachypanchax playfauri* (Play fair panchax) of Seychelles, *Rasbora heteromorpha* of Malaysia, *Rivulus cylindraceus* of Cuba, *R. dorni, R. harti, R. ocellatus, R. urophlhatmus, R. xanthonotus* and *Rasbora heteromorpha*.

Egg Buryers

These fishes bury their eggs in soil before the advent of summer season. Before rain, the water gets dried up and along with it the fishes also die. But the eggs remain buried and dormant in the bottom soil. During rainy season the eggs hatch out and grow to perpetuate the race. Such fishes are found in S. America/Africa where water is available only during rainy season. Though they are some of the most beautiful of all aquarium fishes, their breeding in aquaria is hampered because of the above mentioned breeding habit and long incubation period. Example of egg buryers are given below.

Arnold's Lyre Tail *(Aphyosemion arnoldi)*

It is a native of Nigeria and it grows to 2" length. It has a bright blue body with irregular dots and streaks of carmine red. One variety has oliver green dorsal fin having carmine dots, anal and caudal fins are orange. Another variety has predominantly blue fin.

Breeding

Tanks of 5-10 gallon capacity can be used for breeding. Glass bottom tanks are preferred as it helps in locating the eggs easily. Add 4 to 5 inch depth seasoned water. Bottom of the tank can be provided with 1″ layer of willow moss or spanish moss or fluffy nylon knitting wool. Some floating plants are given to cut down light. Sides of the tank can also be pasted with black or dark green paper. After the tank is made ready, release the mature male and female fish. They breed in the dark corners of tank readily accepting the bottom as a substrate for mud into which, they usually bury the eggs. When the spawning is observed, remove the floating plants, and then the bottom plants carefully and gently shaking to release the nonadhesive eggs.

Collection of eggs for incubation: Prepare a series of clean, glass jars each having 1″ deep water. Collect clean rain water, add peat moss and allow the same to stand till pH reaches 5.8-6. This is considered good. Eggs are transferred from the breeding tank to these glass gars. Once breeding has commenced, a given brood fish may continue to lay eggs every day for a month or more. Breeders can be fed on live daphnia. During summer, incubation period varies from 5-6 weeks while during winter it may take 2½ to 3 months. Soon after hatching, the hatchlings are picked up by glass tube and tranfered to a rearing tank where infusoria is provided as feed. Within a few days they will grow to a large size when Artemia nauplii are given as feed.

Other species of Aphyosemion that are egg buryers are *A. filamentosum*, *A. gardneri*, *A. sjoestedti*, (Blue gularis), A. *occidentalis* which are all found in African continent. Blue gularis is tthe largest of all attaining a length of 5½″. They all can be bred to get seed following the method described for *A. arnoldi*.

Cynolebias adolfi

It is a native of Brazil. Male is bright blue with dark bars (9-12nos) vertically running through the body. Fins are dark blue with black margin. The female is light brown with no vertical bars. They also lay their eggs in pond bottom buried in the peaty mud. After the pool is duried up, the eggs hatch out in the next rainy season.

For spawning, a setup used for *A. arnoldi* may be used for this fish also. But after collecting the eggs, eggs are incubated in between 2 layers of moist peat moss (each layer of ½″ thick) kept in a covered petridish. Petridish with eggs is kept in a dark place at constant temperature of 75-78°F. Examine the eggs after 6 weeks, if eyespots are visible place the same in shallow aquarium having the same water as the parent had. After the hatching the fry can be yared as earlier.

Other species of Cynolebias are *C. bellotti* (Argentine pearl fish), and *C. nigripinnis* of Argentina.

Nothobranchius guentheri

Native of east coast of Africa, attains 3″ size. Male is blue with golden border at the edge of anal fin. Tail is bright red with brown edge. Dorsal and anal fins are yellowish green with red markings, rest of the fin bright blue with white edge. Female is green with colourless fins.

Breeding can be done as in *C. adolfi*.

Other species of the genus are *N. orthonotus* of E. Africa and *N. rachovi* of *E. Africa*.

Egg Stowers

Fishes that lay eggs inside the fresh water mussel such as bitterling *(Rhodeus sericeus)* are included in this group (Fig. 7.2) Bitterling is a native of Central and Southern Europe with its breeding season between April and June. It grows to 3″.

Male is distinguishable from female in its colour pattern. During non breeding season, they are silvery with a horizontal stripe on the lower half of the body. During spawning season, the stripe becomes bright blue, belly and anal fin of the male are bright red; outer half of the tail fin is also yellow, dorsal fin yellow with green markings. The female shows horizontal stripe but all other colours are pinkish. When the female exhibits bulging abdomen and male exhibit colour pattern noted above, the pair can be introduced into a breeding tanks (aquarium) prepared with sandy bottom, and provided with green plant. The freshwater mussels are also introduced into the tank. The mussels burrow and remain buried in the sand with its exhalent and inhalent siphons projecting out. The female, when ready to spawn developes a pink tube (ovipositor) which is introduced into the inhalent siphon of the mussel. As she lays about 40 eggs, the male that stands by releases the sperm that is inhaled by the mussel along with stream of water and it fertilizes the eggs present in the mantle cavity of mussel. Thus fertilization takes place inside the body of mussel. The eggs remain inside the mussel for 4-5 weeks. After the young emerges out along with exhalent waters it feeds on infusoria and later on brine shrimp nauplii.

Ornamental Fishes that Lay Eggs and the Same are Guarded by Either Male Parent or Female Parent or Both but the Developing Eggs are not in Direct Contact with the Body of the Parent

Included in this group are ornamental fishes that exhibit parental care but the eggs are not carried by the parent fish. They have been classified into 5 types as noted below (Axelrod and Vorderwinkler, 1986):

(a) Egg anchorers,

(b) Egg splashers,

(c) Egg scoopers,

(d) Egg hiders, and

(e) Nest builders.

Egg Anchorers

Definite spawning sites (surface of stones, wooden piece etc.) are sought and prepared by both the parents or either of the two. As the female lays the eggs on substrates male releases the sperm and fertilizes the egg. As parent fish keep a watch on the eggs, the fanning movement of pectoral fin causes water circulation over eggs which ensures adequate oxygen for the developing eggs. In spite of their parental are, some times the parent fish may turn cannibalistic and eat away its own eggs. To avoid this, soon after egg laying, the parent fish can be separated from the tank and provide mild aeration in the breeding tank near the place where eggs are laid. Common examples along with their breeding habit are described below.

Angel Fish, *Pterophyllum scalare* [Fig. 7.2]

Being one of the attractive fishes among aquarium fishes because of its graceful moment and shape, breeders have developed different strains of the fish exhibiting different colour patterns by this time. Though it is a native of Amazon basin, it is distributed all over the world as it is in high demand. It grows up to 5 inches. It is a cichlid fish having lateraly compressed body with fins exhibiting unusual feature such as existance of thread like filaments arising from the top and bottom rays of the tail fin, first fin ray of dorsal and ventral fins and also of pectoral fins. Strains with silvery body with black vertical bars; pure black strains, strains with silvery colour in the front half and black colour in the posterier half; gold coloured, albino coloured etc have been developed. Sexes are difficult to be distinguished externally but the female has a larger vent and the tube extending from the vent during spawning is twice the diameter of that of male.

Breeding can be done by introducing a mature pair into an aquarium provided with enough of aquatic plants. The pair selects usually a stiff plant stalk or leaf and after cleaning they attach the eggs in rows to the surface. The eggs are fanned by the parents and incubation period is about 4 days. The hatchling is attached to the leaf by a sticky thread attached to its head. Yolk sac is absorbed within 3-4 days and by this time the thread also disappears. The fry then drop to the bottom. Initially, the fry

Fig. 7.2: Veil Tailed Angel Fish.

can be fed with infusoria and later brine shrimp nauplii can be given. It is better to remove the parents soon after hatching to avoid loss of fry due to cannibalism.

Brown or Common Discuss *(Symphysodon aequifasciata axelrodi)*

This fish is also a native of Amazon basin and it grows to 9″. This is also one of the most attractive fishes for aquarists. It has a laterally compressed body of round shape. Body has deep blue and brown colour with 3 vertical bars a central bar of which is darker colour.

Male and female differ in the colour pattern-many wavy bright blue lines around the head, back and belly region are seen in male but female has fewer wavy lines.

They breed exactly like angel fish. But after the yolk sac is absorbed, they feed on the mucus secreted by the parents skin for about one month. After one month only, they shall feed on other live feed like brine shrimp nauplii.

Aequidens curviceps

It is also a cichlid fish, a native of Amazon which grows to 3″. The male is distinguishable from female in its colouration. In male, body sides are blue becoming green in the region of belly and upper part of the body has 3 distinct bars. Tail and anal fins are yellow, rest of the fins being blue with red edge. Females are brighter in colour. The dorsal and anal fins are less pointed than male.

A small tube begin to protrude through the vent of both sexes when they are fully mature. The pair will select a site for spawning and clean the same meticulously. The female swims with the vent close to the surface of spawning site and lays eggs on it. The male follows close behind shedding its milt upon them. After the spawning act is over, one of the parents take a position, close to the eggs fanning them with the pectoral fin movement. At times it may take the eggs in the mouth to keep it clean. Incubation period is 4 days. Parents scoop out depression to keep the hatchlings. If the hatchlings go too far as they start moving, parents pick them up in their mouth and spit them back. As the parent may turn cannibalistic, it is advisable to separate the parents soon after spawning is over.

Oscar/Peacock Cichlid *(Astronotus ocellatus)*

It is a native of eastern Venezuela and Amazon basin which can grow to 1 foot. Breeding habits are similar to that of other cichlids described above.

Egg Splashers

The splashing Tetra *(Copeina arnoldi)* come under this category. They are found from Brazilian Amazon to Venezuela and it grows to 3″ size. Upper lobe of the tail fin is elongated and edged with black. Dorsal fin which has a black triangular spot is longer in males than females.

These fishes have the habit of laying eggs on overhanging leaf or stone which it reaches by jumping. The eggs are kept moist by splashing water at them by the parents swimming below it. Incubation period is 2-3 days. The hatchlings fall into the water and lead independent life.

These fishes can be bred in tanks of 10 gallon capacity containing aged, soft and neutral to slightly acid water. Temperature can be between 82° to 84°F. The tank should be kept covered because of its jumping habits. As the fish accepts substitutes for overhanging leaf substrates such as green frosted glass with rough side down can be given about 2" above water surface. The mature male and female can be introduced into the water. The pair shall leap out of water under the spawning site. The female lays the egg and attach to the spawning surface when the male fertilizes them. The parents fall back into the water after the sex act. This is repeated until the female is depleted. After this the male start splashing water up at the eggs till they hatch. Incubation period is 2-3 days after hatching they fall into the water.

Egg Scoopers

Usually, the male choses the site for egg laying and scoops out a depression by swimming in a circle. Any particle of debris or sand which cannot be removed in this manner are picked up by the mouth and deposited outside the depression. After the above work is over, the male coaxes the female to the site and the eggs are laid by the female. This is guarded by the male until the young hatches out.

eg. *Copeina guttata* (Red spotted copeina). It is a native of Brazilian Amazon. The body colour is bluish grey becoming white on the belly. The male has six rows of red dots on the body horizontally which are absent in female.

Etroplus maculatus (orange chromide) of India has similar breeding habits but some times they place their egg on the side of a rock instead of shallow pit.

Egg Hiders

These fishes are similar to egg anchorers but differ only in one aspect. They prefer to hide their eggs out of sight but the egg anchorers place the eggs in the open. They search for crevices between rocks or scoop out a hole by the side of the rock. Both the parents or either of the two guard the eggs.

Ornamental fishes such as Apistogramma, Nannacara, and Pelvicachromis - commonly called dwarf cichlids are of this category.

Dwarf cichlids are easily bred in aquaria of 2-5 gallon capacity at an average temperature 78 to 80°F. Breeding tank (aquarium) is provided with aquatic plants, gravel/rocky bottom. If a flower pot is put with its open end away from light it is easily chosen by the female for egg laying. Usually, it is the male that prepares the spawning site by scooping out the hole. The male shall coax the female to spawning site and as she deposits the egg in group, the male fertilize the same. Soon after this the female get aggressive towards the male, the male is bitten and butted and chased away. The female takes care of the egg and hatchling. As the female develops cannibalistic tendencies and eataway the eggs/hatchlings she can be separated soon after spawning. Incubation period is 4-5 days and the hatchlings remain at the bottom for about 3 days till the yolk sac is fully absorbed. Initially the hatchlings can be fed on infusoria and later live rotifer can be provided. Other species such as *Apistogramma agassizi*, of Amazon basin, *Nannacara anomala* (golden eyed dwarf cichlid), *N. taenia* (Lattice dwarf cichlid) a native of Amazon basin, *Pelvicachromis pulcher* of West Africa are fishes of similar breeding habits.

Nest Builders

These fishes prepare a nest for keeping the eggs safely till hatching. The nests are either prepared out of air bubbles produced from the mouth or nests are built out of twigs of plants, roots, floating objects etc.

Bubble Nest Builders

Some fishes create a heap of froth on the surface of water by blowing bubbles which are covered with mucus by its mouth. The nest made of bubbles float on the surface of water into which the female lays the egg and fertilized by male which watches the eggs till hatching.

Siamese Fighting Fish, *Betta splendens* (Fig. 7.3)

It is an anabantoid fish which is a native of Thailand attaining a length of 2½″. Two male fishes, if kept together, shall fight with each other. They have accessory respiratory organs like other Anabantoid fishes. As they are attractive fishes, several strains have been produced having different colours such as red, blue, green, albino, black, white with red fin etc. through breeding and hybridization.

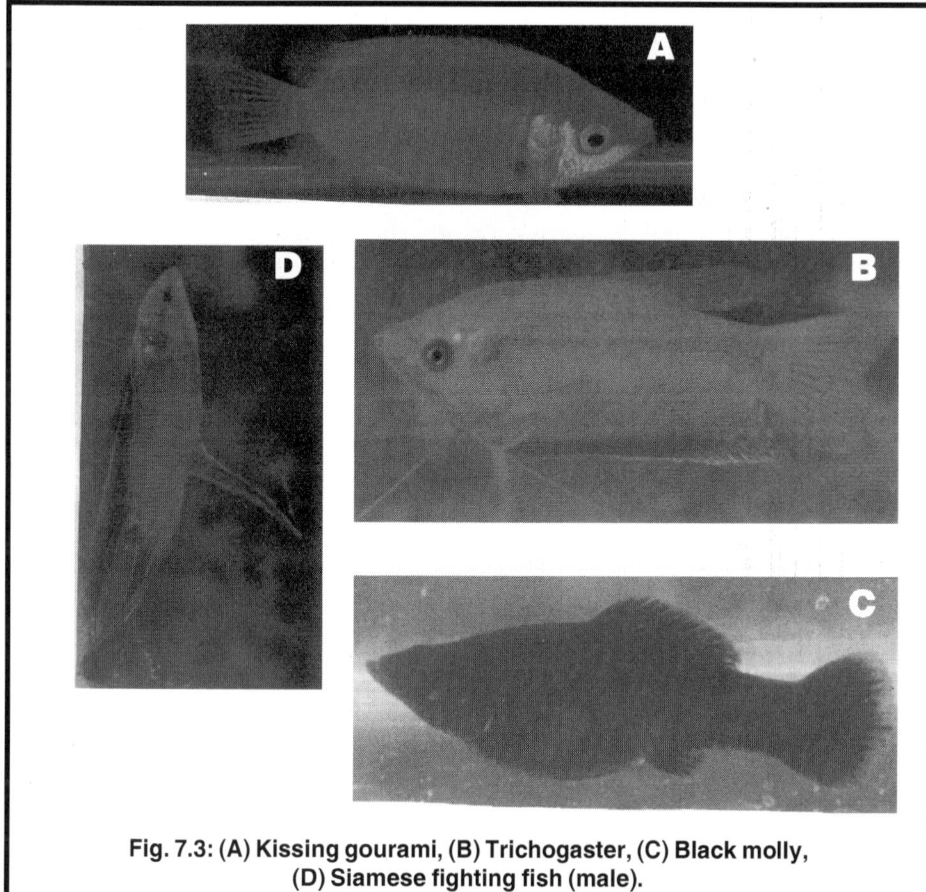

Fig. 7.3: (A) Kissing gourami, (B) Trichogaster, (C) Black molly, (D) Siamese fighting fish (male).

Breeding

Male is grown individually in separate containers to avoid infighting and damage. Upon reaching sexual maturity, keep one male in a large aquarium. Male is distinguishable from females by their larger and longer fins and brighter colouration. The male shall soon start producing bubbles from its mouth which float on the surface. A large number of bubbles floating together form the bubble nest. At this time a fully mature female is introduced into the Aquarium. The male shall guide her to a position below the nest he has prepared and then shall wrap his body around the female's body in such a way that their vents are close together (Fig. 7.3). This 'nuptial embrace' may be repeated till the female releases the eggs which is immediately fertilized by the male. As the eggs fall slowly to the bottom, the male leaves the female and catches the egg in his mouth. Then he moves up and blows the eggs into the nest and stand guard over them. At each nuptial embrace 10-40 eggs are released and the act is repeated till all the eggs are released by the female. At this point, the male shall drive away the female. Hence at this time it is better to remove the female from the breeding tank to avoid injury to the female. The incubation period is 30-40 hours and on hatching, the hatchlings remain in the bubble nest. If any egg during incubation or any hatchling after hatching falls down from the nest, the male shall pick it up and replaces them in its proper place. Within 3-4 days hatchling becomes free swimming. After this the male shall leave the nest. The hatchling, soon after the yolk sac is absorbed shall start feeding and then infusoria may be provided as food. Mild aeration of the tank water also shall be required. This can be stopped after 3-5 weeks when the fry develops accessory respiratory organs.

In addition to the Siamese fighting fish mentioned above, fishes such as giant gourami *(Colisa fasciata)*, thick lipped gaurami *(Colisa labiosa)*, dwarf gourami *(Colisa lalia)*, the gourami, *Osphronemus goramy*, Kissing gourami, *Helostoma temminckii* [Fig. 7.4(a)] pearl or mosaic gourami *(Trichogaster leeri)* snake skinned gourami, *Trichogaster pectoralis* [Fig. 7.4(b)] threespot gourami *(Trichogaster trichopterus)*, croaking gourami *(Trichopsis vittatus)* round tailed paradise fish *(Macropodus chinensis)*, the paradise fish *(Macropodus opercularis)* (Fig. 7.5) and armored cat fish *(Callichthys callichthys)* build bubble nest and their breeding habits are similar to that of Bettas. Croaking gouramis and armored cat fish make bubble nests under leaves of plants. Gouramis lay floating eggs and hence they do not drop down while spawning. Incubation period is also short (only one day) and for the fry to become free swimming it may take only one week.

Nest Fabricators

These fishes collect twigs/roots of plants, debris etc. and weave them into a tunnel shaped nest and the same will be attached to plants or other objects found in water. The mucus secreted by the fish works like cement holding different materials used for nest making together. When the parent fish is in the tunnel, head and tail project from it at both ends. Stickle backs belong to this category.

Four spined stickle back *(Apeltes quadracus)*

Males are distinguishable from female by their red ventral fin during spawning season. They are brackishwater fishes that can be acclimatised to fresh water gradually.

Breeding

It is advisable to keep the breeding tank (aquarium) in a cool place and provide enough of adequate plants and aerators it is the males that construct the nest (females do not participate in nest making). After the nest is completed, the male leads the female to the nest- the male squeezes in along with the female. The female lays the egg and male fertilizes it. This process may be repeated till female lays all eggs. After this, the male drives out the female and watch the eggs. Hatching takes place after 3-4 days. The hatchings are fed on infusoria and later on brine shrimp nauplii.

The three spined stickle back *(Gasterosteus aculeatus)* has also similar breeding habits. Male is distinguable from feamle in breeding season. Normally, sides of the abdomen are silvery with light pink tinge but in the breeding season, the side becomes metallic blue. Ventral side become bright red. Females have more silvery abdomen.

Egg Carriers

Fish belonging to this category carry their eggs around with them during incubation. This is performed in either of the following ways:

(a) The female swims around with the eggs attached to the vent. Later they are brushed off to a plant or any other object *eg.* cuban Killie. *(Cubanichthys cubensis)* and Japanese medaka *(Oryzias javanicus)*

(b) Eggs may be carried in the mouth of the male parent or female parent. These fishes are called mouth breeders.

Mouth Breeders

Mouth breeding betta *(Betta brederi)*. This is found in India. Males have longer fins than female. The fins are reddish with golden edge.

During nuptial embrace, the eggs released by the female are taken on the anal fin of male when it is fertilized by male. The female picks them up in her mouth and then spits them one by one to her mate. The male catches it in his mouth. Incubation period is 4-5 days. After hatching, the fry escapes from the mouth.

Haplochromis philander

It is a native of Congo attaining a length of 4″. Anal fin is red with blue dots but in male there is a red spot at the tip of anal fin.

For breeding purpose, a larger aquarium is preferred for this fish to suit its swimming habit. Usually the male and female swim around in a circle for some time. Then the female drops a few eggs which is duly fertilized by the male which follow her. As the pair makes the next circular movement, the female pauses to pick the eggs in her mouth. This habit is repeated till the female depletes all her eggs. Incubation period is 10-14 days during which time the female does not fed any thing. Soon after hatching the hatchlings come out of the mouth.

H. multicolor is an Egyptan mouth breeder, a native of Nile river with similar breeding habits.

Parana Mouth Breeder *(Geophagus balzani)*

It is a fish that has both egg anchoring and mouth breeding habit. First the fish spawns on the surface of flat rocks carefully cleaned by both parents. Later *i.e.* after one day, the female picks up the egg in her mouth and keeps them for about 7-8 days. After the hatching, young ones are released into the water. But the female continues to protect her young for a long time. As soon as the mother fish give signal to the young one regarding impending danger by wriggling movement of fin, the young flee back to her mouth and take shelter there.

Nyasa Golden Cichlid *(Pseudotropheus auratus)*

This cichlid fish, a native of Africa grows to 5″ size and starts breeding on attaining 3″ size.

The male is usually having 2 longitudinal dark stripes at the middle and dorsal part of the body. During spawning season, male develops dark brownish black colour except for a white lateral stripe and a light brownish band at the base of dorsal fin. The female is bright golden yellow, with black lateral stripes. Another black longitudinal stripe is also found on the dorsal fin of female. The tank should be provided with hiding place for the fish such as rocks and plants. During courtship, the male chases the female, eggs laid by the female are taken in the mouth by the female itself. There is prolonged incubation period of 23 to 30 days. After the hatching, the young remain in the mouth for 10 more days, for the yolk sac to be absorbed. During incubation period, the female does not feed.

In addition, fishes like chocolate gourami *(Sphaerichthys osphromenoides)* a native of Malaysia and various species of Tilapia are also mouth breeders.

Ovoviviparous/viviparous Ornamental Fishes (Live Bearers)

Live bearers are those fishes that give birth to young ones. In case of these fishes, fertilization is internal and the development of eggs take place inside body of the mother fish.

Common live bearers of value as ornamental fishes are guppies *(Poecilia reticulata)*, black molly *(P. sphenophs × P. formosa)* (Fig. 7.5.c) and lyre tail sword tail *(Xiphophorus helleri)* and Platy *(Xiphophorus maculatus)*. In all these fishes, lower part of anal fin of male is modified to a thick rod like structure called gonopodium which is meant for inserting into the genital aperture of the female during mating and the spermatozoa are injected into the reproductive organ of female. The female stores the sperms in her body for a considerably long period and the same is utilized for fertilizing the ovum. Once the ovum is fertilized, it undergoes development inside the body of female. Developing embryo can be seen as a dark spot infront of anal fin which is known as "gravid spot". It may take about 4 weeks to complete the development and hatched out young ones are released through the genital aperture of female. The number of young ones released at one time may vary from species to species - Sword tail may deliver up to 200 numbers, mollies up to 100, guppies from 6 to 60.A newly born young one may be ¼″ long, may have their fin formed in the normal shape. They are capable of swimming and searching for food and protection.

In most cases, the female has the habit of eating its own new born young ones and hence before delivery the female is to be shifted to a breeding trap kept inside a breeding aquarium. The breeding trap has perforated bottom so that when young ones are born, they can escape into the surrounding water as a result female can not reach the young one and eat them. It is better to put aquatic plant at the bottom so that young can take refuge in the plants.

Guppies are hardy fishes that can withstand adverse conditions such as poor water quality, low oxygen, diseases etc. They are of diverse colour and fin pattern. Males are smaller (4cm) than female (6cm) and they are more colourful. They are prolific breeders and four generations can be raised in one year. Among the hybrids produced through selective breeding of guppies are Green-lace guppy, scarf tail guppy, veil tail guppy, and sword tail guppy. Black mollies are hybrids of *P. sphenops* and *P. formosa*. Male can grow to 10 cm and female up to 11cm. Sword tails of different colours-gold, green, albino and red are available. Platy fish are smaller, growing up to 4cm size. These are also found in various colours such as red, gold, blue, black etc. Many other fishes belonging to the family poecilidae are of importance as ornamental fishes such as *P. latipinna* (sail fin molly), *P. velifera* (green sail fin molly) *P. vittata* (spotted linia) and *Gambusia affinis* (mosquito fish) etc.

Chapter 8

Breeding and Seed Production of Shell Fish (Crustaceans)

TIGER PRAWN (*Penaeus monodon*)

Introduction

Among the several penaeid prawns available in the Indian coast *viz.*, (a) various species of Penaeus (*P. indicus, P. merguiensis, P. japonicus, P. semisulcatus, P. monodon* etc.), (b) various species of Metapenaeus (*M. monoceros, M. dobsonii, M. brevicornis* etc.) and (c) Other penaeids such as *Parapenaeus* sp., *Parapenaeopsis* sp., *Trachypenaeus* sp. etc., *Penaeus monodon* or tiger prawn (Fig. 8.1) is the most favoured candidate species for culture. It has a fast growth rate, attains large size, has high resistance to handling stress and hence exhibits high survival rate during culture. It has also good consumer preference in the domestic as well as export market thereby fetching maximum unit value. Hence its production through extensive/modified extensive/semi intensive/intensive culture systems in the coastal areas became popular in India and other countries like China, Taiwan, Thailand, Indonesia, Philippine, Equador etc. This was accompanied by the establishment of monodon hatcheries all along the coast by private companies as well as governmental organizations. Before describing the hatchery production of seed, an account of the natural spawning and seed collection from natural grounds are presented below.

Reproductive Biology

Age and Size at First Sexual Maturity

Age at first breeding of monodon is not clearly established. It is reported that a combination of age and body size influence the onset of sexual maturity.

According to Motoh (1981) sexual maturity is defined as the minimum size at which spermatozoa are found inside the terminal ampullae of the vas deferens in male and inside the thelycum of female. Based on this criterion, the males of 37 mm carapace length and 35 g body weight and female of 47 mm carapace length and 68 g body weight are found sexually mature among wild stock where as among farm reared shrimps male weighing 20 g and with carapace length of 31 mm and female weighing 41.3 g and carapace length of 39 mm were found sexually mature (Motoh, 1981). But according to Primavora (1988) male of wild as well as farm reared stocks mature on attaining 40 g body weight and 38.5 mm carapace length but female wild stocks matured at 63 g body weight and 46 mm carapace length and pond reared females matured on attaining 40 g weight and 41 mm carapace length.

Male Reproductive System

There are a pair of testes located in the cephalothorax above the hepatopancreas. The testis is transluscent, composed of six lobes, all the lobes are connected on the inner margin leading to vasdeferens. Vasdeferens has 4 regions such as a poximal region which is short and narrow, a medium portion which is thick and larger, long narrow distal part and a muscular portion called as terminal ampoule which opens at the base of the coxa of 5th pereioped. The spermatozoa produced by the lobules of testis are aggregated and stored in a bag secreted by the wall of vas deference. This is called as spermatophore. Spermatophores are stored in the terminal ampoule. Spermatozoa of tiger prawn are small in size (3 microns) and lacks flagellum. Each spermatozoa consists of a central spherical body and a short nonflagellated spike Sperms are not capable of swimming. It adheres to the egg at spike end.

Female Reproductive Systems

There are a pair of ovaries which are partly fused extending almost the entire length of a mature specimen. Each ovary consists of an anterier lobe, lateral lobes located dorsal to the hepatopancreas and an abdominal lobe lying dorso-lateral to the intestine. The oviduct originate from the sixth lateral lobe, leading to the external genital aperture located at the base of the coxa of third pereiopod.

Ovarian Development

Five stages of ovarian development are discernible based on changes in colour, size, shape and texture of the ovary in course of development (Fig. 8.2).

Stage I: Immature Stage

The ovaries are thin, transparent and unpigmented. The inner lining of the ovarian capsule is lined by a layer of epithelial cells called germinal epithelium. Oogonia develop from this layer. At this stage ovaries are not visible through the dorsal exoskeleton. Outer to the germinal epithelium is the connective tissue layer and the outer most layer is the outer epithelium.

Stage II: Early Maturing Stage

Externally, ovaries are flaccid, white to light yellow to yellow green in colour. Through the exoskeleton, it is discernible as a linear band. The oocyte grows to 170 microns in diameter. Yolk granules are also formed in the cytoplasm of the oocyte.

It is at this stage that inhibitory hormones secreted by the x-organ-sinus gland complex located in the eye stalk prevent further ovarian growth when females are

Fig. 8.1: Mature Tiger Shrimp, *Penaeus monodom* (Fabricius).

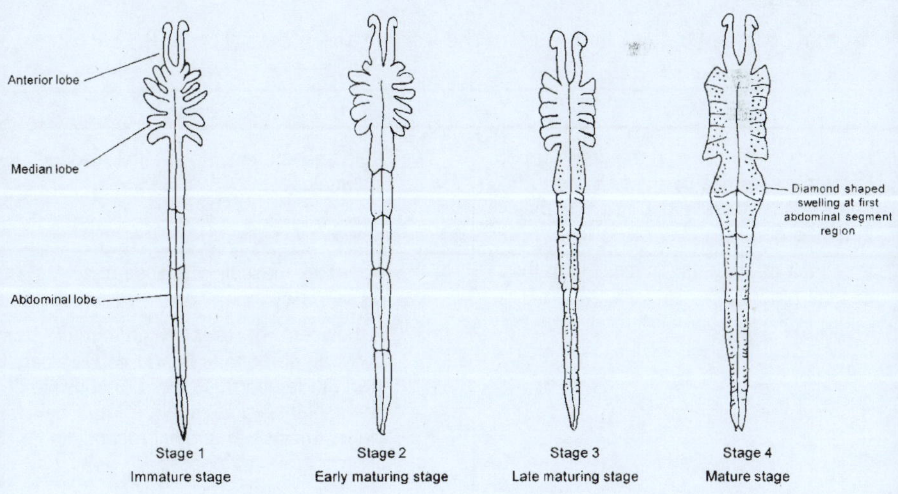

Fig. 8.2: Stages in the Ovarian Development of Tiger Shrimp.

held in captivity in tanks. Captive conditions stimulate the secretion of ovary inhibitory hormones. Unilateral eyestalk ablation is practiced at this stage to remove one of the x-organ sinus gland complex which produces ovary inhibiting hormone so that block in ovarian development is removed.

Stage III: Late Maturing Stage

The ovaries are visible to the nacked eye as a thick, dark, linear band through the exoskeleton. Anterior and middle lobes are fully developed. Maturing ova are opaque due to the accumulation of yolk in their ooplasm. The ovary is also slightly expanded in the first abdominal segment.

Stage IV: Mature Stage

Ovaries are dark green in colour due to the accumulation of carotinoid pigmentation. In the first abdominal segment it assume a distinct diamond shape which is called saddle. The oocyte develop cortical granules which are filled with a jelly like substance destined to form part of egg shell membrane after ovulation. Oocytes are released into the oviduct during ovulation. Ovulation is followed by oviposition *i.e.* release of eggs from oviduct in to the water.

Stage V: Spent/Recovery Stage

After extrusion of eggs, the ovaries revert to immature condition. It appears as a thin linear band. In case of partial spawning oocytes of either abdominal/lateral/ anterier lobe may remain unreleased.

Sexual Dimorphism

A male can be distinguished from the female as follows (Fig. 8.3).

Male	*Female*
1. Presence of petasma in the first pleopod. Petasma is formed by the fusion of endopodites of first pleopod along their inner margin.	Petasma is absent in the first pleopod in female.
2. Presence of appendix masculina in the second pleopod	Appendix masculina is absent
3. Thelycum is absent	A thelycum is present in the thoracic sterna between pereopods IV and V. It is meant for storing the spermatophore. It consists of 2 lateral plates and a median plate. The lateral plates enclose a seminal receptacle for storing spermatophores.
4. Gonopore is located at the base of fifth walking leg.	Gonopore is located at the base of 3rd walking leg.
5. Smaller in size	Larger than male of the same age of group.

Mating

Upon reaching sexual maturity, the female is inseminated by a male each time she molts. *i.e.* the hard shelled male mates with a freshly moulted female. This facilitates

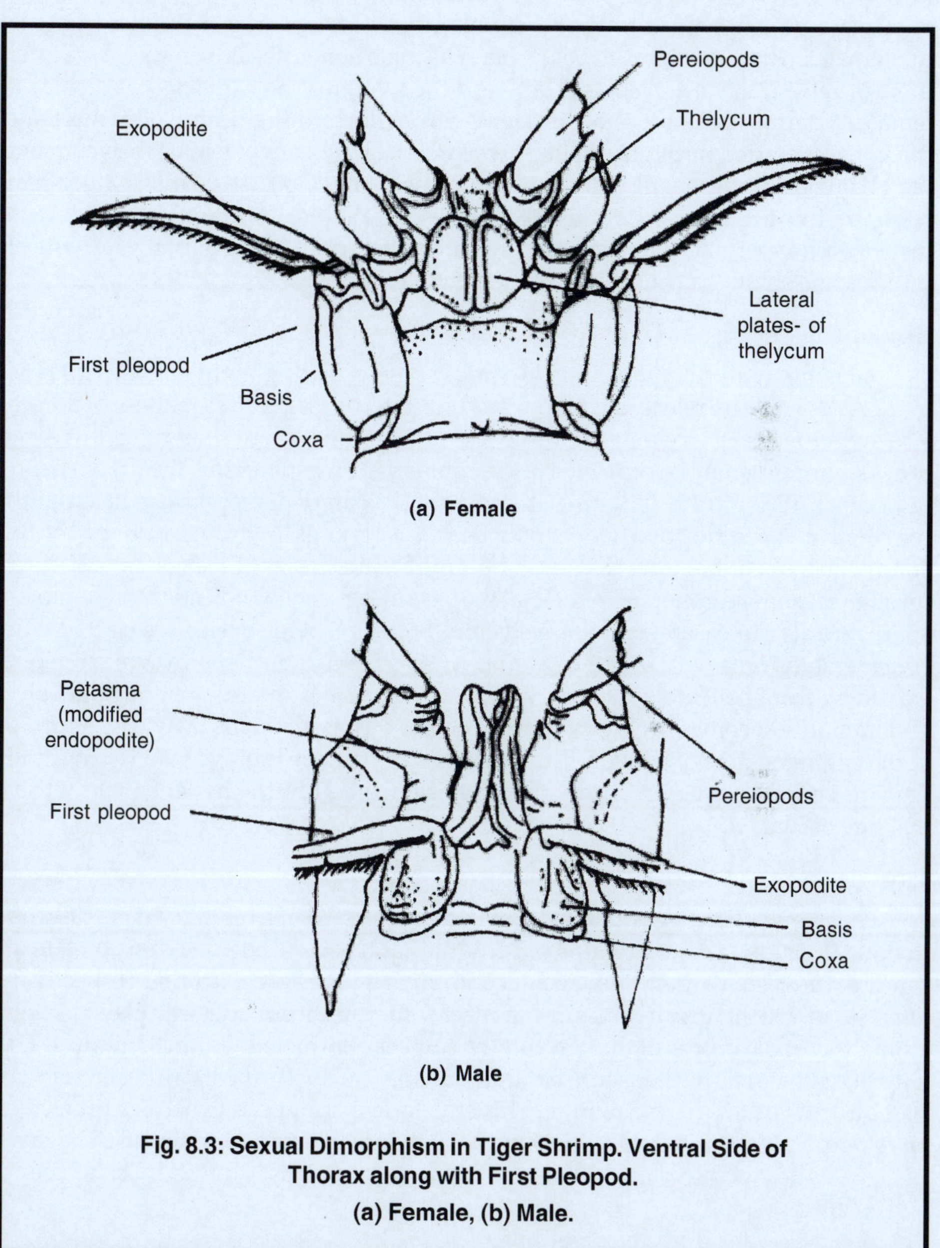

Fig. 8.3: Sexual Dimorphism in Tiger Shrimp. Ventral Side of Thorax along with First Pleopod. (a) Female, (b) Male.

the insertion of spermatophore from the male into the seminal receptacle of the thelycum of female. (In species of prawns having open thelycum, mating usually takes place between hard shelled male and hard shelled female). Generally mating takes place at night immediately after female moults. One or more males swim parallel to a female positioning himself below her. The male bends its body in such a way that there is ventral to ventral contact between the two, then he turns parpendicular to female. A pair of spermatophore is released from the terminal ampullae of male and the same is inserted through the slit between the lateral plates of female thelycum into the chambers underneath with the help of petasma. Courtship and mating may continue from ½ hr to 3 hours and requires parallel swimming for considerable distance. The spermatophore in the thelycum can be easily seen externally by vertical, milky white streaks on both sides of thelycum.

Spawning

A female with fully developed ovaries spawns during night between 10 P. M. and 2 A. M. Eggs are released while swimming in the water near the bottom of the sea. The swimming activity may continue for a few hours before spawning, but some times swimming may be only for a few minutes. During spawning, the prawn bends its body posterior to 4th abdominal segment and 3 pair of pereopods are held tightly together and are flopped vigorously. Release of eggs is brought about by the contraction of caphalothoracic and abdominal muscles as the eggs are extruded through the female gonopore located at the base of 3rd pereiopod. Side by side, stored sperm are also released from the thelycum. Both eggs and sperms are ejected with considerable force, where the eggs appear as greenish and sperm whitish cloud extruding from beneath the female as she swims. As the sperms are non flagellated and immotile, it comes in contact with the ovum by passive collision with the surface of the egg at the time of ejection. The entire process may lasts only about two minutes or up to 4 minutes at times. The fecundity ranges from 2 to 10 lakhs per female with an average of 5 lakhs.

Egg

Each newly laid egg is 250-300 mm in diameter and of irregular shape. The first meiotic division takes place after 2-5 minutes and second meiotic division 8 to 14 minutes after spawning. First and second polar vodies are extruded respectively after each division. After it comes in to contact with water, it extrudes jelly like substance from the cortical crypts of the egg and the same is completed within 8 minutes. The jelly like substance hardens into a tough transparent shell - the hatching envelope - by 12 to15 minutes post spawning. This envelope is supposed to act as a protective envelope, acting as a shield against pathogenic organisms and toxicants. The male nucleus unites with female nucleus and the zygote nucleus is formed. The fertilised egg is called zygote.

Cleavage

The first cleavage is holoblastic and equal and it takes place 30-60 minutes after spawning, though some eggs may remain without division even upto sixty minutes. However, by 90 minutes, all the fertilised eggs will undergo at least the first cell division. On the other hand, the unfertilized eggs either has not divided or has divided assymetrically. At four cell stage it is easy to distinguish fertilized egg from an unfertilized egg.

Determination of Fertilization Rate

Successful fertilization depends on several factors such as egg quality, sperm quality, correct sperm count per spermatophore and health of brood stock. Physical dimension of spawning tank also affects the fertilisation rate by not allowing optional mixing of sperms and egg.

Rate of fertilization can be determined by microscopic observation. Fertilized eggs exhibit symmetrical pattern of cleavage where as unfertilized eggs exhibit assymetical cleavage. One hour after fertilization, the fertilized eggs will have 4-8 cells where as unfertilized ones will have only two cells. There is rapid cleavage after first cleavage in fertilized eggs while in unfertilized eggs erratic cell division is seen after 1st cleavage.

Flow cytometry is another method used for monitoring fertilization rate. The technique depends on incubating a sample of eggs in the presence of a DNA flourescent dye. The intensity of flourescene is dependant on amount of DNA-unfertilized egg being haploid has half the amount of DNA compared to the fertilized diploid egg. Thus hundreds of eggs can be pumped through a flow cytometer and the relative proportion of fertilized and unfertilized eggs can be determined.

The zyote undergoes rapid mitotic division and at the end of 2nd hour 6th (64 cell stage) or 7th (128-cell stage) cleavage is completed. A layer of cells is formed at the periphery of the egg which is termed as blastoderm. At the end of 4 hours, a secondary envelope is noticed within the external hatching envelope. Process of gastrulation sets in is followed by tissue differentiation. By 11-12 hours, naupliar form is discernable with a single reddish pigmented spot representing naupliar eye. Appendages also get developed. First the rudimentary antenna appear at the middle part of the embryo followed by appearance of root of mandible below and root of antennule above it. Embryo shows signs of movement within the envelope. Around 12 to 13 hours after spawning hatching takes place.

Nauplius Larva

A newly hatched nauplius larva is 0.34 mm in size and dark yellow in colour. It is oval in shape having 3 pairs of appendages - a uniramous antennule, biramous antenna and a mandible. The appendages are locomotory in function, antenna being most active, antennule coming next. They are attracted to light. When at rest, it keeps its dorsal side down, remaining in water in a perpendicular position with 3 pairs of appendages slanting upwards.

The nauplius is a nonfeeding stage as it utilizes the reserved nutrients of the egg stored in it. As it grows, it moults once in 4-6 hours. There are six naupliar stages (Fig. 8.4) each stage is formed after the moulting of previous stage. At each moult size of the body increases along with increase in structural details. The distinguishing features of six naupliar stages are listed in Table 8.1. After moulting, colour of the larva becomes whitish and semitransparent. Reddish brown specs appear on appendages. The duration of the nauplius stage is 36-48 hours.

Table 8.1: Distinguishing Features of Six Naupliar Stages of *P. monodon* (modified from Silas, E.G. *et al.*, 1978).

	Nauplius I	*Nauplius II*	*Nauplius III*	*Nauplius IV*	*Nauplius V*	*Nauplius VI*
1. Size (length–mm)	0.29 to 0.32	0.31 to 0.32	0.31–0.34	0.34–0.36	0.36–0.41	0.46–0.53
2. Furcal formula	1+1	1+1	3+3	4+4	6+6	7+7
3. First antenna	Inner lateral seta two numbers-distal one longer than proximal. Terminally 2 long and one rudimentary stae. Outer distal margin has one long seta	Outer terminal and outer lateral setae are comparatively shorter Rudimentary seta of stage I develops into a seta. Another setal rudiment added to outer distal portion	Outer terminal seta is shorter than inner terminal. Outer lateral seta is short. Another seta added proximal to inner lateral seta	Outer distolateral seta lost. Inner terminal seta is longer	Minute seta is added on outer lateral margin. Proximal half exhibits faint segmentation	2 more setae added to outer margin
4. Second antenna	Exopodite has 5 long setae along inner distal margin. Endo-podite with 2 short inner lateral setae and terminally 2 long setae and a minute setal rudiment	On exopodite, 4th seta from proximal end is bifurcated	Exopodite has six plumose setae and a setal rudiment. Setal rudiment of endopo-dite has grown into short non plumose setae	Exopodite exhibits faint segmentation. Setal rudiment of stage III develops into seta and another 2 setal rudiments are developed one terminally, another proximally in inner margin. The endopo dite has longer, inner terminal seta and another setal rudiment developed terminally	Exopodite with outer most seta longer but still non plumose	Exopodite develop 2 more setae. Endopodite with 4 terminal setae out of which 3 are long and one short. Another short seta is added to base
5. Mandible	3 long setae are present on expodite and also on endopodite	Same as in stage I	Same as in previous stage	Base of mandible is swollen	Same as in stage IV	Basal swelling more prominent. Cutting edge of mandible seen
Duration	3–4 hours	3–4 hours	4–5 hours	5–6 hours	10–12 hours	15–24hours

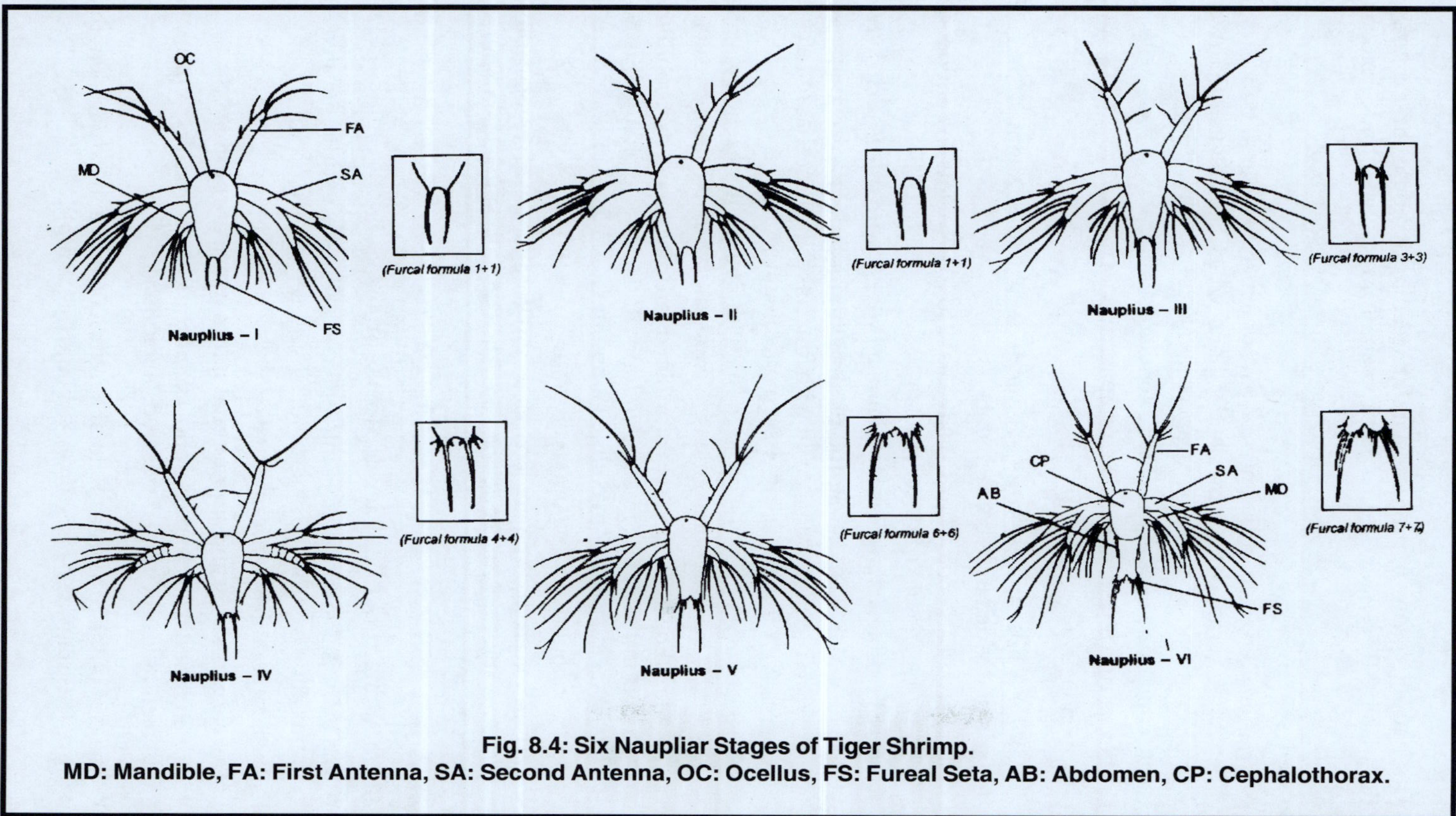

Fig. 8.4: Six Naupliar Stages of Tiger Shrimp.

MD: Mandible, FA: First Antenna, SA: Second Antenna, OC: Ocellus, FS: Fureal Seta, AB: Abdomen, CP: Cephalothorax.

Protozoea

As the nauplius grows to protozoea, it starts feeding. It moults three times, there by distinguishing three stages such as protozoa I, protozoa II and protozoa III. (Fig. 8.5)

Protozoea -I

(i) It has an elongated body anterior part of which is covered by a carapace. Behind the carapace there is a slender thorax which is six segmented. Average total length is 1.06 mm and carapace length is 0.47 mm.

(ii) The compound eyes at the anterior end are sessile.

(iii) The rostrum and spines on the carapace are not developed.

(iv) Appendages

(a) First antenna is three segmented, proximal segment is further subdivided into five segments

(b) Second antenna: The exopodite has 9 to 10 segments and endopodite is two segmented, both bear setae

(c) Mandible: It has incisor and molar process, with free teeth in between.

(d) First Maxilla: It has a protopodite with 2 segments-proximal segment with six and distal segment with four setae. Expodite has four feathery setae; endopodite is three segmented with setae.

(e) Second Maxilla: The protopodite with five endites each having setae. Endopodite is four segmented. Exopodite with five feathery setae.

(f) Maxilliped: The first two maxillipeds are developed, protopodite with two indistinct segments, endopodite is four segmented, expodite unsegmented, 3rd maxilliped is absent.

(g) Pereiopods and pleopods are absent.

(h) Telson is not distinctly separated from last abdominal segment, uropods are absent. Setae of telson 7+7. This stage lasts for 36-48 hours [Fig. 8.5(a)].

Protozoea – II

After moulting, protozoea I develops in the protozoea II.

(i) The body is distinguishable into anterior carapace, middle six segmented thorax and posteror abdomen which is five segmented.

(ii) The compound eyes became stalked

(iii) Develops rostrum and supraorbital spines which are bifurcated

(iii) Appendages: All apendages are of same the same pattern as in protozoea I but mandibles get more developed. (Table 8.2).

(iv) The size of the protozoea II: total length:1. 06 mm and carpace length 0.72 mm. This stage also lasts for 36-48 hours [Fig. 8.5(b)].

Table 8.2: Distinguishing Features of Three Stages of Protozoea of *P. monodon* (modified from Silas, E.G. *et. al.*, 1978).

Sl. No.		*Protozoea I*	*Protozoea II*	*Protozoea III*
1.	Size (mean total length and mean carapace length in mm)	1.06, 0.47	1.06, 0.72	2.28, 0.79
2.	Eyes	Sessile	Becomes stalked	Stalked
3.	Rostrum and supraorbital spines	Absent	Rostrum and bifid supraorbital spines appear	Rostrum is long, supraoital spine is simple not bifid
4.	Abdomen	Abdomen is unsegmented	First five abdominal segments are demarkated	Six abdominal segments.- first five segments with dorsomedian spine, 5th and 6th with posterolateral spines, 6th has an additional ventro lateral spine
5.	Appendage (a) First antenna	Three segmented, proximal segment further subdevided into 5 segments	Almost same as in stage I	Five basal segments fused into one
	(b) Second antenna	Expodite has 9-10 segments, endopodite is 2 segmented boh bearing setae	Almost same as in Stage I	Almost same as in previous stages
	(c) Mandible	Mandible has incisor and molar process and free teeth in between	Mandible asymmetrical-left mandible with 5 free standing teeth, right with one free standing tooth molar and a concavity in the molar surface, having a number of crenellate teeth in the rim. Molar surface of left is not concave, but having a number of toothed ridges that fit into the concavity of the right	Right mandible with 2 and left with six free standing teeth
	(d) First maxilla	Protpodite with 2 segments, proximal segment with 6 setae, distal with 4 setae. Exopidite has 4 feathery setae, 3 segmented endopodite has setae in each	Almost same as in previous stage	Almost same as in previous stage but more setae appear

contd...

Table 8.2–contd...

Sl. No.		*Protozoea I*	*Protozoea II*	*Protozoea III*
	(e) Second maxilla	Protopodite with 5 endites each bearing setae. Endopodite is 4 segmented. Expodite has 5 feathery setae.	Same as in previous stage	Same as in previous stage but more setae
	(f) Maxillipeds	First 2 maxillipeds are developed protopodite with 2 indistinct segments endopodite is 4 segmented, exopodite unsegmented 3rd maxilliped is absent	Same as in previous stage	1 st and second same as in previous state but more setae appear. 3rd maxillipeds appear
	(g) Pereiopods	Absent	Absent	5 pereiopods appear
6.	Telson and uropod	Telson is not distinctly separate from last abdominal segment uropods are absent. Telson with 7 + 7 setae	Almost as in stage I	Uropods appear. Telson is separate from last abdominal segment and it has 8 pairs of setae
7.	Duration	36–48 hours	36–48 hours	36–48 hours

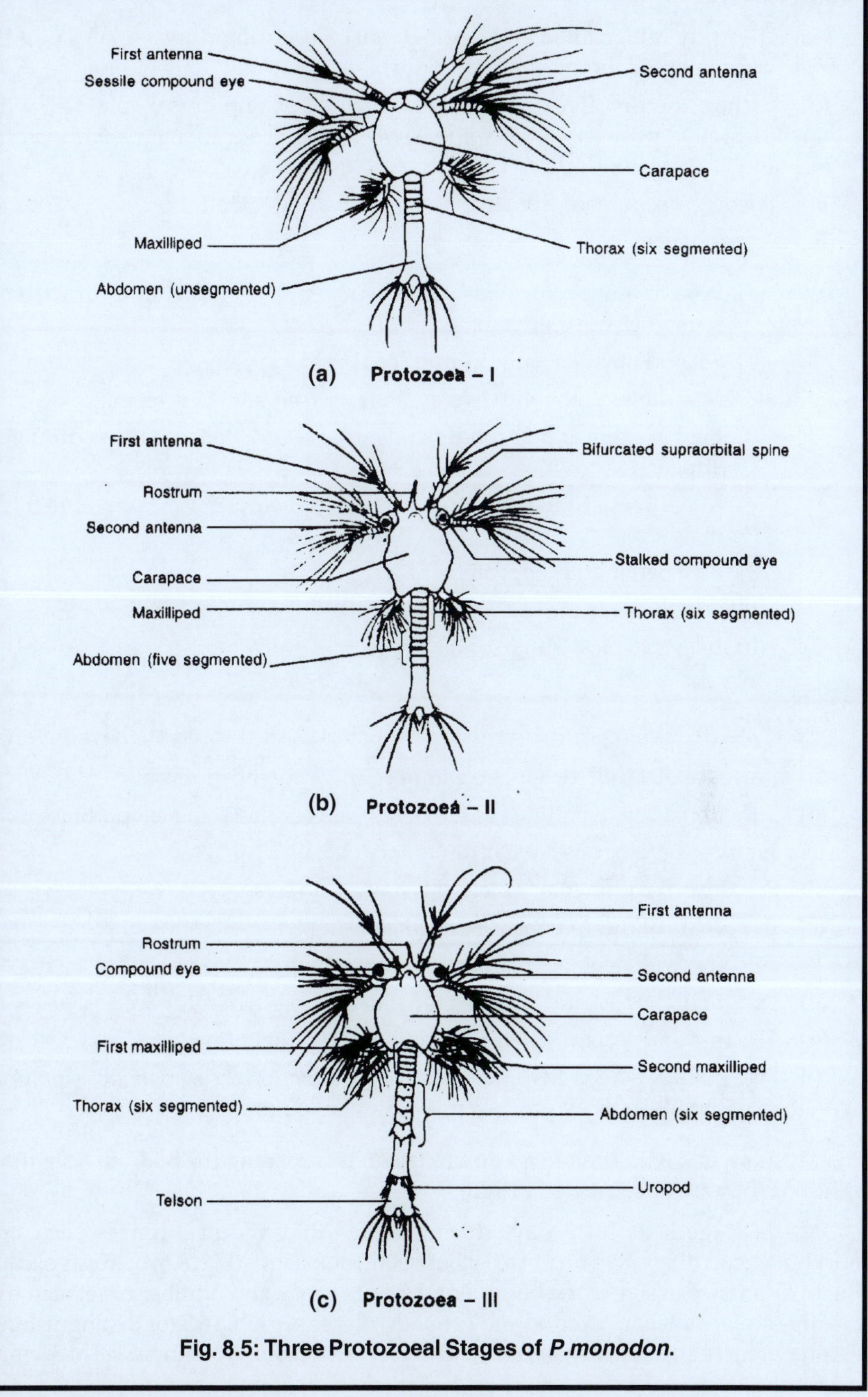

Fig. 8.5: Three Protozoeal Stages of *P.monodon*.

Protozoea – III

(i) The body is elongated and distinguishable into anterior carapace, middle six segmented thorax and posterior six segmented abdomen.

(ii) Each of the first five abdominal segments develop dorsal spine, 5th with additoinal postero lateral spine, sixth with the additional postero-lateral and ventro lateral spine.

(iii) Telson get separated from the last abdominal segment.

(iv) Appendages

(a) First antenna : Five basal segments (five divisions of the proximal segment) are fused into one.

(b) Second maxilla: Same as in stage II

(c) Mandible: Right with two and left with six standing teeth.

(d) First Maxilla and Second maxilla : same as in stage II but more setae appear.

(e) Maxillipeds: First two maxillipeds develop more satae and third maxilliped appear.

(f) Five pereiopods appear.

(g) Uropods appear.

(h) It lasts for 36-48 hours [Fig. 8.5(c)].

Mysis

Protozoa III develops into mysis stage which is characterised by the following

(a) Presence of a well developed carapace covering the thorax also

(b) Functional 3rd maxilliped and 5 pereiopods each having exopodites

(c) 1st three pereiopods have rudimentary chela

(d) Presence of rudimentary pleopods having no setae.

(e) Exopodite of the antenna is scale like, losing its segmentation.

(f) Telson is well developed and notched medially.

(g) Length of the rostrum is more than half the length of carapace

(h) The first and second antenna are olfactory in function

(i) Five pairs of pereopods are locomotory in function assisted by three pairs of maxillipeds. It measures 3.28 to 4.87 mm in length.

Mysis swims with their head down, telson up, keeping the body in a slanting position. They are not attracted to light.

Mysis stage lasts for 3 days during which they moult 3 times. Clear cut morphological differences do not exist between 3 substages (Fig. 8.6) of mysis except due to difference in size of the body, length of pleopods and number of setae in the antennal scale, maxillary exopod and uropod. (please see table 8.3 for distinguishing features of three stages of mysis.) At the end of 3rd stage, mysis attains a size of 3.1 mm to 4.5 mm.

Table 8.3: Distinguishing Features of Three stages of Mysis in *P. monodon* (modified from Silas, E.G. *et. aL*, 1978)

Sl. No.		*Mysis I*	*Mysis II*	*Mysis III*
1.	Size, mean total length and mean carapace length (mm)	3.79,1.18	4.16,1.39	4.24, 1.4
2.	Carapace	Carapace is well developed having a toothless rostrum and spines such as pterygostomial, hepatic and supraorbital	Rostrum is well developed, without teeth rarely a minute tooth may be found. spines as in stage I	Rostrum with one tooth in some. Spines as in previous stage
3.	Appendage			
	(a) First antenna	Three segmented with minute inner flagellum and a larger outer flagellum, setae present	Larger, inner flagellum 3/4 length of the outer	Inner and outer flagella more or less equal in length
	(b) Second antenna	Exopodite unsegmented with 11 setae, and endopodite with 3 shorter plumose outer distolateral seta	Exopodite with 19 plumose setae, and a prominent distolateral spine. Endopodite unsegmented with 2 short setae	Exopodite with 22-23 plumose setae, prominent distolateral spine, endopodite is 2 segmented
	(c) Mandible	Right mandible with 3 and left mandible with 7 standing teeth. No mandibular palp	Teeth as in previous stage, but mandibular palp is developed	Teeth as in previous stage mandiblular palp becomes longer
	(d) First maxilla	Exopodite is found	Maxilla without exopodite	With 13 setae on distal endite
	(e) Second maxilla	Exopodite with 10 setae	Exopodite with 16 setae	Exopodite with 22 setae
	(f) Maxilliped I	Exopodite with 12 setae	Exopodite with 12 or 13 setae	Exopodite with 12/13 setae, gill rudiment on protopodite
	(g) Maxilliped II	Outer lateral setae added to first and second segment of endopodite	Penultimate segment of endopodite indistinctly divided into 2, distal segment with six setae, exopodite with 4 terminal setae	Same as in previous stage

contd...

Table 8.3–contd...

Sl. No.	*Mysis I*	*Mysis II*	*Mysis III*
(h) Maxilliped III	Well developed, with 5 segmented endopodite, exopodite with six setae on tip	Almost same as in previous stage	Endopodite longer than exopodite
(i) Pereiopods	First 3 pereiopods have 2 segmented endopodite with incipient chela and 4 long setae at the tip, exopdite with 7 to 8 setae. 4th and 5th pereiopods have endopodite tipped with 3 terminal setae, exopodite with 7 or 8 setae	First 3 pereiopods with chela on endopodites, endopodite is 2 segmented but distal segment is indistinctly divided into 3. 4th and $5t^{h}$ pereiopods have no chela on the endopodite	First 3 pereiopods have 5 segmented endopodite having distinct chela. 4th and 5th pereiopods have no chela on endopodite. Pereiopods are locomotory assisted by maxillipeds
(j) Pleopods	Minute pleopod buds appear in abdominal segments	Pleopods short unsegmented	Pleopods become long, 2 segmented without setae
(k) Uropod	Exopdite with 16-17 plumose setae, endopodite with 16 long plumose setae, exopdoite has prominent spine distolaterally	Endopdoite with 20 plumose setae, exopodite with 18 plumose setae	Endopodite with 22 plumose setae, exopodite has 20 plumose setae.

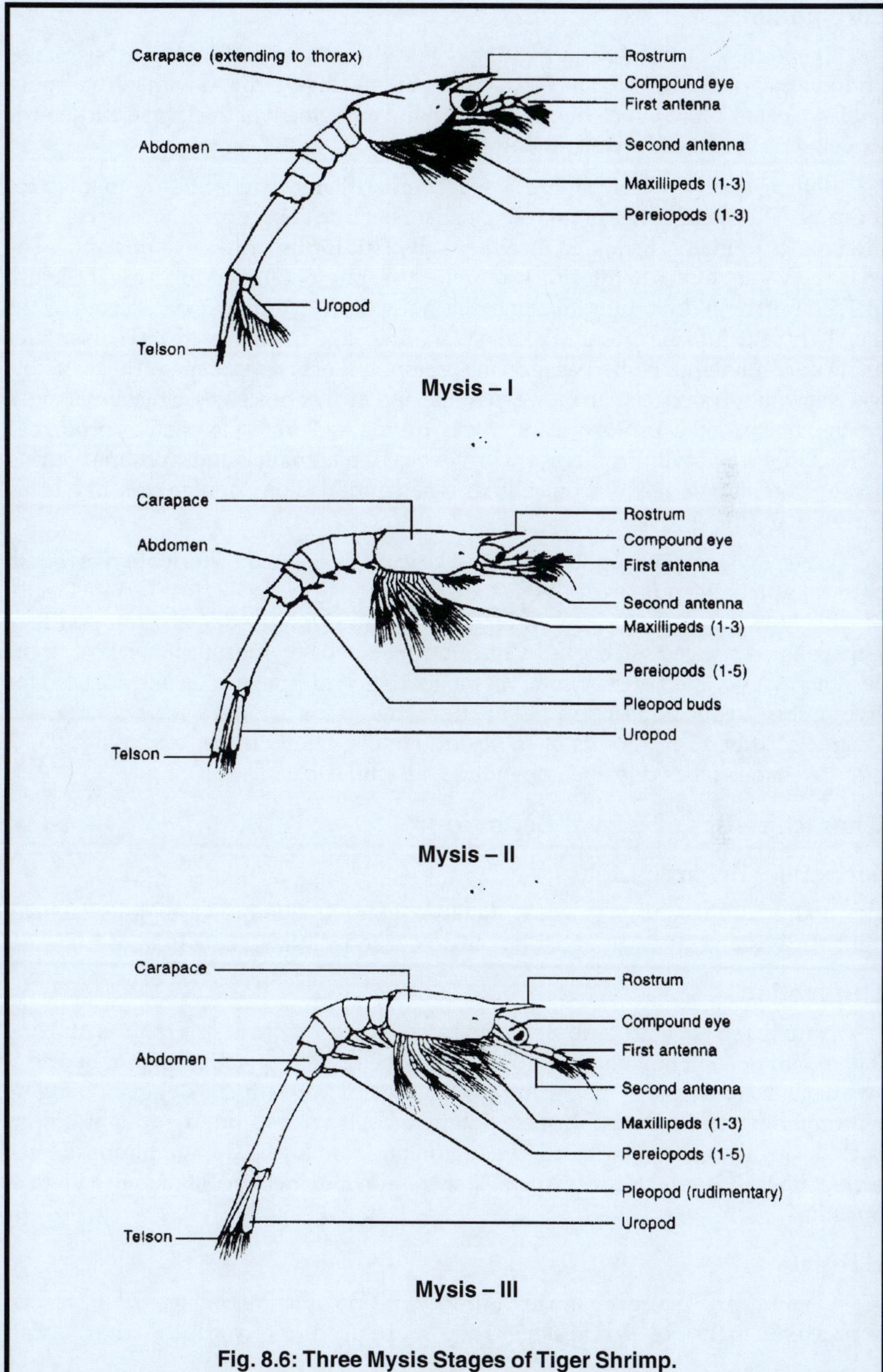

Fig. 8.6: Three Mysis Stages of Tiger Shrimp.

Locomotion

Exopodites of thoracic appendages help in swimming during mysis stage. Endopodites of pereiopods are also natatory in this stage. Uropod with telson from a tail fan like structure which helps in the jerking movement of the larvae backwards enabling them to escape from predators.

Both protozoea and mysis are filter feeders as their mouth parts are adapted for the same. The maxillae has numerous close set setae on the endopodite. Each seta has fine setules pinnately arranged on either side. This forms a filtering apparatus. The space between setules of filtering setae allow the phytoplankton to go in. The brush like setae of the endopodite of first maxilla sweep the phytoplankton collected by the maxillary filter into the mouth. The maxillary filter is supplemented by a coarser filter made up of numerous barbed setae on the protopodite of first maxilliped. The mandible is weak with serrated teeth in between molar and incisor processes. Unicellular algae are the choice food of protozoea and mysis. protozoea I prefers isochrysis, protozoea II chaetoceros and cylindrotheca and protozoea III platymonas and spirulina. During mysis stage, larvae feed on unicellular algae and also on zooplankton like brine shrimp nauplii.

When mysis metamorphose into post larvae (Fig. 8.7), a transformation of mouth parts take place. Mandibles develop sharp cutting edge in place of lose serrated teeth. Maxilla loses the filtering setae, endopodite become highly reduced. Protopodite of 1stmaxilliped develop stiff bristles. Chleae on pereopod become functional. Post larvae become carnivorous. Later five pairs of pleopods get fully developed and are used for swimming. Uropods assist in balancing. Pereiopods are used for walking and grasping. After 10-12 moults, it settles down and creep on the sand. After 20-22 moults, shape of the body and appendages resemble that of adult.

Characteristics of Larval Behaviour

Attraction Towards Light

Both nauplius and protozoeal stages exhibit attraction towards low intensity light but 3rd stage mysis and post-larvae do not show attraction to light.

Locomotion

Nauplii remain with ventral side up in water. They swim in short spurts utilizing the antennule, antenna and mandible. Protozoea is a very active swimmer and it swims with dorsal side up. During mysis stage, it is relatively sluggish, hanging with anterior end pointing oblequely downwards. It hovers around, suddenly jumps back flexing the abdomen. Post-larvae swim horizontally because of plumose setae on pleopods. It is pelagic for 4-5 days. They rest on the sides of container in a vertical position.

Habitat

In nature, the larvae remain in the bottom layers upto mysis stage. They come to surface only in the post-larval stage. They gradually drift towards the coast carried

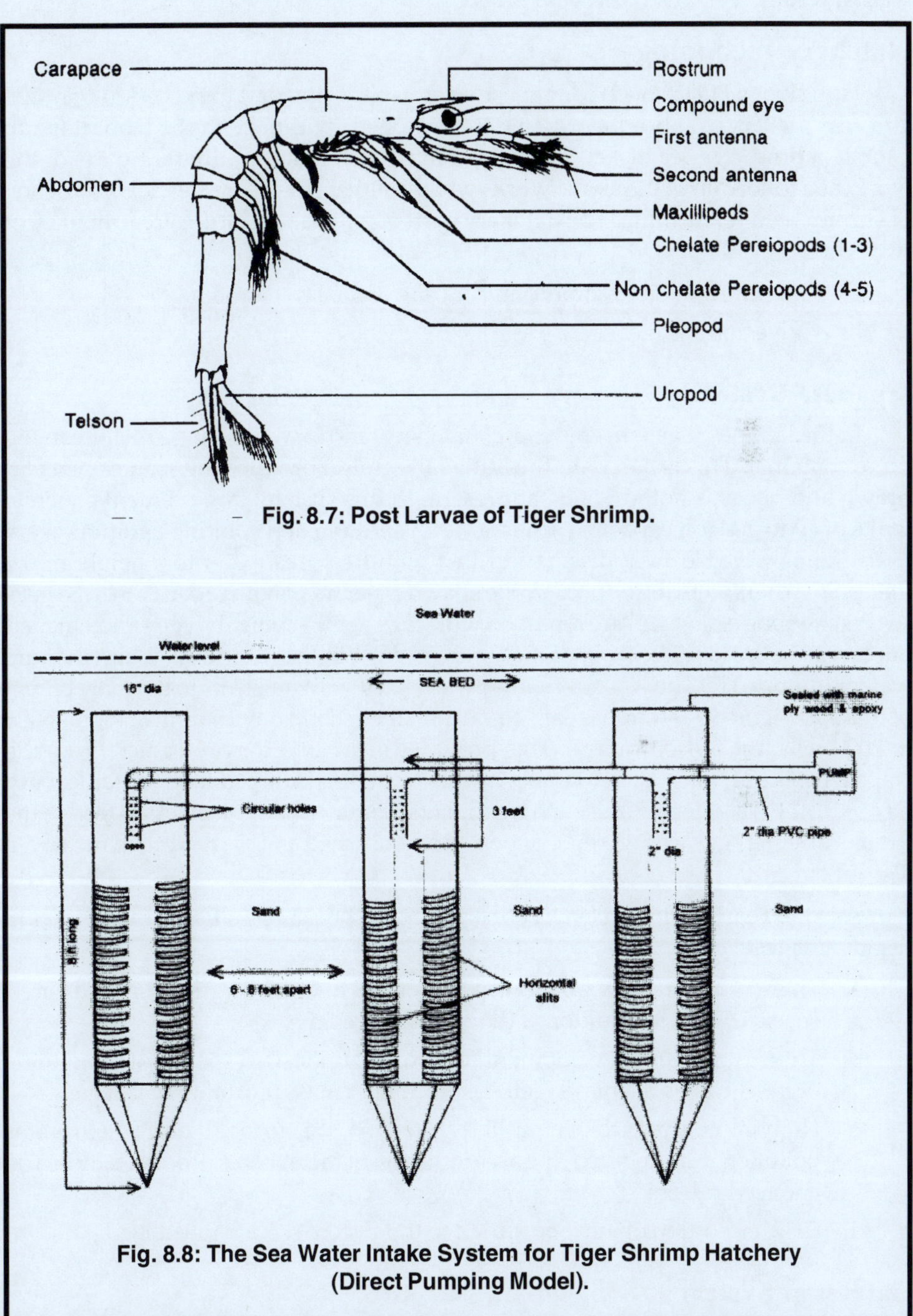

Fig. 8.7: Post Larvae of Tiger Shrimp.

Fig. 8.8: The Sea Water Intake System for Tiger Shrimp Hatchery (Direct Pumping Model).

by water current and are found in swarms. They become euryhaline and are found in coastal water creeks, estuaries and lagoons

Hatchery Technology

Hundinaga (1935, 1942) a Japanese scientist, was the first to successfully breed and rear the larvae subsequently under controlled conditions in the laboratory. In course of time, a series of developments have taken place in different parts of the world as a follow up of the above works which helped in the commercial production of shrimp seed for farming. The hatchery system for prawn culture are primarily of two types.

(a) Japanese system or community culture or fertilized system,

(b) Galveston system.

Japanese System

In the Japanese system, spawning, hatching, and larval rearing are done in the same container. This is based on the initial attempt to produce the seed of penaeid prawn and hence is not in vogue at present. In this system, large cement concrete tanks of 60 to 200 ton capacity provided with aeration and rotating agitators were used. Tanks were cleaned, dried and filled with fresh sea water to a height of 0.4 meter. Spawners were introduced @ one spawner per m^3 of tank capacity in cage nets. After spawning, spawners are removed with cage nets, leaving the eggs and hatched out larvae in the tank. Rearing of the larvae is done in the same tank wherein more water is pumped in. The water is regularly fertilized with nutrients to promote bloom of diatom. Vigorous aeration is carried out. On the second day, diatoms bloom (5000 to 2000 cells/ml concentration). This provides ideal food for protozoea. From the stage I mysis to 4th day of post-larvae, fresh sea water is pumped in to the tank every day untill the water level is raised to two meters. For mysis, supplementary feed in the form of artemia eggs were given. For post-larvae crushed and washed clam meat was also given.

As stated earlier, this system is not practiced at present because of the following disadvantages.

(a) There is uncertainty due to lack of control over the production of desired species of phytoplankton at the appropriate time.

(b) There is frequent development of bloom of undesired species of planktonic organism such as dinoflagellates leading to mass mortality of larvae.

(c) High initial strength of nauplii resulted in poor survival rate, due to water pollution resulting from the accumulation of metabolites produced by larvae and feed residue.

(d) Greater portion of the food added to the system remain unutilized

Galveston System

This system was developed in the Galveston laboratory, USA (Cook and Murphy, 1966)In this system, rearing of brood stock, spawning hatching, larval rearing and live feed culture are done separately in separate containers. Spawning is carried out in small indoor plastic pools. Newly hatched nauplii are transferred to larger (2000 litre capacity) plastic pools @ 50 nauplii/litre of water. Simultaneously separate cultures of mixed plytoplankton predominated by Chaetoceros and culture of rotifer (Brachionus) and cladoceran (Moina) were undertaken in the lab.

Sea water of 32 ± 2 ppt were pumped into large containers where it is filtered through 60 μ mesh nylobolt cloth before use in the operation. From second day after spawning *i.e.* last nauplius stage onwards, 200 litres of mixed phytoplanton culture, predominatly Chaetoceros (2 lakh cells/ml) is pumped into culture tank after reducing equal quantity of water from it every day as protozoea feed on phytoplankton. From mysis, in addition to the above feed, frozen Brachionus @ 100 rotifer/larva/day is given. When it metamorphose in to postlarva feeding on diatom is discontinued and Moina @ 20 per larva per day is given. Throughout the rearing period, vigorous aeration of water is provided. Constant check is made on the quality of sea water and slow exchange of 15 to 25% of water is made after the larvae reaches mysis stage.

The Galveston system prescribed the following precautions for optimization of larval production:

Water Quality

Water should be clean unpolluted sea water free of suspended impurity and planktonic organisms. This is ensured through filtration of seawater. The fresh sea water is allowed to settle and then filtered through suitable mesh cloth filters or sand filters or through mechanical filters. Sea water is drawn either from the sea or from a well sunk in the intertidal zone to a depth of three meters. The sea water intake system using subsoil filter (Fig. 8.8) usually consists of cylindrical PVC tubes of 6″ dia with conical closed end having three rows of horizontal slits. Another PVC tube of 2″ dia is fixed three feet deep from the opening of outer PVC tube and has circular holes. Three such pipes are joined together to the same supply canal using elbow and T-joints. Each structure is wound by 25-30 layers of 20-40-mesh nylon net. This whole unit is burried ten feet deep in the sea bed at about 10-15 meters away from the coast. Water is pumped to a reservoir where it is treated with 8-10 ppm chlorine and later dechlorinated with sodium thiosulphate. The water is filtered through a sand filter and 3 μ catridge filter and taken to second reservoir. There again it is filtered through a sand filter and 1 μ catridge filter. Before taking to larval rearing tanks water is treated with 10ppm EDTA. Then again it is filtered through 1 μ catridge filter and then released to larval rearing tank. Cook and Murphey (1969) has recommended use of EDTA @ 1 g for every 100 litre of rearing medium to enhance the quality of sea water to minimize larval mortality.

Temperature

28°C ± 2 is found most suitable. Lower temperature retard the development of egg and larvae. Average range of 24°C to 32°C is found suitable for development of penaeid larvae.

Salinity

Range of 27 ppt to 34 ppt is suitable.

Light

Role of light in the larval rearing is disputed. According to Murata and Shiguano (1979) adequate light is required to successful rearing as post-larvae reared outdoors were healthier. But Cook and Murphy (1966) opined that light is not essential for larval rearing as they could rear larvae in dark. Lumare *et al.* (1971) observed that light intensity below 1000 lux inihited normal development of protozoea to mysis in case of *P. kerathurus*.

pH of Water

Higher pH caused mortality of larvae. Hence pH should not exceed 8.2–8.5.

Ammonia and Nitrite

Accumulation of ammonia (above 0.1 mg/litre)and nitrite (above 6.5 mg/litre)is toxic to larvae.

Dissolved Oxygen

Optimum DO level has to be maintained through continuos aeration

Modern Shrimp Hatchery

A model shrimp hatchery should have the following essential units:

(a) Broodstock rearing/maturation unit

(b) Spawning unit

(c) Hatching unit

(d) Larval rearing unit

(e) Artemia cyst hatching unit

(f) Algal culture unit

Broodstock Rearing/Maturation Unit

The basic requirement for seed production is healthy brood stock of desired size so that it will yield maximum number of viable eggs. For shrimp hatcheries, brood stock is obtained from wild habitat *i.e.* sea as the rearing of brood stock in farmed conditions is not found to yield desired results. Males of a minimum average size of 200 mm total length and weighing over 65 gs (6-8 months old) and females of average total length of 240 mm and weighing between 90-150 g (9-12 months old) are reported to give good results. From the collection site to the hatchery site, breeders are transported in bag containing 40-50 litre of sea water @ 4-6 numbers per bag. After introducing the breeders fill the bag with oxygen and seal. After the breeders are brought to the hatchery, they are to be acclimated to the hatchery conditions by transferring the breeders to tubs with sea water used in hatchery. Care must be taken to do this slowly in case there is considerable difference in salinity and temperature.

After acclimation, breeders are treated with 100 ppm formalin for about ½ hr for disinfection. Incase breeders exhibit some sign of disease such as lying on one side/abnormal pigmentation/lesions in the exostkelton a dip treatment in antibiotic (50 ppm oxytrtracyclin/erythromysin/prefuran) for 10 minutes can be given. Such disinfected, healthy breeders are then transferred to maturation tanks.

Maturation Tank

Maturation tanks [Figs. 8.9(a,b)] are usually circular, 4 meter in diameter and 1.25 meters in height made of black fibre glass sheet. At the bottom of the tank, a layer of gravel of 10 cm thick is given over which a synthetic permeable cloth is placed. Above the synthetic cloth sheet a 5 to 10 cm thick layer of sand is provided. In the gravel layer, perforated plastic pipes fitted to PVC tube of 90 mm thick are concentrically arranged. This PVC pipe is connected to 90 mm dia vertical pipe by the side of the tank into which fresh sea water supply pipe of 40 mm is fitted. The drain pipe of 90 mm dia. is located at the centre of the tank and the same is kept in a PVC pipe of 250 mm dia. The water enters the tank through the inlet pipe, vertical tube and then through the perforations of the concentric pipe, then through the gravel, synthetic permeable sheet, and sand. Facilities are usually provided for aeration at the inlet point through air stones to ensure maximum DO in the rearing tank water.

Maintenance of Broodstock: Factors Affecting Maturation

Water Quality

For proper maintenance and healthy growth of brood stock, clean, clear seawater free from pollutants is the primary requirement. Water level in the maturation tank may be kept between 60 to 100 cm height, and a constant water flow @ 10 titre per minute is also necessary. Water is to be constantly aerated so that DO shall be at saturation level.

Salinity

Sea water, salinity range of which is between 30-36 ppt is ideal for brood stock maintenance. According to some scientists (Posadas, 1986)*P. monodon* can mature and spawn even at lower salinities such as 15 and 25 ppt but full seawater salinity is required for incubation and hatching. It has been observed by some workers that female collected from ocean have higher maturation rate and fertilization rate when compared to those collected from brackish water (22 to 28 ppt).

Temperature

Desirable temperature range is between 21°C and 31°C.

pH of the Water

It should not be less than 7.3 as it affects the calcification of cuticle and normal moulting process in prawn. pH within the range of 8 to 8.5 in ideal.

Light

Reduction in the intensity of light to 10 to 40% of the natural day light (100 Lux

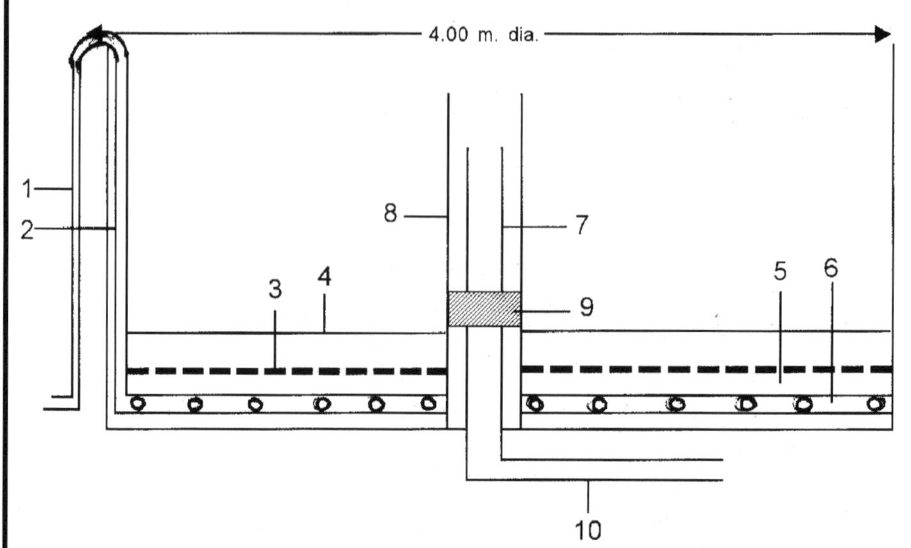

Fig. 8.9 (a): Maturation Tank (Cross Section).

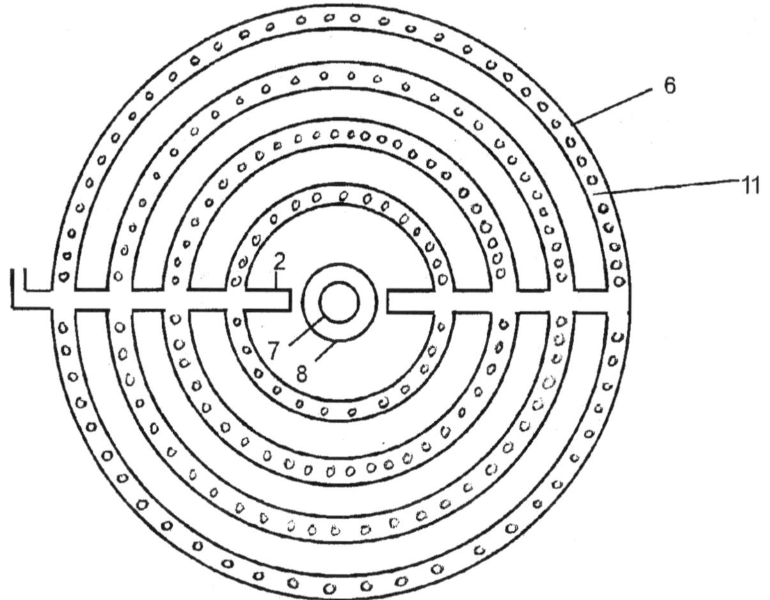

Fig. 8.9 (b): Concentric, Perforated PVC Pipes Embedded in the Gravel at the Bottom of Maturation Tank (Front View).

1. Water inlet pipe 40 mm PVC, 2. PVC pipe 90 mm, 3. Filter sheet, 4. Sand layer, 5. Gravel layer, 6. Concentric PVC pipe, 7. Removable waste pipe 90 mm PVC, 8. 250 mm PVC pipe, 9. Plastic screen, 10. Outlet pipe 90 mm PVC, 11. Perforations on the concentric pipe embedded in gravel.

–dim light) is reported to have beneficial effects on growth and maturation of gonads. For this purpose, fluorescent lights covered with dark blue acrylic sheets are used in maturation division. Photoperiodic regime can be 12 hour light-12 hour dark.

Food and Feeding

Meat of squid/mussel/clam are found most suitable for brood stock. Feeding @ 12-15% of the body weight have been reported to yield good results. Among the several pelleted feed tested, diets containing 50% protein and 10 % PUFA (such as arachidonic acid, eicosopentaenoic acid, docosohexaenoicacid) @ 2% of the total bodyweight are stated to enhance gonadal maturation. Feeding can be done 4 times a day.

Stress

Stress conditions such as overcrowding, handling, or poor water quality retards gonadal maturation and even causes regression of developed ovary. Stress due to over crowding can be avoided by restricting the stocking density @ 4 shrimp per square meter with male to female ratio at 1:1 or 1:2.

Sanitation

The maturation tank should be cleaned thoroughly using detergent and disinfected using bleaching powder. All glasswares, PVC, floor of the area, underwater torch etc. should be disinfected using 100 ppm chlorine water. Moulting of prawn, dead prawns if any etc. must be removed using scoop nets with long handle.

Health of Brood Stock

Proper gonadal maturation and high fecundity depends mainly on the health of the brood stocks. Injured or diseased breeders do not yield viable eggs of desired quality. The diseases are to be prevented by prophylactic treatment and controlled by therapeutic treatment.

Prophylatic Treatment

(a) Application of antobiotics: Chloramphenicol/oxytetracycline/erythromycin is applied at a dose of 4 ppm after stopping the water exchange for 2-3 hours. This treatment is to be done consecutively for 3 days.

(b) To prevent the appearance of protozoans and fungus 50 ppm formalin is added to static water for 2 hours.

In case any diseases appear, increase the water exchange rate and resort to dip treatment in 500 ppm formalin or 50 ppm antibiotic for 10 minutes

Induction of Maturation through Eyestalk Ablation

Panouse (1943) observed that removal of eyestalk of shrimp *Palaemon serratus* led to ovarian development. Ever since this discovery several workers have tried to induce gonadal maturation in prawn through ablation of eyestalk. Usually, unilateral eyestalk ablation is recommended as bilateral eyestalk ablation is reported to give adverse effects.

An important neuroscretory system located in the eyestalk *viz.* X-organ –sinus gland complex produces neurohormones noted below which regulate important physiological functions related to reproduction in addition to several other hormones controlling a number of other physiological activities.

(a) Ovary inhibiting hormone, and

(b) Moult inhibiting hormone.

If the eyestalks are ablated, it deprives the body of these horomones. As the ovary inhibiting hormone is meant to inhibit the maturation of oocytes, once this gland in removed along with eyestalk, the oocytes undergo maturation quickly. If both eyestalks are ablated, all the physiological activities controlled by the other neurohormones of the x-organ – sinus gland complex (such as movement of pigment in chromatophores, regulation of calcuium during formation of cuticle, production of inhibiting hromone meant to inhibit the secretion of moulting horomones by Y-organ affecting water regulation during ecdysis, inducing hyperglycaemia in blood to combat stress, regulate lipid metabolism and protein synthesis in hepatopancreas) shall be halted there by adversely affecting normal physiology and growth of brood stock. Hence instead of removing both eyestalks, unilateral eyestalk ablation is practiced generally.

The Procedure of Eye Stalk Ablation

It may vary as indicated below:

(a) Eye stalk ablation using scissors (cutting). It is a simple method but it may cause bleeding and subsequent mortality

(b) *Electrocauterisation* : Removing the eyestalk using eletrocautery appartus. This seals the cut instantly and avoids blood loss.

(c) *Incision-Pinching* : An incision is made on the eyeball with a sharp blade, squeeze the content of the eyeball outwards holding it between thumb and forefinger, crush the eyestalk 2-3 times to destroy the tissue.

After the unilateral eyestalk ablation, each female is double tagged (one tag is put around the remaining eyestalk with a ring of coloured elastic silicone bearing a label and another label with a number glued over the cephalothorax). After the moulting label around the eyestalk remain while the other one is on discarded carapace. After ablation, place the female in a tank containing antibiotic 10-20 ppm for half an hour and then transfer to maturation tank. Females are examined under water with water proof hand light every day for examining the ovarian development. Some females become gravid three days after eye stalk ablation. Some may take longer duration even upto 2-3 weeks. A female which is ready to spawn will have dark green, granular ovary having a diamond shaped swelling on the first segment of abdomen. The gravid female is dipped in 500 ppm formalin for 5 minutes rinsed in sea water and then taken to the spawning tank.

Latency Period

The interval between eyestalk ablation and spawning is referred to as latency period. The latency period depends on several factors such as stage in the moult cycle, age and source of the brood stock.

Stage in Moultcycle

It is advisable to do the ablation during intermoult stage. If ablation is done during premoult, moulting will follow immediately and latency period shall be longer *i.e.* 2-4 weeks Ablation during post –moult leads to mortality because of excessive loss of haemolymph and added stress on female

Age

Sub adult stage of monodon takes longer time - 40 days to mature and 69 days to spawn (Hillier, 1984)

Source of Collection

Adult monodon collected from brackish water lake had a latency period of 30 days whereas brood stock collected from offshore regions of Indian ocean had a latency period of only 4-5 days

Sperm Quality

If normal sperms are more than 50% sperm quality is considered good.

Ablated shrimp mature and spawn repeatedly three to five times during inter moult peroid and may remain productive for 50-60 days.

Spawning Unit

Spawning Tank

It is made up of black fibre glass sheet and fabricated into cylindro-conical shape (Fig. 8.10). The outlet pipe is fitted to the centre of the cyclindroconical tank and the water inlet can be from the top of the tank above the mouth. A perforated plastic sheet is kept at the junction of conical and cylindrical part on which the spawner can rest. Through the perforations, the eggs released by the female can sink to the conical part but the females cannot enter into the conical part as it has the habit of eating its own eggs. Usually one spawner is kept in one tank but the number can be increased if the size of the prawn is less. Spawning occurs usually in the night and the eggs are collected through the outlet into an egg concentrator [Fig. 8.10(b)] which is also fabricated from 200 mm PVC pipe and water outlet is guarded by 100 micron mesh net cloth which will retain the eggs.

Counting of Eggs

The eggs are then taken to a graduated bucket to know the total volume of eggs. In order to assess the total number of eggs a known volume of eggs is taken into a petridish and the number of eggs is counted with the nacked eye. Repeat the same with at least 2-3 more samples and the average is taken from three to four data. The total number of eggs in the bucket can be calculated using the formula:

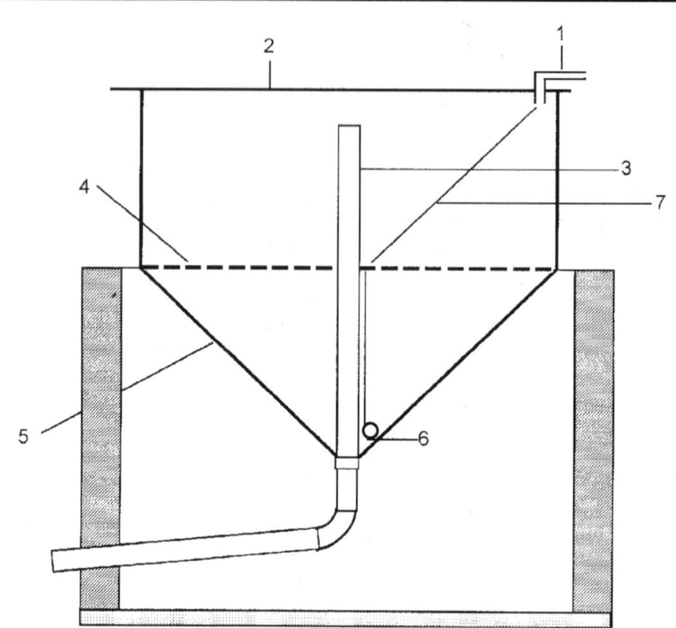

Fig. 8.10 (a): Spawning Tank (Cross Section).

1. Water inlet pipe, 2. Black cover, 3. Water outlet pipe, 4. Perforated plastic plate, 5. Conical bottom of tank, 6. Air stone, 7. Air pipe.

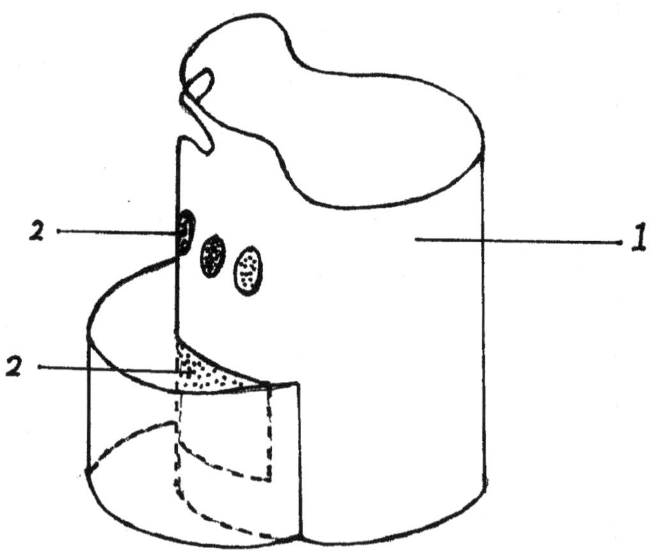

Fig. 8.10 (b): Egg Concentrator.

1. PVC pipe - 250 mm (closed at bottom), 2. 100 μ mesh net.

$$\text{Total number of eggs} = \frac{\text{No. of Eggs in a Known Volume}}{\text{Known Volume}} \times \text{Total Volume of Eggs}$$

After assessing the total number of eggs, the percentage of fertilization is also assessed as follows :

Observe the eggs under compound microscope. Unfertilized eggs appear as opaque, dark brown spheres with unsymmetrical cell division (*i. e.* 2-3 big cells and several small cells) where as fertilized eggs exhibit symmetrical cell division and sometimes having rudimentary appendages and setae. Fertilized dead eggs containing embryo at different stages of development can also be encountered. Lack of spermatophore in the thelycum is the main cause of non- fertilization. Lack of proper nutrition to mother prawn particularly during last few days of ovarian maturation has been also reported to be the cause of production of non viable eggs in *P. monodon.*

A healthy female tiger prawn weighing 60 to 130 g yields 1 lakh to 6 lakh eggs per spawning with 60% viable eggs. Wild females have a fecundity of 2 to 10 lakh eggs depending on the size of the prawn average being 4 lakh eggs per female. Ablated female of 90 to 150 g spawn 1 to 5 lakh eggs with an average of 2 lakhs eggs/prawn. It has been reported that fecundity and quality of eggs decrease with repeated spawning. As ablated females tend to show a decline in fecundity, hatching rate and egg viability during successive spawning in a given molt cycle, tiger prawn brood stock are replaced 6-8 weeks after ablation so that only the first or utmost second spawn in a molt cycle are harvested as later spawnings maybe less vialable. Various workers have undertaken comparative studies on the reproductive performance of ablated and unablated female tiger prawn (Hillier, 1984; Emmerson, 1983). Ablated females spawned more number of times in a molt cycle than unablated ones. But the number of eggs in the first spawn is found higher in unablated when compared to ablated which indicated that ablated ones spawn mainly from rematuration.

Hatching Unit

Eggs collected from the spawning tank are transferred to the hatching box (Fig. 8.11.) which is made up of 315 mm dia PVC pipes closed at the bottom by a PVC disc. At the middle of PVC pipe one 100 micron mesh screen is fixed. The inlet pipe may be of 40 mm PVC pipe, attached to the lower part of the hatching box so that water flows through the mesh screen and finally passes through the outlet which can be a 20 mm PVC pipe. The outlet opens in to a chamber with 48 micron seive which will work as a nauplii concentrator. The eggs are placed on the 100 μ mesh screen when the nauplii hatch out, they swim to the surface, and are borne away by the over flow into the harvesting tank with seive. The incubation period is 8-10 hrs depending on temperature.

In certain hatcheries, the fertilized eggs are transferred to hatching tanks contianing 300 litre sea water. Before transferring eggs, tanks are washed in running filtered sea water (one micron level using catridge filters) for 10 minutes. The water in the tank is treated with 10 ppm EDTA and 4 ppm chloramphenicol or 1 ppm

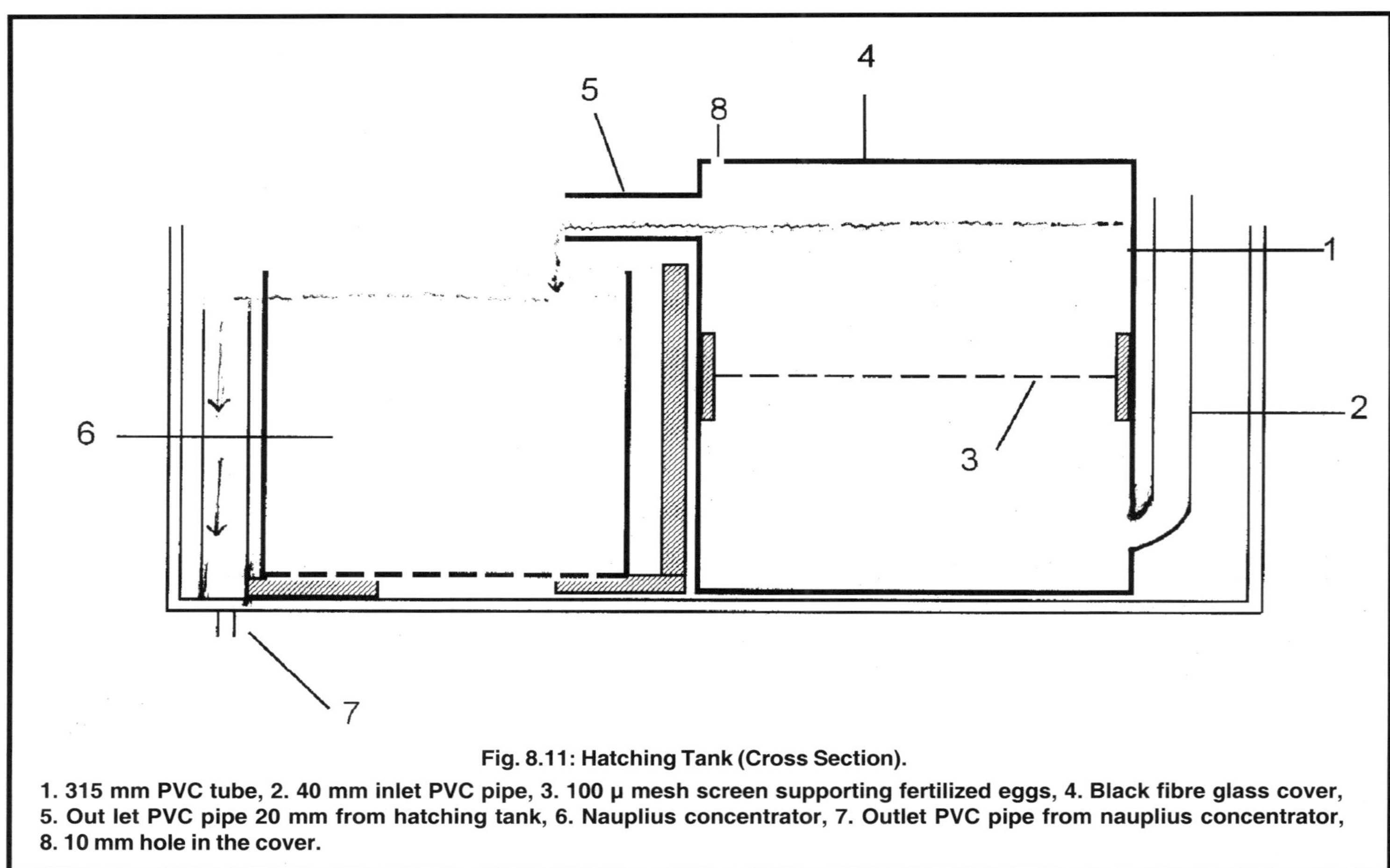

Fig. 8.11: Hatching Tank (Cross Section).

1. 315 mm PVC tube, 2. 40 mm inlet PVC pipe, 3. 100 µ mesh screen supporting fertilized eggs, 4. Black fibre glass cover, 5. Out let PVC pipe 20 mm from hatching tank, 6. Nauplius concentrator, 7. Outlet PVC pipe from nauplius concentrator, 8. 10 mm hole in the cover.

erythromycin or prefuran or oxytetracycline. Aeration is done gently in the tank. At the desired temperature (28-32°C) and salinity (29-34 ppt) nauplii hatches out from the egg within 12 hours. It takes 36 hours for the nauplii to undergo six moults (Nauplii I to VI) and become protozoea I. During this stage no feeding is required as it grows on the stored nutrients derived from eggs.

Quantitative Evaluation of Nauplii

Collect a known voume of water with nauplii from the harvesting bucket by mean of a pipette and count the number of nauplii in the known volume. Repeat this at least three times. Average of 3 data are taken, Total number of nauplii can be worked out by multiplying the number of nauplii in known volume with the actual volume in the harvesting bucket.

$$\text{Total number of nauplii in the bucket} = \frac{\text{Mean Number of Nauplii in the Sample}}{\text{Volume of the sample (ml)}} \times \text{total volume (ml) of water in the bucket}$$

Hatching rate is also calculated as follows:

$$\text{Percentage of Hatching} = \frac{\text{Number of Nauplii in the Given Volume}}{\text{Number of Eggs Stocked in the Given Volume}} \times 100$$

Evaluation of Quality of Nauplii

Before harvesting the nauplii for release into the larval rearing tanks, it is necessary to ensure the quality of nauplii. This can be ascertained by observing the movement and behaviour of nauplii. If the nauplii are inactive exhibiting lethargic swimming and no phototaxis it maybe due to physical deformities such as incompletely formed appendages/setae or due to accumulation of dirt/ occurrence of diseases. In such cases it is better to reject the whole lot.

Harvesting of Nauplii

Before harvesting stop the aeration in the hatching tank. Remove the cover of the hatching tank and keep the light on for a few minutes. The nauplii swim up to the lighted area of the tank and the same is siphoned off immediately to the harvesting bucket taking care to avoid siphoning of egg shells and unhatched eggs in to the harvesting bucket.

Larval Rearing

Tanks of different shapes (cylindroconical to parabolic) and size are used for rearing larvae in different hatcheries. These tanks are placed in room located away from other section to avoid cross contamination. The tank should have facilities for aeration and water exchange. The shape of the tank should be such that it should facilitate uniform aeration, uniform distribution of food materials, least scope for settlement of organic detritus on the interior of the tank and easy siphoning of feed residues, faecal matter and organic detritus.

Prior to stocking the tanks with sixth stage nauplius, it should be thoroughly cleaned and disinfected. Disinfection is done by 200 ppm chlorine water for 8-10 hours and thoroughly scrubbing with detergent. It should be washed with freshwater and drained for atleast 24 hours. After this, filtered sea water is released. Prophylactic treatment with fungicide and antibiotic is given one hour before stocking. Algae is added at a density of 100000 cells/ml just before stocking with nauplii @ 100000 nauplii/ton. The larvae should be acclimated to the larval tank water slowly for 10-20 minutes and it should be released slowly at different points in the tanks in small quantities

Rearing of Protozoea

Certain hatcheries prefer feeding of stage I of protozoea with isochrysis at the rate of up to 80000 to 100000 cell/ml of rearing media. Chaetoceros can be fed from stage I to stage III. Stage III can be fed on platymonas. Three days old chaetoceros and four day old isochrysis and platymonas are recommended. During this stage, 2 ppm treflan @ 10ml/m^3 as antigufal treatment is given. Starting from day 3, Furazolidone 0.2-0.3g/m^3 is recommended once in every 2 days. During late stage I water level may be increased from 6 ton. to 8 tons and from stage II onwards it may be increased to 8 ton. From stage III onwards water exchange is started. To start with at stage III 30% to 50% of water exchange in recommended using 250 micron mesh screen. Algae is fed 2 times a day once in the morning immediately after water exchange and next at 4PM.

Rearing of Mysis

Rearing media level maybe 10 tonnes and 70% of water is exchanged every day using 350 micron filter. From stage I mysis *i.e.* 6th day of stocking, freshly hatched artemia nauplii @ 1 nauplius/4 ml is given as feed in addition to unicellular algae such as chaetoceros/platymonas @ 100000 cells/ml. During stage III mysis Artemia nauplii may be fed @ 1 nauplius/2 ml. Antibiotic treatment may be @ 0.4 to 0.5gm/m^3 once in every 2 days and antigungal treatment can be 2 ppm treflan @ 15 ml/m^3. Algal feeding can be 3 times a day once in the morning after water exchange, another at 4 PM and another at 12 noon.

Initial Indoor Rearing of Post-larvae

As mysis grows to post larvae, *i.e.* generally on eleventh day to fifteenth day of rearing water level can be 10 tonnes and water exchange can be 100% using 350µ filter. Algal feeding can be reduced to 60000 cells/ml. Artemia nauplii may be increased to 1 nauplius/ml. Usually larvae are harvested at PL3 stage for rearing at post-larval section of the hatchery Treflan @ 20 ml/m^3 and furazolidone @ 0.5 g/m^3 (in every 2 days) is recommended during this period.

Usually, drugs are dissolved in distilled water or fresh water and added to the rearing medium at aeration points immediately after water exchange and feeding.

As larval rearing is the most sensitive component in a hatchery operation utmost care must be taken to avoid primary stress factors like bad water quality pollution, overcrowding, undernourishment etc.

Very often microenencapsulated feeds are given as supplementary feed to the larvae. For protozoea stage 5.30 µ is suitable, for mysis 40-90 microns and for postlarvae 90-150 micron size are recommended.

Artemia Unit

Freshly hatched artemia nauplii is the choice food for shrimp larvae from mysis stage onwards. Hence in a shrimp hatchery appropriate facilities are created to ensure adequate supply of artemia nauplii daily during the period of hatchery operation. For a brief account of the culture, life cycle and use of artemia cysts in shrimp hatchery please refer to the chapter on culture of live feed organisms

Algal Culture Unit

Shrimp larvae from the period of transition from sixth naupliar stage to protozoea start feeding on microalgae voraciously and hence for their survival and growth adequate quantity of the same is to be made available in a hatchery. Hence algal culture unit is essential in a shrimp hatchery. For details of microalgal culture refer to the chapter on culture of live feed organisms.

Out Door Rearing of Post-larvae of Shrimp

For stocking the grow out ponds 20 day old post larvae (PL-20) are usually supplied from a hatchery. Hence the 3 day old post-larvae (PL-3) reared in the indoor larval rearing tanks are to be further reared in outdoor larval rearing tanks taking all precautions noted below to avoid mortality

(a) Separate post-larval rearing tanks,

(b) Abundant supply of sea water to maintain running water and freshwater for cleaning the tanks,

(c) Facilities for aeration of rearing tanks, and

(d) Facilities for feeding the larvae.

Larval Rearing Tanks

These are reinforced concrete tanks of large size (20 tons). It may be eliptical in shape with sides gently sloping to the centre of the floor and provided with duck mouth inlet pipes kept slanting to maintain a circular movement of water and outlet facility is guarded by fine mesh sieve to prevent the escape of post larvae

Water Exchange

Clean, clear, filtered pollutant free sea water is utilized for the larval rearing. Water exchange is done to ensure water quality through the removal of metabolic wastes, feed residue etc. Siphoning of faecal matter, feed residues etc. settling at the tank bottom is done after turning off the aeration and running water facilities.

Aeration

Gentle aeration is to be done in the tanks to ensure optimum level of oxygen in the rearing medium

Feeding

PL-3 to PL-6 stage artemia nauplii is given at the rate of 2 nauplii/ml, 3 times daily

PL7 to PL12 – 3 nauplii/ml, 3 times daily

PL13 to PL20 – 3 nauplii/ml, once daily

While adding artemia nauplii care must be taken to avoid cyst shell as it may bring bacteria into the rearing medium.

In addition to Artemia nauplii, artifical feed is also given such as microenencapsulated feed and egg custard. Microencapsulated feed is dissolved in water, stirred and allowed to stand for 5-10 minutes. Before feeding scum from the top containing broken capsules may be removed. The feeding is done @ 0.5 g per ton. The egg custard is prepared as follows:

Eggs	—	36 nos.
Cod liver oil	—	75 ml
Yeast	—	75 ml
Beef liver	—	130 g
Polychaete worm	—	150 g
Vitamin	—	20 drops
Flesh of squid/prawn	—	200 g
Milk powder	—	75 g

Mix all ingredients in a mixer. Cook the same in a pressure cooker for 1 hr. Take required quantity and seive through the screen (500 μ) which gives fine granules. Wash the granules till all the fat content is removed. Mix it with water and broadcast in the tank. Egg custard is fed from PL8 onwards as shown below :

PL8 to PL 10 – 15 g/feed, 2 times daily

PL11 to PL 15 – 25 g/feed, 3 times daily

PL15 to PL20 – 40 g/feed 5 times daily

Prophylactic and Therapeutic Treatment

On the first day of post-larval rearing antibiotic and antifungal drugs are added as a prophylactic measure in the rearing medium. Thereapeutic treatments are done as and when needed as per the disease conditions.

At present hatchery technology for over 31 species of shrimp including twentyfour penaeid species and seven metapaenid species has been standardised. Out of this, hatchery for 11 species *viz., P. monodon, P, japonicus, P. indicus, P. orientalis, P, stylirostris,*

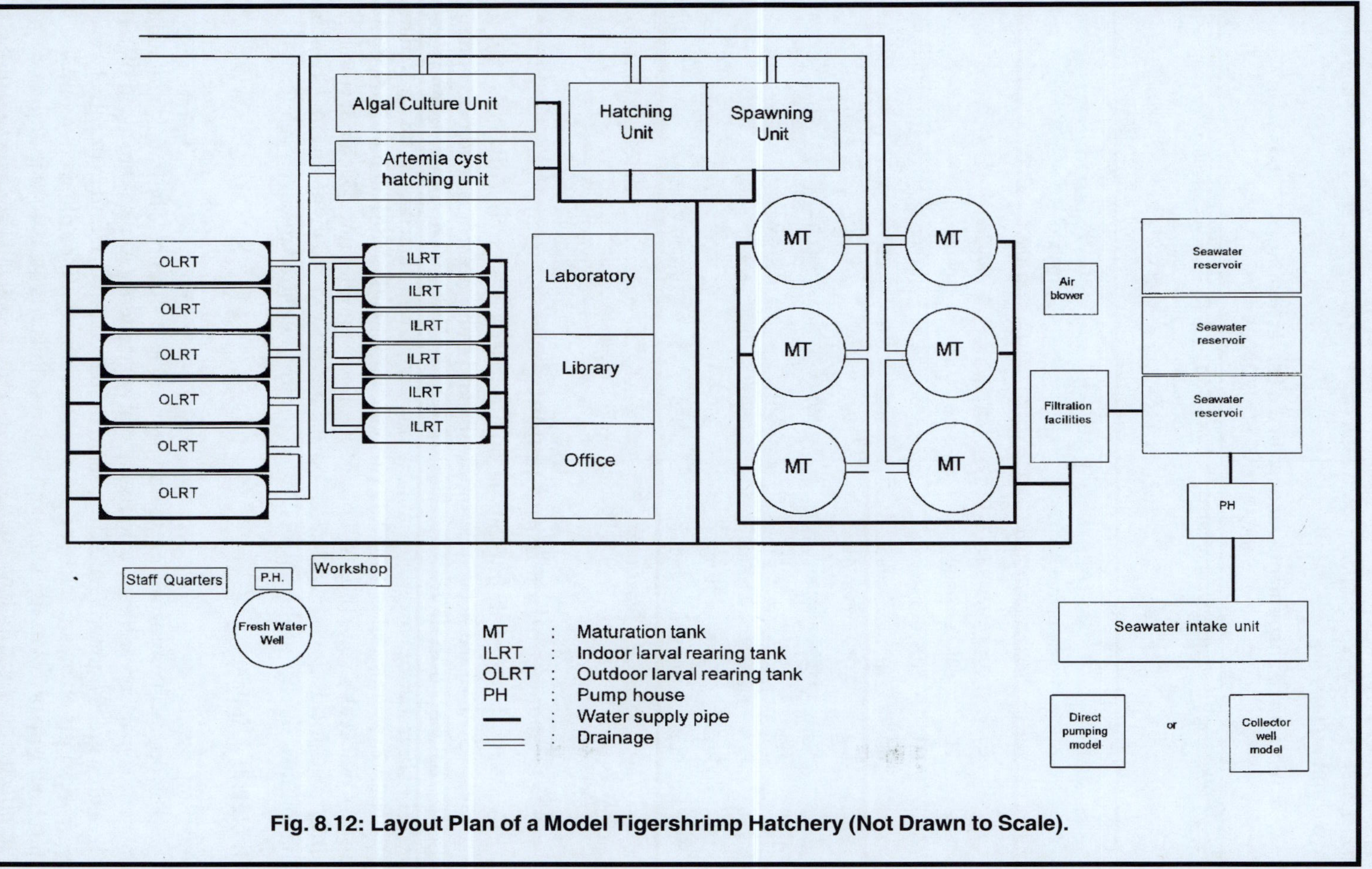

Fig. 8.12: Layout Plan of a Model Tigershrimp Hatchery (Not Drawn to Scale).

P. merguiensis, P. aztecus, P. duorarum, P. *setiferus, P. vennamei, M. ensis*- are established on a commercial scale.

Table 8.4: Management Practices in the Larval Rearing of *P. monodon* (Joshua, K *et al.*, 2000)

Day	*Substage*	*Water Management*	*Mesh Size of Screen (micron)*	*Feed Management*	
				Algal Feeding (cells/ml)	*Artemia Nauplii (no. /ml)*
1	N6/Z1	Fill 60% of tank	100	100000	—
2.	Z1	make upto 80%	100	100000	—
3.	Z 2	make upto 100%	100	100000	—
4.	Z3	30% exchange	250	100000	—
5.	Z3	50% exchange	250	100000	—
6.	M1	70% exchange	350	100000	0.25
7.	M1	70% exchange	350	100000	0.25
8.	M2	70% exchange	350	10000	0.25
9.	M3	70% exchange	350	100000	0.5
10.	M3	70% exchange	350	100000	0.5
11.	PL1	100% exchange	350	60000	1
12.	PL2	100% exchange	350	60000	1
13.	PL3	100% exchange	350	60000	1

N6 :sixth naupliar stage, Z1: Protozoea 1, Z2: Protozoea 2, Z3 :Protozoea 3, M1: Mysis1, M2: Mysis 2, M3 :Mysis 3, PL1: Post larva 1, PL2: Post larva2, PL3 :Post larva 3.

Larval Diseases and its Management

Mass mortality of shrimp larvae due to diseases are frequently reported in hatcheries. Such diseases occur sporadically and cause heavy loss and devastating effects in hatchery operations. As prevention is always better than cure, strict measures are to be observed with regard to sanitation and hygiene in a hatchery. While designing a hatchery care is also taken to isolate various components of the hatchery so that any disease outbreak in one unit will not contaminate another unit. Bacterial/fungal/ viral/ parasitic diseases have been reported in the larval stages of shrimp. Some of the common disease of shrimp larvae and control measures are given below:

Bacterial Diseases

Necrosis

This occurs in protozoea, mysis and post larvae of shrimps. The initial symptoms are noticed in the protozoea as a liquifaction of gut contents. As algae are incompletely digested, a green liquid fills the gut instead of shaped faeces found in healthy larvae. Necrosis starts on an antennae or uropod in prolozoea III or pleopods in mysis. It starts at the tip of the appendages and progress towards the base of the appendage. The affected area gets melanized. As the larvae gets weakened, its setae are broken, appendages are opaque, necrosed appendages hold back old carapce during moulting. Several bacterial species such as *Vibrio, Pseudomonas* and *Flavobacterium* are implicated in this disease.

Neerosis also affect internally. Necrosis at the region of the mouth causes the larvae to stop feeding. Hepatopancreas and stomach become greyish and hence called, grey stomach diseases.

Protozoea is more prone to the necrosis. Chloramphenicol and furazolidone at the doses shown below are used as preventive and curative drug for the disease at the doses noted below:

Age of Larvea	*Chloramphenicol*		*Furazolidone*	
	Preventive	*Curative*	*Preventive*	*Curative*
Day 3	2 g/m³	Double the dose of preventive for 2 days	0.2 g/m³	Additional 0.1 g for 2 days
5	3 g/m³	-do-	0.3 g/m³	-do-
7	4gm/m³	-do-	0.4 g/m³	-do-
9	5 g/m³	-do-	0.5 g/m³	-do-
11	5 g/m³	-do-	0.5 g/m³	-do-
13	5 g/m³	-do-	0.5 g/m³	-do-

Filamentous Bacterial Disease

It affects zoea, mysis and post larvae. If examined under the microscope, it is seen as fine, colourless thread like growth on the body surface and gills. This interferes with normal locomotory process and moulting. It may choke the gills affecting respiration.

Treatment with prefuran 1 ppm, erythromysin 4-6ppm, Malchite green 0.075 ppm for one day. Potassium permanganate 2.5 to 5 ppm for four hours.

Luminous Bacterial Disease

It is caused *by Vibrio alginolyticus/Vibrio harveyi.* Infected larvae exhibit continuous greenish luminescence when observed in total darkness. It may cause sometimes 100% mortality of the larvae. This can be prevented by using filtered, ultraviolet irradiated water in hatchery. It is better to discard the affected stock after treatment. Treatment can be by any one of the following:

(Chloramphenicol – 15 ppm; Furazolidone – 3/10 ppm; Malachite green –0.75 ppm; Neomysin – 8/15 ppm' formalin –25 ppm; Erythromycin –2-4 ppm)

Vibriosis

It is caused by some species of vibrio *(V. parahaemolyticus, V. alginolyticus, Pseudomonas* sp., *Aeromonas* sp. It affects haemolymph and mid gut gland. During initial stages larvae exhibit yellow, vermillion/red colour. Later the larvae show white turbid colour. Treatment by prefuran 1 ppm, furazolidone 2-5 ppm.

Fungal Diseases

The common fungae that attack the shrimp larvae are Lagenidium, Sirolpidium, and Haliphthoros. Infected larvae/ post-larvae appear whitish, become week and eventually die. It causes heavy mortalities, even up to 100%. Mysis stages are most vulnerable to fungal attacks.

Treflan containing Trifluralin 0.1 mg per litre is effective in controlling the disease. Recommended dosage for nauplii, protozoea, mysis and post larvae are 10 cc/m^3, 20 cc/m^3/30 cc/m^3 and 40 cc/m^3 respectively.

Viral Diseases

Viral diseases appear mainly during post-larval and juvenile stages. 2 groups of viruses are implicated in post-larval disease (1) Baculovirus (2) Parvovirus or picornavirus group. Baculovirus group affect the midgut and hepatopancreas where as parvovirus group (IHHNV –infectious, hypodermal and haematopoetic necrosis virus) affect ectodermally derived tissues such as cuticular epidermis, hypodermis lining the foregut, antennal gland, nerve cords and mesodermally derived tissues such as striated muscles, connective tissue, haematopoetic tissue, mandibular organs heart and gonads. This may cause high mortality in post larvae.

No effective treatment has been found for the viral infections.

Larval Fouling by Protozoans Like Ciliates

Large scale infestatation of some cilliates like Zoothaminium, Vorticella etc on the gills, eyes etc. of larvae lead to hypoxia and death. Though they are not directly pathogenic, they can cause mortality in larvae. Malachite green 0.0075 ppm and formalin 10 ppm are effective treatments.

Diseases Due to Environmental Stress/Nutritional Disorders

Blakc Gill Disease

Gills exhibit reddish/brownish/black discolouration and atrophy. In advanced stage, the gill become totally black. It causes physical deformities, loss of appetite and mortality. Some times, secondary infection by bacteria and fungus attack the dying cells. Methylene blue 8-10 ppm/ prefuran 1 ppm/malachite green 0.0075 ppm is recommended.

Muscle Necrosis

Adverse environmental condition such as low oxygen, temperature, salinity shock, over crowding, severe gill fouling etc. cause this disease. Opaque white areas appear in the abdomen. Edges of uropod blackens followed by erosion. Later liquid filled boils appear at the tip of uropods. This may lead to secondary infection by bacteria later. Reducing the stocking density, adequate feed and adequate water exchange are the suggested preventive measures.

Black Stomach Diseases

This is developed when old chactoceras culture is used to feed the larvae which lead to incomplete digestion of algae. Presence of black plug in the stomach obstructing the digestive tract is seen. Therapy consists of monitoring the algal quality. Change water and add fresh algae. Feed the larva with other unicellular algae like Isochrysis.

SEED PRODUCTION OF GIANT FRESH WATER PRAWN *(Macrobrachium rosenbergii)*

Introduction

Macrobrachium rosenbergii commonly called as scampi, is the giant fresh water prawn. About 125 species of Macrobrachium are distributed in the tropical and subtropical region of the world out of which 49 species are of fisheries importance and 15 species are of cultural significance. They are found in inland fresh water areas including lakes, rivers, swamps, canals, ponds and also in estuarine areas. They require brackish water during initial stages of life cycle and hence they are found in water bodies that are directly or indirectly connected to the sea. Species that grow to large size are *M. rosenbergii M. americanum and M. carcinus. M. rosenbergii* [Fig. 8.13 (A)] are found in the coastal belt extending upto 200km interior from coast in the South/South east Asian region, northern Australia and western pacific inlands. *M. americanum* is found in the western coast of America while carcinus is found in those connected with Atlantic, Rosenbergii is now farmed in several countries such as Thailand, Taiwan, Hawaii, Mauritis, Costarica, Indonesia, Israel, Malayasia, Mexico, Phillippine, Zimbabwe and India. Specimen of rosenbergii weighing 654 to 1000gm have been reported. Because of its fast growth rate and consumer preference, it is considered as an ideal candidate species for monoculture or polyculture along with major carps. Another species of macrobrachium commonly available in India is *M. malcomsonii* commonly called as Indian river prawn [Fig. 8.13(B)].

The giant freshwater prawn is noctural in habit, found hiding under shades and shelters during day time. They are found in extremely turbid condition being an omnivorous bottom feeder. The food includes aquatic insects and their larvae, algae nuts, grain, seeds, fruits, small molluscs, crutaceans and carcasses of fish and other animals. They are found in varying salinities such as 0-20ppt and temperature varying from 25°C to 34°C. Under culture conditions it attains a growth of 40-50gm in five months. Males grow faster than females

Reproductive Biology

Age at First Sexual Maturity

Scampi has been reported to attain first sexual maturity on reaching a length of 140 to 150 mm and at 35-40gm weight

Sexual Dimorphism

They exhibit distinct sexual dimorphism as shown in Fig. 8.14.

Courtship, Mating, Spawning

Copulation (mating) takes place between hard shelled males and mature soft shelled females which have just completed their premating moult. Within a few hours after premating moulting, females become receptive to male. The time period between

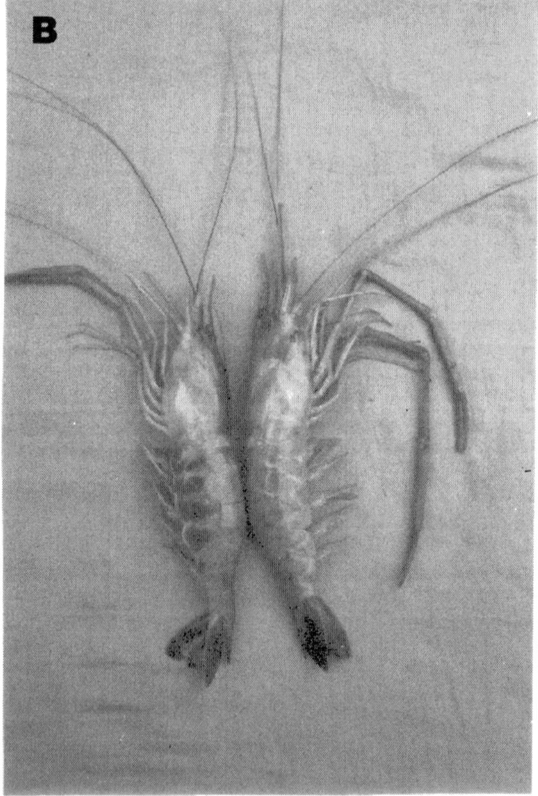

Fig. 8.13 (A): The Giant Fresh Water Prawn *Macrobrachium rosenbergii* (scampi), (B) The Indian River Prawn, *Macrobrachium malcomsonii.*

moulting and mating varies from 1.2 to 21. 8 hours. (average 9.1 hours). Pre copulatory behaviour involve attempts by male to contact females by its antenna and claws. Once the contact is established the male turns the female upside down, clears her thoracic region by his pereopods and male gonopore is brought close to sperm receptacle of female. The transfer of spermatophore takes only a few seconds and it is accomplished by first two pleopods of male. The spermatophore adheres to the sperm receptacle because of the glutinous nature of attachment matrix.

Table 8.5: Distinction Between Male and Female (Fig. 8.14).

Male	*Female*
1. Mature males are larger, second walking leg much larger and thicker.	Smaller in body size, the second waling leg thinner and shorter.
2. Cephalothorax is larger, but abdomen is narrower.	Cephalothorxa is smaller, abdomen is broader.
3. The endopodite of second pelopod has appendix masculina in addition to appendix interna.	Appendix masculina is absent in the endopodite of second pleopod.
4. Genital pores are at the base of 5th walking leg.	Genital pores at the base of 3rd walking leg.
5. Pleura of the abdomen is shorter.	Pleura of first, second and third abdominal segments are longer, and it forms a brood chamber in which eggs are carried till hatching.
6. Ventral side of the first abdominal segment has a lump -a hard point- at the centre which can be felt with finger.	No lump or hard point at the ventral side of 1st abdominal segment.
7. Reproductive setae are absent on the ventral side of thorax and pleopods of mature male.	Reproductive setae appear on the ventral side of thorax and pleopods of mature females such as ovipositing and ovigerous setae. Ovipositing setae appear on the coxa of last 3 pairs of pereopods, and posterior margin of sperm receptacle area, and the pleopods. It helps in guiding and propelling eggs during spawning Ovigerous setae serve to anchor the eggs to pleopods.
8. No sperm receptacle in the thoracic sterna between last three pairs of pereopods.	There is an unspecialised sperm receptacle on the thoracic sterna between last three pairs of pereopods.

Usually sexually active male may build breeding depression at the bottom of culture ponds, several females may be found around such depression which is occupied by a male. Usually male provides protection to female during their period of vulnerability immediately preceding and following premating moult and then copulate with each female. After mating the female may spend several hours in the prespawning preening behaviour. Chelae of first pereopod is used for cleaning and arranging oovipository and ovigerous setae before spawning.

Male deposits semen in a gelatinous mass on the ventral side of thoracic region between walking legs. Within a few hours after copulation, the female flexes its abdomen tightly below the thorax and first two pairs of pleopods are extended so

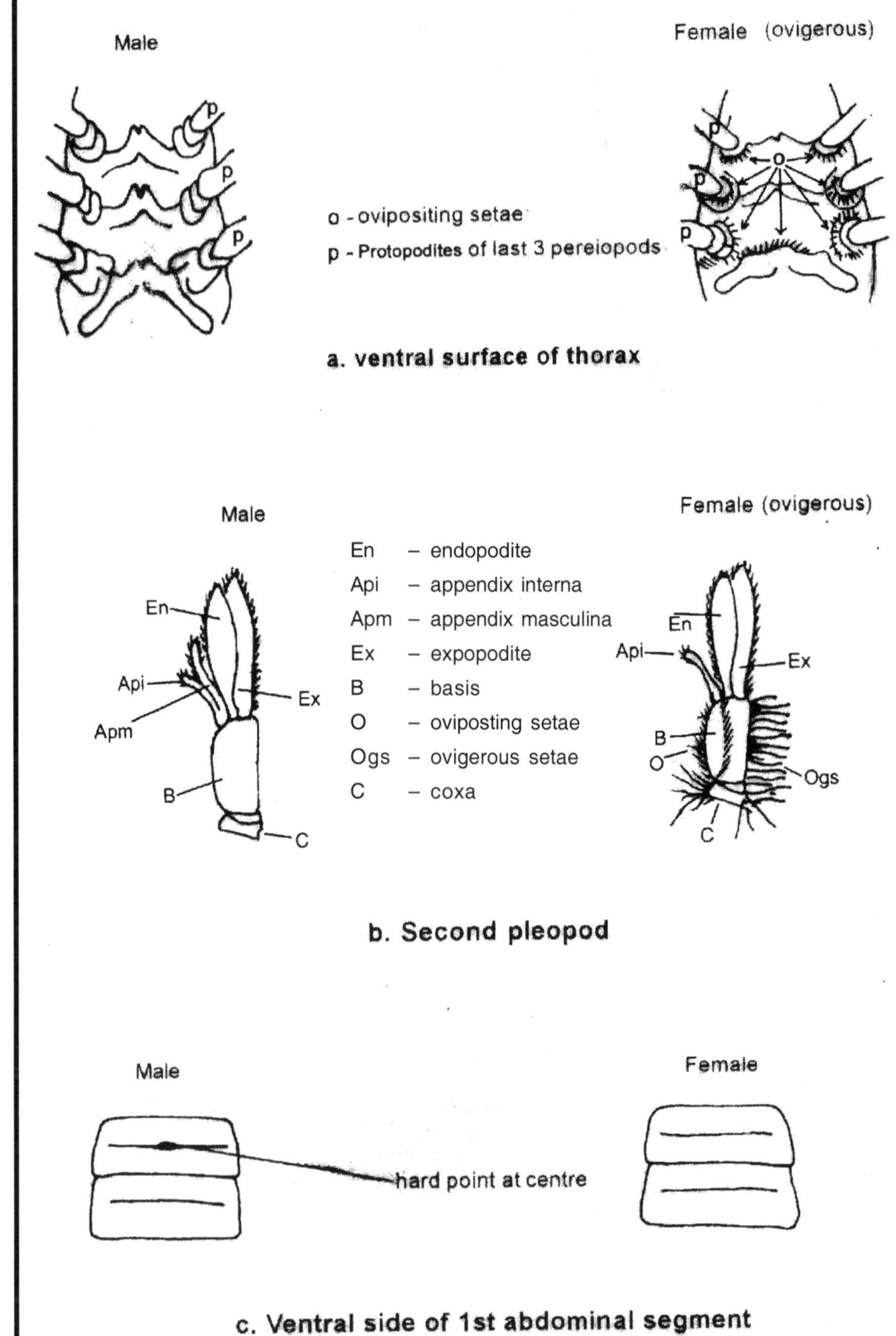

Fig. 8.14: Sexual Dimorphism in Macrobrachium.

that it will overlap sternum and form a floor for the passage way of eggs. The eggs laid by the female are fertilized by the sperm attached to the underside of thorax. The fertilized eggs are transferred to the ventral side of abdomen (brood chamber) by ovipository setae and the same is attached to the ovigerous setae of 1st four pleopods and kept in place by a membrane. This membrane is formed by the hardening of cementing substance which may take a few minutes. Movements of the pleopods helps in aerating the eggs. Fecundity varies from 20000 to 100000 eggs. Adult females carrying eggs on the ventral side of abdomen are called berried females (Fig. 8.15). The eggs are slightly eliptical (0.6 to 0.7 mm in the long axis) bright orange in colour. 2-3 days before hatching, it turns grey black in colour.

Breeding Season

Mating and reproduction are reported throughout the year in farm reared ponds. Though there are seasonal peaks related to the temperature. Minimum number of berried females are noticed during December and January. In India in natural waters the breeding season of Hooghly estuary stock is reported to be from December to July with the peak being March to May. In Kerala backwaters it is from July to December with the peak in October and November. In Andhra Pradesh (Kollerulake and Godavari system), it is throughout the year with peak in August to October.

Reproductive Systems

Male posesses a pair of fused testes, located on the mid dorsal side of the cephalothorax below carapace. Arising from it is the vasdeferens on each side located anterior to heart extending as a tube along the poster lateral margin of the cephalothorax terminating in the enlarged ampulla or terminal ampulla which opens by gonopore at the coxa of 5th pereiopod. Each terminal ampulla contains half of the spermatophore. On ejection through the gonopore the spermatophore halves fuse along the median margin to form complete spermatophore. Each half of spermatophore contains a sperm mass in which non motile sperms are embedded in a dence fibrous matrix.

Female: A pair of ovaries are located dorsal to the stomach and hepatopancreas in the cephalothorax cavity. When it is fully ripe, the ovary extends from behind the eyes to first abdominal segment. From each ovary, the oviduct arises laterally and it opens on the coxa of 3rd pereopod.

Maturity stages of the ovary can be seen through carapace through external examination from the colour of the ovary and size in relation to carapace cavity. 4 stages such as immature (neuter period), early maturing, maturing and ripe (mature) are distinguishable in the ovarian development. During immature stage (neuter period) ovary is very small transparent seen in the posterior most region of carapace cavity. Oocytes are 0.064 to 0.128 mm in diameter, spherical with large nucleus. In the early maturing phase, ovary grows to occupy ¼ to ½ of the total carapace cavity. Colour is yelowish due to light deposition of yolk in the cytoplasm. Oocytes are 0.191 to 0.447 mm in diameter and nucleus is not large. During late maturing phase ovary occupies ¾ of the carapace cavity, ovary is light orange in colour, due to heavier deposition of yolk in the cytoplam, and ova size range from 0.319 to 0.574 mm. Ripe

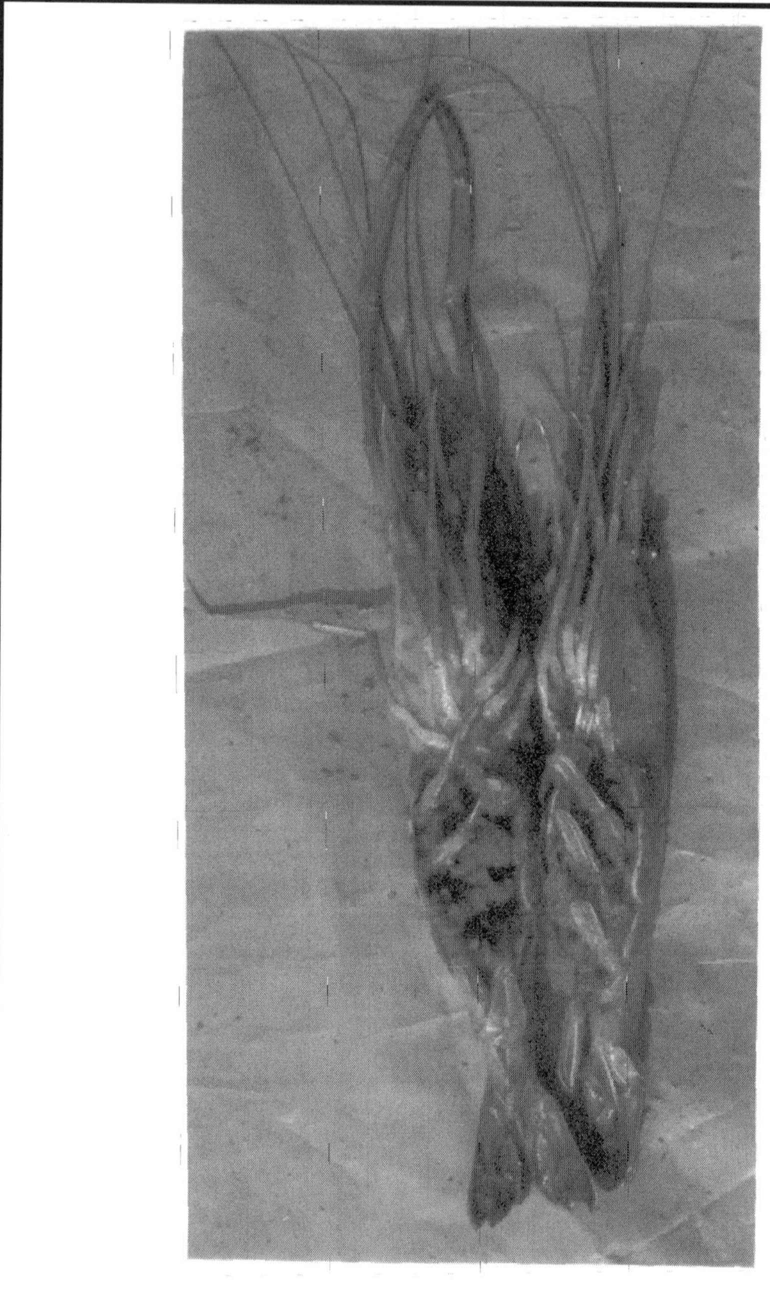

Fig. 8.15: Berried Female of *Macrobrachium*.

ovary occupies the entire carapace cavity, colour turns dark orange, ova opaque due to heavy yolk, oocyte diameter varies from 0.4468 to 0.7761 mm. Development of ovary takes place within 15 to 20 days.

Once the eggs are extruded, fertilised and the embryonic development proceeds, side by side, the ovary also undergoes development and attain ripe (gravid) condition just before the completion of embryonic development and hatching of laid eggs in the brood chamber. Hence a given prawn breeds repeatedly, atleast 3 times in a breeding season, under favourable condition. The neuter period in an individual female prawn also gradually increases with the number of times it has bred during a season. The incubation period varies from 15-21 days (19 days normally) depending on temperature. The freshly hatched larvae are called zoea which require salinity from 11-13 ppt for survival and growth.

Development of Larvae

The newly hatched zoea larva is about 1.92 mm in length. It has the zoeal characteristics such as a body distinguishable into cephalothorax and abdomen. The cephalothrax is covered by a carapace having the dorsomedian rostrum. All cephalic and thoracic appendages are present. The segmented abdomen is without pleopods. The larvae passes through 11 zoeal stages (Yuno and Soo, 1969) within a period of 23-32 days depending on temperature, food and water quality to become the postlarvae exhibiting adult features. The distinguishing features of 11 zoeal stages is represented is a tabular form below (Fig. 8.16)

Table 8.6: Identifying Characters of Larval Stages of Prawn (Macrobrachium)

Larval Stage	*Age (days)*	*Total Length mm*	*Distinguishing Features*
I	1	1. 92	Sessile eyes
II	2	1. 99	Stalked eye
III	3	2.14	Uropod present
IV	4-6	2. 5	Two dorsal rostral teeth uropod biramous with setae.
V	5-8	2. 8	Telson narrow, elongated
VI	7-10	3. 75	Pleopod buds appear
VII	11-17	4. 06	Pleopods biramous and bare
VIII	14-19	4. 68	Pleopods with setae
IX	15-22	6. 07	Endopods of pleopods with appendix interna
X	17-24	7. 05	Three to four dorsal rostral teeth
XI	19-26	7. 73	Teeth on half of the upper dorsal margin
Post Larvae	23-27	7. 69	Teeth on upper and lower margin of rostrum, adult behaviour

Hatchery Technology

The following are the different units of a freshwater prawn hatchery:

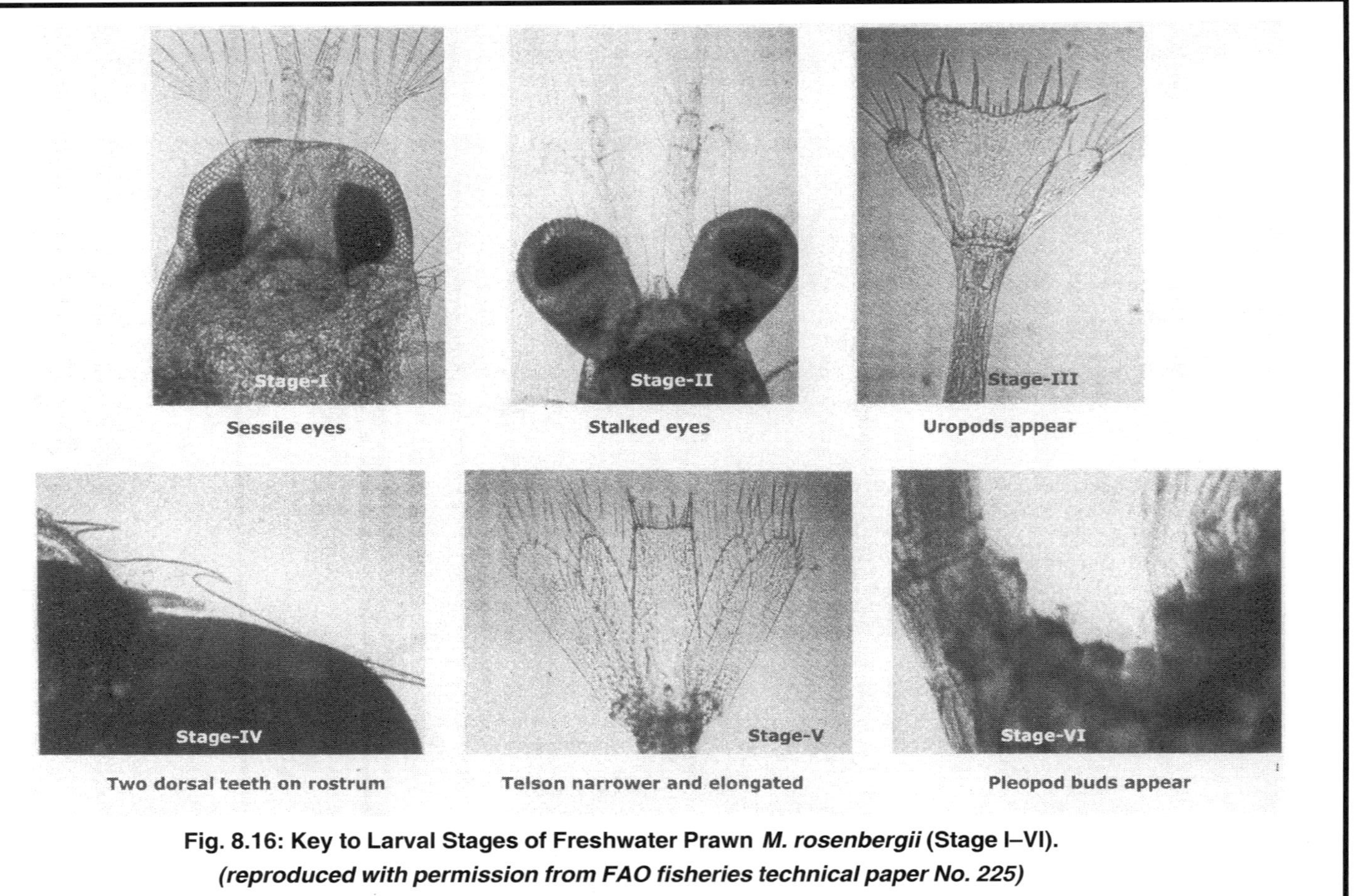

Fig. 8.16: Key to Larval Stages of Freshwater Prawn *M. rosenbergii* (Stage I–VI).
(reproduced with permission from FAO fisheries technical paper No. 225)

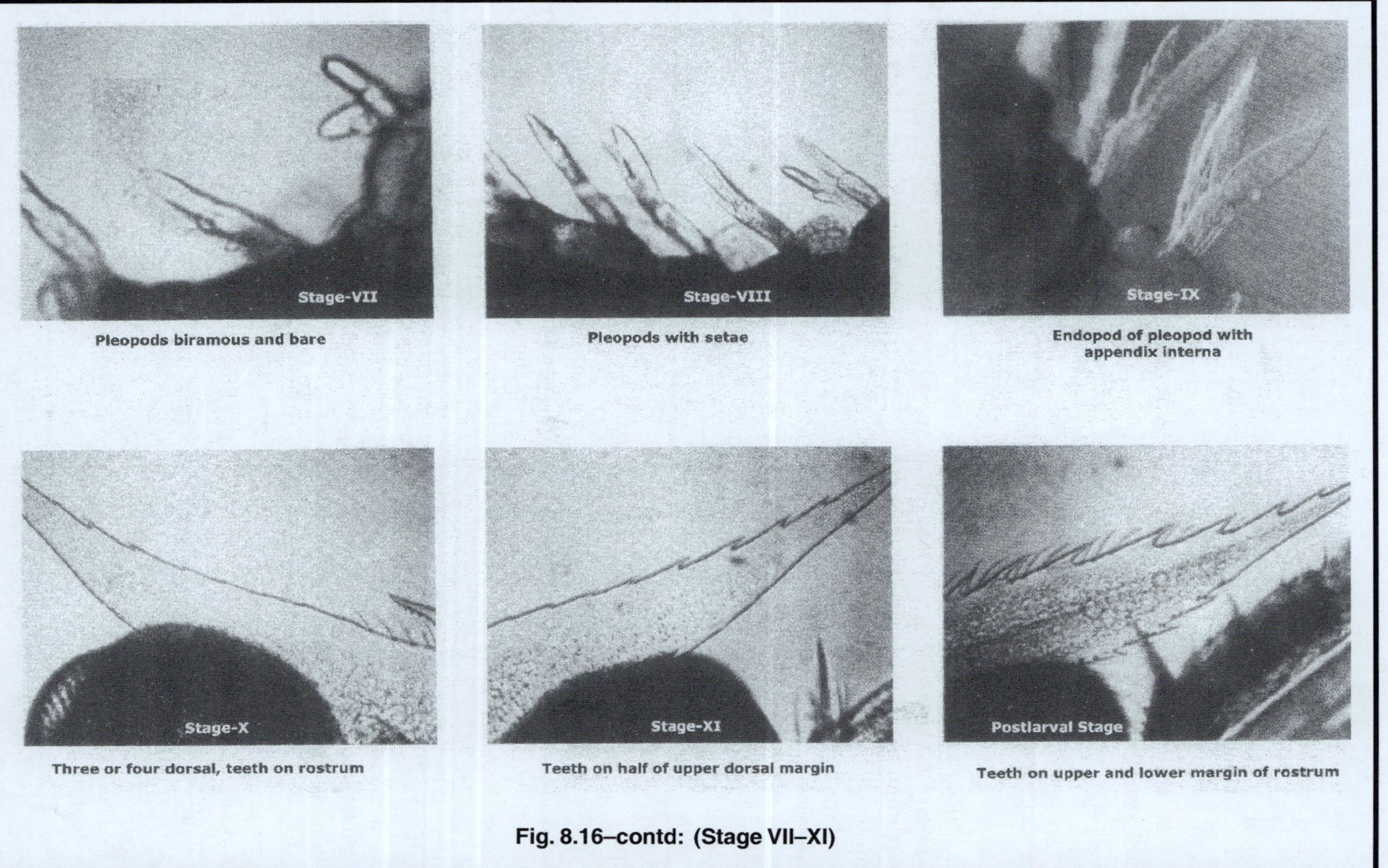

Fig. 8.16–contd: (Stage VII–XI)

(a) Brood stock rearing unit,
(b) Incubation and spawning unit,
(c) Larval rearing unit,
(d) Artemia unit, and
(e) Post larval holding unit.

In addition to the above infrastructural facilities such as storage tanks for freshwater, saltwater and mixed water, air blowers for aeration, generator as a standby during power failure etc. are required.

Maintenance of Brood Stock

For developing a brood stock early juveniles are stocked in the ponds at a rate of 3-4/m^2 and adults @2 nos/m^2. The sex ratio can be 1:4 (male:female) For complete harvesting of the pond, rectangular, drainable ponds are advisable in prawn farming. Size may vary from 0.2ha to 1.6 ha with an average depth of 0.9m (minimum 0.75 and maximum 1.2m). The bottom of the pond should be smooth with the slope from water intake towards the drain (slope of 1.500 for larger ponds and 1.200 for smaller ponds). The bundh should have minimum 60cm free board, high enough to protect the pond from flooding. The internal slope of the bundh can be 3:1 and external slope 2:1. The water can be preferably fresh though culture has been successfully done in saline water (salinity range 10ppt to 25ppt) Growth rate is lesser in harder waters, the total hardness can be above 40ppm and below 150 (preferably 100) ppm $CaCO_3$. pH can be 7 to 8.5 optimum temperature 29°C to 31°C (it can grow in the range 18°-34°C) temperature below 14°C and above 35 C is lethal. It must be pollutant and predator free.

For hatchery purpose, healthy and active berried females are chosen. Though berried female may be available from natural grounds, its transportation to hatchery in healthy, undamaged condition is difficult. Hence it is advisable to maintain a brood stock in production ponds near hatchery

For a target of 25 million post larvae per year, it is necessary to have a stocking tanks of 1.5 ha.

The prawn can be fed on a pelleted feed containing 35-40% proten which can be prepared from groundnut oil cake (60%) fish meal (20%)deoiled rice bran (18%) and vitamin and mineral mix (2%). They are fed 3-4% of the total body weight about 3 times per day. If pelletted feed are not available they can be fed on a variety of item such as rice and rice products, tapioca, trash fish, molluscs, prawn waste etc. Prawn attain maturity within 4 to 7 months depending on feed, temperature and other ecological conditions.

Incubation and Hatching Unit

The berried females bearing dark grey coloured eggs in the abdomen are utilized for incubation and hatching. Grey coloured eggs contain advanced stage embryo that will hatch out within 2-3 days. As orange coloured eggs will contain embryos at the early stages of development, hatching will be delayed. If more than one berried female is to be taken in a given tank care must be taken to take females having eggs at the same stage of ripeness so that tanks will have larvae of the same age. This will lessen mortality due to cannibalism. It will also maintain the same level of feeding activity in the tank. Catching, handling and transportation of berried females have to be done carefully to minimise egg loss/damage. The berried females are first disinfected by placing them in freshwater containing 0.3ppm copper sulphate or 15 to 20 ppm formalin and released into the larval rearing tank having 5ppt salinity at the rate of 5 numbers/m^2 or maintained in a hatching tank. Egg hatching takes place usually during night time. Normally, under Indian conditions, 500 larvae/gm body weight of brood prawn are obtained after hatching. Soon after hatching is completed, mother prawn are carefully scooped out and released into stocking tank after disinfection. Usually, berried females are not fed during 2-3 days of incubation to avoid deterioration of water quality.

Larval Rearing Unit

Factors affecting larval rearing

Salinity

To ensure optimum survival and growth, 12ppt has been reported to be ideal for larval rearing. Rao and Tripathy (1993) reported the following as optimum salinities for different stages of larvae.

Larval stage	*Optimum salinity (ppt)*
I and II	10
III to V	14
VI to VII	10-12
VIII to XI	10

Quality of Larval Rearing Tank

Copper and zinc and their alloys, galvanised steel, bare concrete and oil are toxic to freshwater prawn larva. If concrete is used, inner surface should be coated with several layers of epoxy resin to prevent harmful chemicals leaking out of concrete and to provide smooth surface. It is necessary to age newly constructed tanks by soaking them in several changes of brackish water for several weeks. This will facilitate leaking out toxic materials.

Temperature

Temperature is a critical factor that influence growth and moulting of larvae. Optimum temperature for larval rearing has been shown as 29-31°C. Temperature below 25°C and above 33°C are detrimental for the larval growth, feeding activity and survival.

Dissolved Oxygen

Dissolved oxygen has to be maintained at saturation level by aeration. Aeration also helps in expelling certain amount of ammonia from the water and in even distribution of temperature.

pH

pH has to be maintained at a level of 7 to 8.5, the ideal being 7.At higher pH, un-ionised ammonia (NH_3) increases which is toxic to the larvae, ionized ammonia (NH_4^+) is not very toxic as it can not penetrate through the gill membrane of larvae. Ammonia concentration should not exceed 1.5 ppm of NH_4^+ and 0.1ppm of NH_3

Water Exchange

50% of the water is to be exchanged every day after cleaning the tank, before feeding the artemia nauplii. Provision is to be made for supplying clean, pollutant free filtered water.

Exposure to Sunlight

Exposure to sunlight shall help in augmenting biological purification of water by algae. Direct exposure is harmful for larvae. Hence larval culture units are partially exposed to sunlight to induce growth of algae in the medium. Algal supplement has been shown to have beneficial role by increasing growth rate, survival rate and reducing the duration of larval cycle. Used water can be stored in tanks kept in sun light to promote algal growth which in turn removes the metabolites.

A two phase larval rearing system have been recommended (Rao and Tripathy, 1993). In phase I *i.e.* for rearing larvae from stage I to V cylindro conical tanks of 1000 liters made of fibre glass are used. Inner surface of these tanks should be smooth painted dark green. For rearing larval stage from V/VI to XI rectangular cement tanks with flat or U-shaped bottom are used. Inner surface is painted dark green with epoxy resin. Each tank should have water inlet and outlet facilities. It should be 14m^2 in size with 1 meter depth to hold 70cm deep water. It should be above ground level to facilitate draining by gravitation.

Stocking density of larvae

In the first phase larvae can be stocked at 500-700 numbers per litre where as in second phase it can be 50 to 80 larvae per litre.

Larval Behaviour

The newly hatched prawn larvae (zoea) are planktonic and they swim with tail first and ventral side upwards and head down at an obleque angle. They are attracted towards light. During early larval stages they exhibit gregarious habit, swimming close to the surface of the water in groups where as unhealthy larvae tend to accumulate at the pond bottom. As the larva grows to post larva, there is radical change in behaviour and appearance. Instead of swimming freely in water, they crawl or cling to the tank surface. The post larvae do not need brackish water but it can be maintained in freshwater : First post larva is usually observed twenty days after hatching (between 22 and 27 days) and 90% develop to post larva within next 10 days.

Larval Feeding

Most commonly used live feed for larval rearing is Artemia nauplii. During recent year zooplankton such as Moina, cut pieces of Tubifex worm/Acetes/

Lamellidens and prepared feeds are used as feed for Macrobrachium larvae. In addition, cutpieces of mushroom, soya products, freshwater snail, marine bivalves, squilla, cut pieces of earth worm, fish flesh and hens egg (custard) have been also used as feed. Rao and Tripathy (1993) categorized the former under principal diets and latter as secondary diets. Quantity of Artemia nauplii to be given at different stages vary as noted below (Rao and Tripathy, 1993)

Table 8.7: Feeding Rate of Artemia nauplii to Prawn Larvae

Age of Larva (days)	*Larval Stage*	*Number of Artemia nauplii/larva/day*
3	II-III	5
4	II-III	10
5-6	III-IV	15
7-8	IV-V	20
9	IV-VI	30
10-11	V-VII	35
12	VI-VII	40
13-14	VI-VIII	45
15-24	VII-PL	50
25-30	VIII-PL	40
31-35	IX-PL	30

Feeding cut pieces of tubifex worm is also found an excellent diet.

Two types of prepared feed developed at CIFA for rearing larvae is given below.

Type-I

	Percentage (by weight)
(a) Freshwater mussel meat (foot and gonad)	68. 00
(b) Hen's eggs	20
(c) Skimmed milk powder	10.00
(d) vitamin mineral mix	2

Type-II

(a) Fish flesh (channa)	68
(b) Hen's egg	20
(c) Wheat flour	5
(d) Skimmed milk powder	5
(e) Vitamin mineral mix	2

All ingredients are thoroughly grinded and mixed in a grinder with little water, dough is steamed for 30 minutes, after cooling stored in refrigerator for 4-5 days for feeding.

Mode of Feeding and Frequency of Feeds

Feeding starts from 2nd day of hatching. Prepared feed are usually given in the day time, Artemia nauplii in the night time. Different stages require 50 to 150 μg prepared feed/larva/day and artemia nauplii from 5 to 50 per larva/day.

Table 8.8: Feeding Schedule for Macrobrachium Larvae (Rao and Tripathy, 1993)

Day	*Larval Stage**	*Feeding 7AM*	*tbime & 9AM*	*type 12noon*	*1 PM*	*3PM*	*5PM*	*10PM*
1	I	-	-	-	-	-	-	-
2	II	-	-	-	-	-	AN	-
3	II-III	AN					AN	
4	II-III	AN			PF_1		AN	
5	III	AN		PF_1	PF_1		AN	
6	IV	AN		PF_1	PF_1	PF_1	AN	
7	IV-V	AN		PF_1	PF_1	PF_1	AN	
8	IV-V	AN		PF_1	PF_1	PF_1	AN	
9	IV-VI	AN	PF_1	PF_1	PF_1	PF_1	AN	
10	V-VI	PF_1	PF_1	PF_1	PF_1	PF_1	AN	
11	V-VII	PF_1	PF_1	PF_1	PF_1	PF_1	AN	
12	V-VII	PF_2	PF_2	PF_2	PF_2	PF_1	AN	
13	V-VIII	PF_2	PF_2	PF_2	PF_2	PF_2	AN	
14	VII-VIII	PF_2	PF_2	PF_2	PF_2	PF_2	AN	AN
15	VI-VIII	PF_2	PF_2	PF_2	PF_2	PF_2	AN	AN
16	VI-IX	PF_2	PF_2	PF_2	PF_2	PF_2	AN	AN
17	VI-X	PF_2	PF_2	PF_2	PF_2	PF_2	AN	AN
18	VI-X	PF_2	PF_2	PF_2	PF_2	PF_2	AN	AN
19	VII-X	PF_2	PF_2	PF_2	PF_2	PF_2	AN	AN
20	VII-XI	PF_3	PF_3	PF_3	PF_3	PF_3	AN	AN
21	VII-XI	PF_3	PF_3	PF_3	PF_3	PF_3	AN	AN
22	VII-XI	PF_3	PF_3	PF_3	PF_3	PF_3	AN	AN
23	VII-PL	PF_3	PF_3	PF_3	PF_3	PF_3	AN	AN
24	VII-PL	PF_3	PF_3	PF_3	PF_3	PF_3	AN	AN
25	VIII-PL	PF_3	PF_3	PF_3	PF_3	PF_3	AN	AN
26	VIII-PL	PF_3	PF_3	PF_3	PF_3	PF_3	AN	AN
27-35	IX-PL	PF_3	PF_3	PF_3	PF_3	PF_3	AN	AN

* Due to non-synchronous larval metamorphosis, all larvae in a given stock will not be in the same stage. Hence in a given day of larval development different stages of larvae will be seen together.

AN = Artemia nauplii

PF1- Prepared feed of 230 micron size

PF2- prepared feed of 350 micron size

PF3- Prepared feed of 600micron size

Brine shrimp nauplii are given in the night to ensure the presence of food in the larval rearing medium throughout night. While inadequate feeding leads to stunted growth due to starvation, cannibalism and prolonged metamorphosis over feeding leads to water pollution. The larval rearing tank is to be cleaned every day, by siphoning out feed residues, metabolic waste and faecal matter from the bottom of tank. This is to be done daily after stopping aeration before water exchange and feeding of Artemia nauplii. 25 to 40 post larvae per litre can be harvested within 35 days. Eight cycles of seed production is possible in one year.

Artemia Hatching Unit

The details of artemia hatching unit please see the chapter on culture of live feed organisms.

Airlift Biofiltration System

Biological filtration unit for continous supply of clean pollutant free water is incorporated in a close door system. This helps also in conserving the brackishwater particularly when hatchery is located in inland areas away from seashore (Spotte, 1970)

This is arranged by using 2 fibre glass tanks of about 400 litre capacity. One tank is used for larval rearing and another for a biological filter. The biofilter tank is filled to half the capacity by six layers as noted below.

Layer 1: Two inch thick layer of stone chip or gravel of one inch size.

Layer 2: Three inch thick layer of oyster shells or coral of 1″ size

Layer 3: Two inch thick layer of gravel of 1″ size

Layer 4: One inch thick layer of gravel of ¼″ size

Layer 5: Two inch thick layer of coarse sand of 2-5 mm size.

Layer 6: Two inch thick layer of gravel of 1″ size.

An L-type airlift water discharge pipe is placed 2″ above the bottom of the filter tank. Aeration pipe is fixed at the end of the water lifting pipe. The other end of the water discharge pipe is placed at the centre of the larval rearing tank. Another pipe is used to siphon out excess water from larval rearing tank to biofilter tank. One end of siphoning pipe is placed 2-3″ inchs above the bottom of the larval rearing tank, where as the other end is placed 2-4″ above gravel of biofilter tank.

In this system, the water from the larval rearing tank continuously flows into biofilter tank and as the water flows down the filter, it gets purified by the action of nitrifying bacteria present in the surface of gravel/coral. As ammonia is the main metabolite that has the maximum toxic effect on larvae, it is changed to nitrites by Nitrosomonas bacteria which is again changed to nitrate by Nitrobacter (another nitrifying bacteria). As larvae can tolerate nitrate up to 22ppm, the same water is again lifted to larval rearing tank through airlift system. Organic matter also undergoes mineralisation in this system.

Post Larval Holding Unit

The newly metamorphosed post larvae are gradually acclimatised to freshwater. This is done by drawing out the rearing medium gradually and replenishing with fresh water. Aeration is to be continued. The post larvae are to be weaned from the larval diet and fed on a variety of other feeds. It can be kept in the holding tank for one week or two weeks after which it can be released to grow out ponds.

Some workers recommend rearing of post larvae in nursery tanks till it reaches juvenile stage (25mm) at a rate of 60-100 post larvae/m^2. Freshwater ponds of 400-500 m^2 water area and depth of 1 meter are utilized as nursery tanks. The ponds are carefully prepared so as to make it predator free and suitably manured/treated with lime (200 kg/ha) and cowdung (2500-5000 kg/ha) other water quality parameter should be optimum temperature (28-32°C), pH 7-85, total hardness ($CaCO_3$) 50-100ppm and DO 4-6ppm. The larvae can be fed on pelleted feed @10-20% of body weight daily for 2 months till 1st reads 0.5 to 1.0gm. Submerged artificial shelters are also provided.

Apart from the low intensity nursery system mentioned above, some workers have resorted to intensive nursery system utilizing FRP/cement tanks with a stocking density of 2000-5000 post larvae per m^2 and survival rate of 68-80% within 15-44 days. The post larva were fed freshwater mussel meat and continuous aeration is done. Post larvae of 7-9 mm size shall grow, to 16 - 21 mm in 15 days, to 35 mm in 44 days and to 47 mm in 54 days

SEED PRODUCTION OF MUD CRABS

Introduction

Ebible crabs of commercial importance are *scylla tranquebarica, scylla serrata, Neptunus pelagicus* and *N. sanguinolentus. N. pelagicus* and *N. sanguinolentus* are entirely marine where as the former two species *S. tranquebarica* and *S. serrata* commonly called as mud crabs migrate to brackish water for growth while they are juvinile and the adults migrate from brackishwater to the ocean for breeding and spawning. Mud crabs are of cultural value and hence the seed production technology for the mud crabs is described below.

Scylla tranquebarica is the larger species and it attains a maximum size of 220 mm (carapace width) weighing 2.5 kg whereas *Scylla serrata* is smaller attaining a size of 140 mm (carapace width) and 0.5 kg to 0.72 kg weight *(S. tranquebarica* has 2 spines at the wrist of chelipeds where as *S. serrata* has only one spine in the wrist). While *iS. tranquebarica* are found in open waters S. *serrata* are found in burrows in the intertidal and subtidal regions of mangrove, estuaries and creeks. In India, crabs formed secondary crop along with shrimp in traditional tide fed farm of Pokkali fields of Kerala, Bheries of West Bengal and Khazan lands of Karnataka. In recent years, mud crab production through culture has been practiced in brackishwater ponds either in a monoculture system or in poly culture system along with milk fish/tiger prawn in countries like Thailand, Philippines, China, Indonesia, Malaysia, Taiwan, Srilanka and India. It can be considered as an alternate candidate species for rotation of crops in brackish water ponds, particularly in areas where shrimp farming has collapsed due to white spot disease. They are nocturnal feeders feeding mainly on bottom dwelling, slow moving animals such as bivalve, small crabs and dead decayed animal matter.

Reproductive Biology

Sexual Dimorphism

The sexes are separate and there is distinct sexual dimorphism

Table 8.9: Distinction Between Male and Female

Male	Female
1. Abdominal flap which is folded firmly against the ventral side of cephalothorax is slender and triangular.	Abdominal flap folded against the ventral side of the body is broad and triangular/semicircular in berriedfemales.
2. Appendages are present only on first and second abdominal segments and the same are modified to copulatory organs.	There are four pairs of abdominal appendages present from second to fifth segment and the same are used for carrying the eggs.
3. The claws are comparatively larger.	The claws are smaller.

Size at First Sexual Maturity

The size at first sexual maturity is 120 mm (carapace width) for larger species *(Scylla tranquebarica)* and 83 mm (cw) for *S. serrata*. The early maturing ovary is bright orange where as in mature ready to spawn female it is deep yellow.

Mating Behaviour

Copulation takes place between a hard shelled male and a freshly moulted, soft bodied female. The courtship is initiated by a "premating embrace" between hard shelled male and hard shelled female which lasts for 2-3 days [Fig. 8.17(A)]. Premating embrace is accomplished by the male climbing over the female and holding her by his cheliped and first two pairs of walking legs. When the female is about to moult, the males leave the riding position and helps the female in casting off the old cuticle. A few hours after the precopulatory moulting, actual process of copulation starts. The male turns the soft bodied female upside down using his chelipeds, and climbs on to her ventral side [Fig. 8.17(B)]. The female unfolds the abdomen and holds the male in position. This embrace lasts for 6-8 hours during which time male deposits the spermatophore, in the seminal receptacle of the female. After this the male and female separate [Fig. 8.17(C)].

Spawning

The ova are extruded by the female and the same are fertilized by the sperm stored in the spermatophores. The fertilized eggs are attached to the ovigerous setae of the abdominal appendages. In *S. tranquebarica*, the number of eggs may be 2 to 3 million where as in *S. serrata* it is 0.5 to 2.5 million such females with egg attached to the abdomen are called berried [Fig. 8.17(D)].

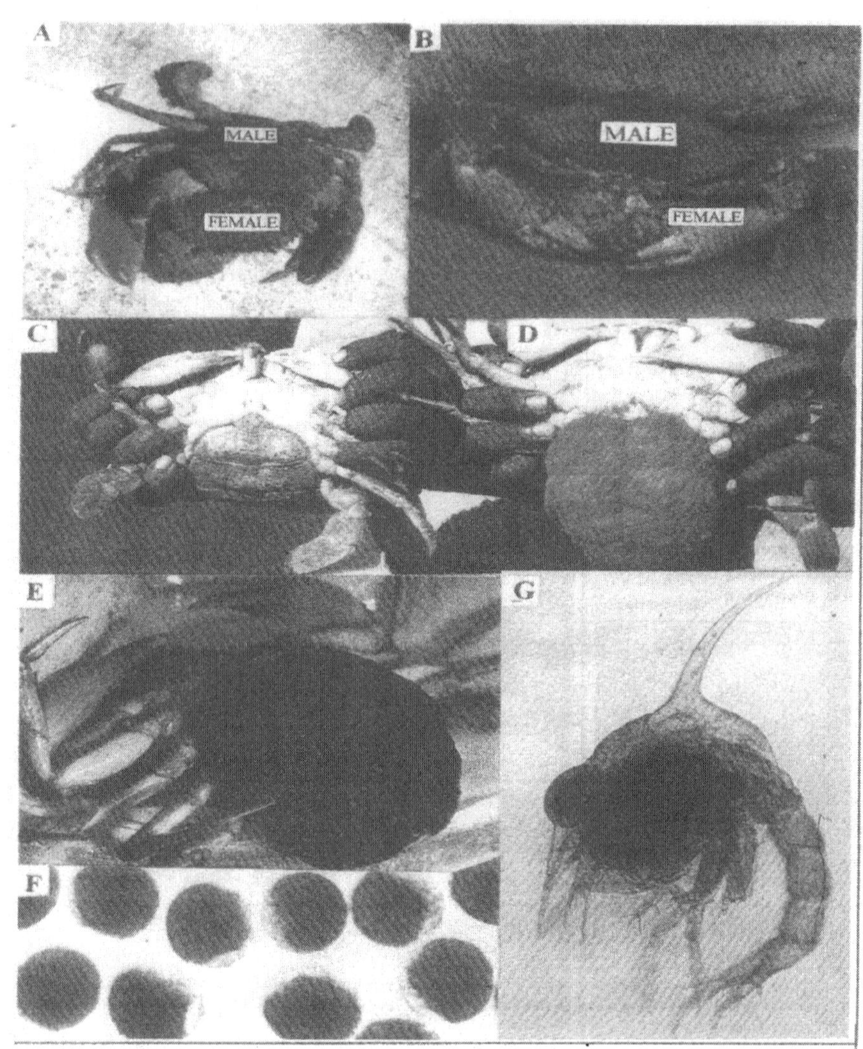

Fig. 8.17: Mud Crab Reproduction.

(a) Premating embrace between hard shelled male and hard shelled female. This lasts for 2 to 3 days, (b) Mating–Hard shelled male with soft bodied female. Male turns the female which has undergone pre mating molting upside down using his cheliped and climbs on to her ventral side, (c) Ventral side of female crab. Note the broad and semicircular abdominal flap folded against the cephalo thorax, (d) A berried female crab showing the orange coloured fertilized egg mass attached to pleopods, (e) Black coloured egg mass towards the end of incubation period just before releasing the larvae, (f) A few isolated developing eggs, (g) Third zoal stage.

(*reproduced with permission from Fishing Chimes, vol. 19 (no. 10,11)*)

Incubation and Hatching

The berried females carry the eggs for 2 weeks, during which period the embryo developes in the egg. The egg change the colour from orange to brown. Just before releasing the larva, the eggs become black [Fig. 8.17(E)]. At the end of incubation zoea larva hatches out [Fig. 8.17(F)].

Breeding Season

Mud crabs breed throughout the year, peak season is observed depending on climate conditions. The season of maximum breeding activity and juvenile abundance in various parts of India is shown in the tabular form below.

Table 8.10: Season of Peak Breeding Activity and Peak Juvenile Abundance in India

Sl. No.	*Region*	*Peak Breeding Season*	*Peak Juvinile Abundance*
1.	Kerala coast	September-February	May to October is Vembanad backwaters
2.	Tamil Nadu coast	September-April	December to May in Pulicatlake, Adyar estuary, Ennore estuary, back water adjacent to Kovalarmetc. April to June and Sept. to October in Tuticorin.
3.	Andhra Pradesh coast	October-February and May-June	December to April and July to August in Kakinada
4.	Orissa coast	November-January	March to June in Chilika lake
5.	West Bengal coast	May-August	November to February in Hoogly Matla estuarine system.

Larval Development

The Zoea larva undergoes 5 successive moults to become the megalopa larva. Thus there are 5 zoeal stages. For each moult it takes 3-4 days. Thus, to develop a megalopa from a zoea it takes 15-20 days. Megalopa develops into juvenile crab after 8-11 days.

A newly hatched zoea larva of crab measures 1.2 mm and it consists of a cephalothorax and a 5 segmented abdomen and telson. The carapace bears 4 spines, one dorsal spine bent backwards, rostral spine bent forwards and 2 lateral spine closely pressed against the sides of the body. The eyes are not stalked. The first antenna is represented as a conical process, second antenna has 2 rudimentary processes, mandible is incipient, first maxilla consist of coxa, basis and endopodite (each with 4, 5 and 6 setae respectively), a second maxilla has coxa and basis bilobed, with 1-1 and 2-3 setae respectively, the endopodite is having 3 rudimentry setae at its tip where as exopodite is with 4 rudimentary setae. The first maxilliped has a short coxa, long basis, and a five jointed endopodite and exopodite. The segmentation of endopidite is not sharply demarkated. Both endopodite and exopodite has 4 rudimentary setae at its tip. The second maxiliped also has coxa, basis and exopodite and endopodite, but the endopodite is 3 segmented, and it has only 3 rudimentry setae. Second and third segment of abdomen has a pair of knobs. Telson is rudimentry and forks are not well defined.

As it passes through 5 zoeal stages, growth in body size, appendages etc take place. Zoea V is 3.5 mm in length. From Zoea II onwards the eyes are stalked (for distinguishing characteristics of larval stages of crab please refer table 8.3 and Figure 8.18).

The megalopa larva has a crab like appearance. The cheliped is well developed. 5 pairs of pleopods are present which becomes locomotory in function. Chelipeds are used to catch prey.

Table 8.11: Distinguishing Features of Five Zoeal Stages and Megalopa of Mud Crab (*Kathirvel, M.* et *al.*, 2000).

Larval Stage	*Distinguishing Features*
Zoea 1	Eyes are sessile. Abdominal segments are 5.Telson is with 3+3 spines.
Zoea 2	Eyes are stalked. Abdominal segments are five. Telson has 4+4 spines.
Zoea 3	There are five abdominal segments.
Zoea 4	Pleopod buds appear in abdominal segments and rudiments of remaining thoracic appendages also appear.
Zoea 5	Setae are present in pleopods. Telson with 5+5 spines. Rest of thoracic appendages develop.
Megalopa	Carpace length is more than the width. Abdomen with five pairs of pleopods. A pair of chelipeds and four pairs of legs are seen.
Crab I	Carapace with nine anterio lateral spines on either side. First pair of cheliped and three pairs walking legs, fifth pairs of legs has paddle shaped dactylus.

CRAB HATCHERY

For the operation of a hatchery for seed production of crab, the following are the component required:

(a) Brood stock development unit,

(b) Hatching unit,

(c) Larval rearing unit, and

(d) Live feed culture unit.

Brood Stock Rearing

Females carrying eggs in the abdomen (berried females) are utilized for producing the seed. The berried females can be either collected from the natural ground or raised in a pond so that breeders shall be available as an when required. If berried females are available from natural ground, only a holding tank shall be required in a hatchery for maintaining the breeders. Berried females with yellow eggs mature to grey eggs within 5-7 days and it can be fed on squid meat twice daily. If brood stock has to be raised, tanks are to be excavated to stock the young crabs (crablings) collected from wild. The size of the tanks may vary from 0.1 ha to 0.4 ha. The central portion of the pond can be kept shallow or slightly exposed which will simulate the natural habitat of *S. serrata* to burrow and live inside burrows It may also serve as a feeding place.

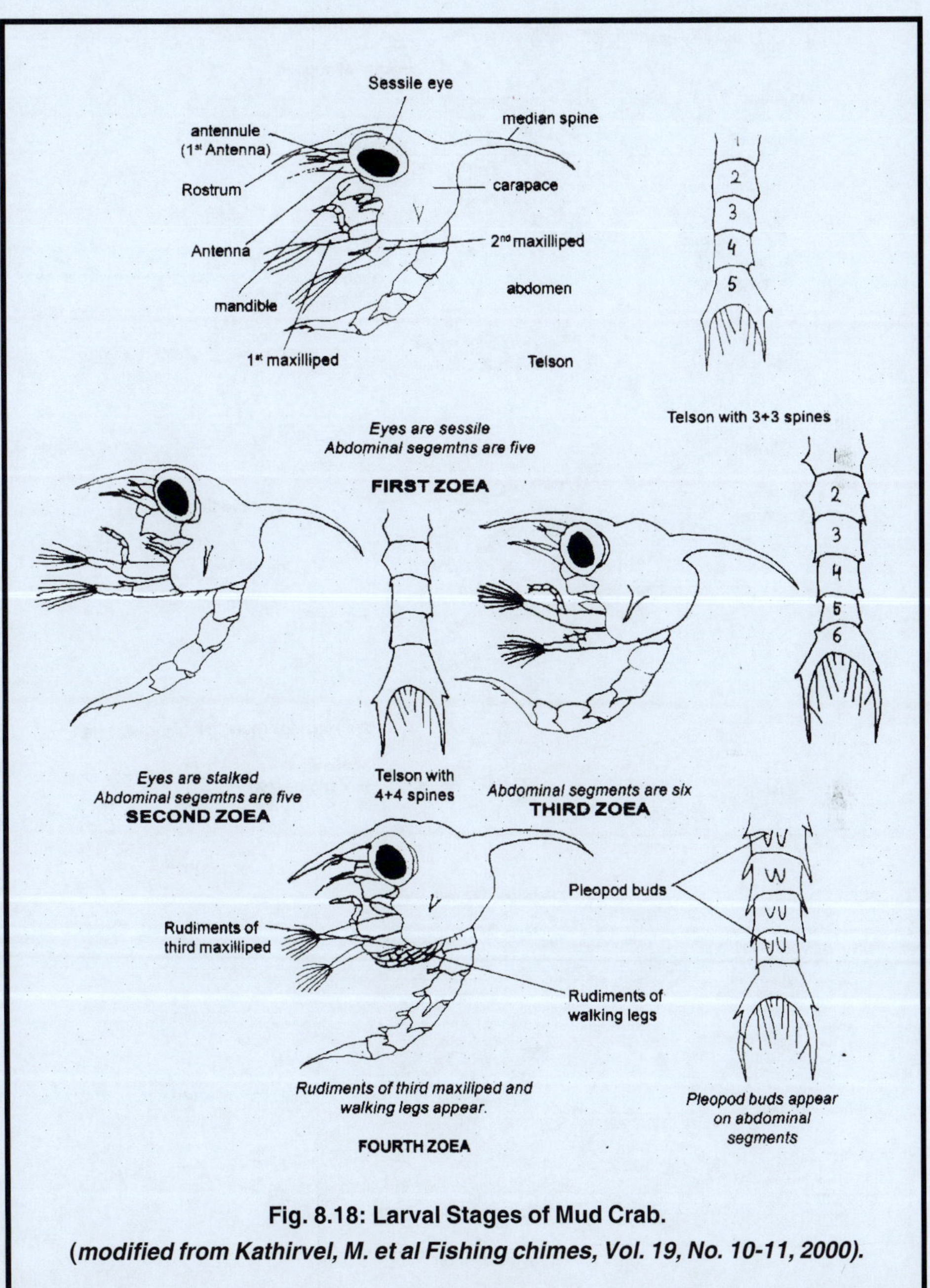

Fig. 8.18: Larval Stages of Mud Crab.

(*modified from Kathirvel, M. et al Fishing chimes, Vol. 19, No. 10-11, 2000*).

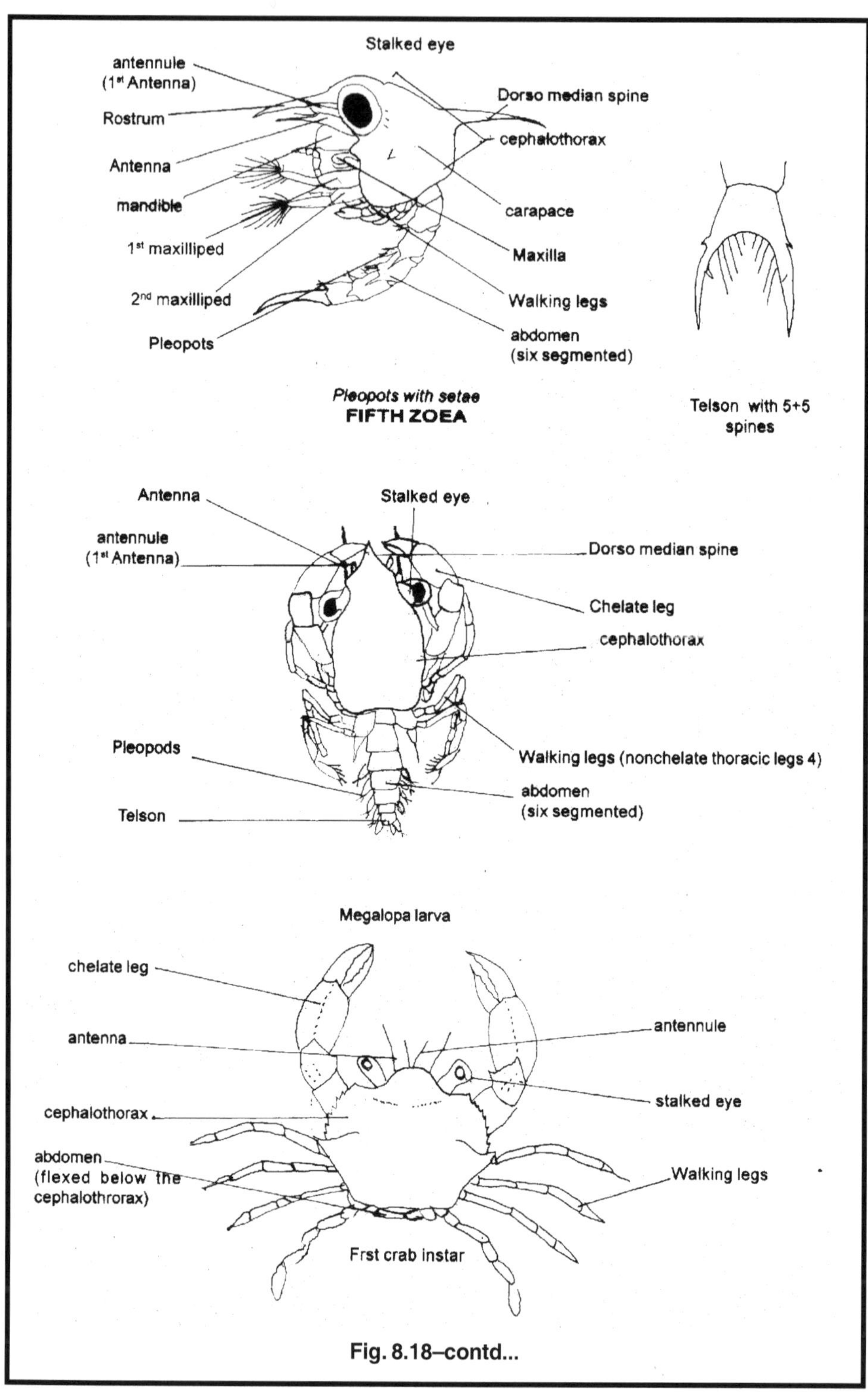

Fig. 8.18–contd...

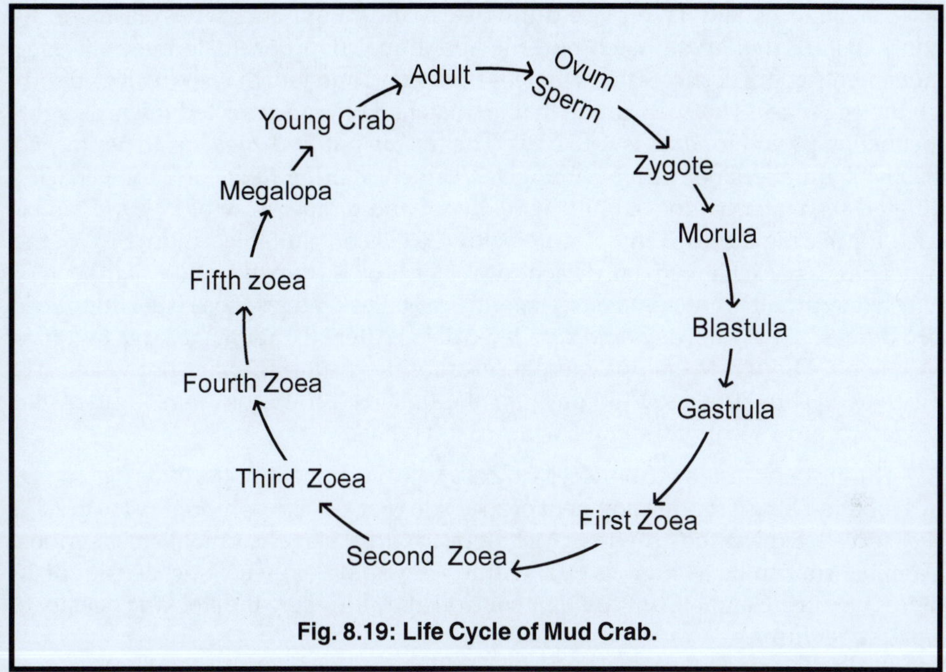

Fig. 8.19: Life Cycle of Mud Crab.

Surrounding the pond, along the inner edge or on the top of bundh fencing is made to prevent the escape of crawling crabs. The fencing of 1 meter height may be made of split bamboo matting/G I chicken wire mesh/nylon netting/absbestos sheet. Inside the pond earthen pipes or PVC pipes or wornout tyres are kept which will be hide outs or shelters which will reduce the fighting among hard shelled crab and prevent mortality of soft crabs. This is required, particularly because of the cannibalistic nature of crabs. Ponds can be constructed in coastal areas with facilities for water exchange by tidal effect through wooden sluice, young crabs of 80-100gm can be stocked @2-5 numbers per square meter. Feeding can be done with trash fish at a rate of 5% of body weight which can be increased to 10% towards the latter part of rearing period of 4 months, when chopped trash fish along with bivalve or gastropod meat is provided. The other desirable parameters for culture are temperature 28-32°C, DO 5-7ppm, depth 0.5 to 1 mtrs, pH 7.5 to 8.5.Periodic sampling can be done to record the growth rate, health and gonadal condition. *S. tranquebarica* can grow from the initial size mentioned above to 400-500 g in 4 months and 800-1000 g in 7 months rearing. From the culture ponds either berried females are collected or to induce gonadal maturation unilateral eyestalk ablatin of females can also be done. Within 15 days from the time of eyestalk ablation, berried females can be obtained for hatchery use. Some workers have recommended eyestalk ablation just after moulting.

Hatching Unit

The berried females are dipped in 10ppm malachite green/methylene blue for 5 minutes as a prophylactic measure and released into 500 litre capacity FRP tanks/ cement tanks covered with black cloth to prevent the passage of light.

Hatching usually take place during early morning hours which lasts for 3-4 hours. Before the larvae hatch out, the abdominal flap of female makes jerking movements in quick succession whereas the second and fourth walking legs tightly jab the egg mass. The zoea larvae that are hatched out are attracted lowards light. Incubation period for crab is 8-15 days. The freshly hatched zoea has to be stocked @200-400 numbers per litre in fibre glass/cement rearing tanks of 2 ton capacity. Filtered sea water having salinity at 30-35 ppt and temperature of 27-29°C can be used for rearing larvae. Temperature below 26 C is not suitable. About 80% of the water has to be exchanged daily. Each zoea stage is of 3-4 days duration, at the end of which they moult to enter into next stage. (Zoea I takes 2-4 days, Zoea II 3-4 days, zoea III 3-5-days, zoea IV 6-7 days and zoea V 5-8 days) After 5th moult, the larva becomes Megalopa. Though Megalopa moults several times, substages are not yet clearly dilienated. Thus it takes 27-30 days for the first zoea to become first crab instar. Megalopa grows to juvinile crabs within 8 to 11 days.

During the transitions from Zoea I to Zoea V heavy mortality (50%-70%) is observed by workers. During the development of zoea V to Megalopa, less mortality (10 to 20%) is observed. During the growth of Megalopa to Crab I stage, cannibalism is serious. Cannibalism can be as high as 60% within a few days at a stocking density of 10 pieces per litre. Cannibalism could be controlled by lowering the stocking density to 5 pieces per litre.

Feeding the Larvae

Rotifers and Artemia nauplii have been shown to be the suitable feed for larvae. Rotifers are preferred during early stages whereas artemia nauplii are given for megalopa to crab stage. During early zoea stage (Z_2 to Z_4) Artemia nauplii swim faster than zoea and hence the zoea larva can't catch the prey *i.e.* free swimming artemia nauplii. Hence one day old frozen artemia nauplii has been recommended as feed for zoal stages and two day old frozen artemia nauplii has been fed to megalopa stage.

Table 8.12: The Feeding Regime Recommended for Larval Rearing of Crab Followed at NAPFRE (National Prawn Fry Production and Research Centre, Malaysia).

Larval Stage	*Morning*	*Afternoon*	*Supplementary feed*
Z_1	*Rotifer 5-10*	*Rotifer 5-10/ml*	-
Z_2	Rotifer 5-10	Frozen Artemia 6	5a
Z_3	Rotifer 10-15	Frozen Artemia 10	7b
Z_4	Rotifer 20-30	Frozen Artemia 15	9b
Z_5	Frozen artemia 10	Frozen Artemia 20	12b
M	2 days old Artemia 10	2 day Artemia 40	15b
C_1	2 days old Artemia 10	2day artemia10	15-20b

(1) In case of rotifer, the number refers to number of live rotifers per ml of water in the culture tank. If live rotifer was not available, frozen rotifers are used.

(2) In case of Artemia nauplii, the number refers to number of Artemia nauplii per larva.

(3) In case of supplementary feed, the number refers to weight of feed in gram given per million of crab larvae.

   (a) feed for shrimp larva imported from Japan having Protein 49%, fat 31%, Ash 5.5% and moisture 4.3%

   (b) feed for giant freshwater prawn containing protein 55%, fat 8%, Ash 7% and moisture 5.5%

Treflan at the dose of 0.7 to 10ppm is used in the culture medium to control fungal growth.

According to Karthirvel, M. *et al*, (2000) feed and feeding rate for larval rearing of crab can be as noted below:

For Zoeal Stages I to III

(a) Algal cultures- *Chlorella* sp. @20000 cells/ml, *chaetoceros* @1-2 lakh cells per ml., *Tetraselmis* @1000cells, *Skeletonema costatum*/*Isochrysis* @5000 cells/ ml.

(b) Rotifer: *(Brachionus plicatilis)* @ 30 –60 numbers per ml

(c) Artificial feed: (shrimp larval feed) @0.5 g. per ton of water

For Zoeal Stages IV to V

(a) Algal culture: *Chlorella* @20000 cells per ml., *Chaetoceros* 1-2 lakh cells per ml. T*etraselmis* @1000 cells per ml.

(b) *Artemia* nauplii 5- 50 numbers per ml.

(c) Artificial feed @0.5 g per ton of water

For Megalopa

(a) Two day old live artemia @ 50 numbers per ml.

(b) Bits of flesh of prawn, bivalve molluscs, squid and fish @150 –200 g. per ton of water

(c) Artificial feed @ 0.5 g per ton of water

First Crab Instar to Tenth and Further Instars

(a) Flesh of prawn, bivalve molluscs and fish @3-5% of biomass

(b) Artificial feed @ 0.5 g/ton of water

During larval rearing high mortality or poor survival rate has been reported by all workers. Survival rate may be as low as 5-15%.

Excess food/faceal matter/dead larvae/moulted shell settled at the bottom of the rearing tank have to be siphoned out daily from the larval rearing tank. Continuous aeration of water is also required during rearing.

Rearing of Post Larvae (Juvenile of Crab)

Rearing tanks for post larva (juvenile crab) is done in tank having 5cm deep sand. They are stocked @5-10 crabs per square meter. They are fed on boiled molluscan meat at a rate of 1-2% of stocked biomass. At the end of 2 months, *S. tranquebarica* grows to 4gm (in weight) and 30 mm carapace width

Live Feed Culture Unit

For details on Artemia cyst hatching, rotifer culture and culture of algae refer to the chapter dealing with culture of live feed organisms.

Chapter 9

Breeding and Seed Production of Shell Fishes Other Than Crustaceans (Molluscans and Holothurians)

Introduction

Molluscans contributed 30.4 % of the global aquaculture production of fin fishes and shell fishes (*i. e.* 10132078 mt. out of the total production of 33310349 mt) in the year 1999 (FAO statistics). Culture of these organisms were done even from the era of Romans. Bivalves such as oysters, mussels, clams, scallops and a few other gastropods such as abalones *(Haliotes rufescens)*, top shells *(Trochus cornatus)* are extensively cultured in mainly in subtropical and temperate countries such as USA, Japan, Korea, France, Spain, Netherlands, Italy etc. Till recent years seed for culture was obtained through collection of spat from natural spawning grounds. During recent years technology has been developed for the seed production of these organisms in controlled conditions and hundreds of commercial hatcheries have been established in several countries. These hatcheries not only cater to the seed requirement of the farms but also provide seed for stock enhancement through ranching the natural water bodies.

OYSTERS

Considering the species wise contribution of fin fish and shell fish to global aquaculture production, oysters top the list. In the year 1999, one species of oyster, *Crassostrea gigas* (Pacific oyster) alone contributed 10% of the total (i. e production of farmed *C. gigas* was 3600459 mt. out of the total production of 33310349mt.). Cultivated oysters belong to two genera such as Crassostrea (cupped oysters) and Ostrea (flat oysters). Edible oysters are cultured extensively in several European countries, America and some tropical countries as it is highly relished as a delicacy. *C. gigas* served on half shell fetches very high price. Other species of Crassostrea of global importance are *C. virginica* (American oyster), *C. angulata* (Portugese oyster), *C. commercialis* (Sydney rock oyster), *C. glomerata* (Auckland rock oyster), *C. plicatula* (Chinese oyster). In India *C. madrasensis* [Fig. 9.1(A)] is the common species found. Common species of Ostrea are *Ostrea edulis* (European oysters) and *O. lurida* (Olympian oysters). Pearl oysters are extensively farmed by Japanese for production of pearl. All oysters are protandrous hermaphrodites, being males first and later changing to female. In Crassostrea change of sex takes place between spawning seasons but in Ostrea it may take place in the same season. In Crassostrea males and females release gametes in to the water where fertilization takes place. But in flat oysters males release sperms in to the water which enters in to the pallial cavity of female along with the inhalent water and fertilization takes place in the pallial cavity. The zygote undergoes development for 8-10 days and when it reaches the veliger stage which escapes out. Rest of the development takes place in the water. Technology of seed production of pearl oyster is described below in view of its importance in pearl culture.

Seed Production of Pearl Oyster

Introduction

Pearl oysters [Fig. 9.1(B)] are of immense value as they yield natural pearls. They are sedentary bivalves found in the ocean, at a depth of 18-22 meters and found attached to rocky ridges, usually 19 kms off shore. A remarkable natural pearl oyster fisheries is found in the area between Capekomerin and Killakarai, Gulf of Mannar, Palk bay and Gulf of Kutch in India. About 28 species of pearl oysters are reported so far. The common species found along the Indian coasts are *Pinctada fucata, P. chemnitzi, P. margaratifera, P. anomioides*, and *P. atropurpurea*. *P. fucata* found in Gulf of Mannar and Gulf of Kutch is most common in India. *P. margaratifera* (Black lip pearl oyster) occur in Andaman and Nicobar islands. *P. chemnitzi* is found in Palk Bay and *P. anomioides* is found in Lakshadeep. As technology for production of cultured pearl has been already standardised, there is need for obtaining more number of oysters for pearl production and the same need is met through production of seed from brood stock under controlled conditions.

Out of the two valves covering the right and left sides of the soft visceral mass, left one is larger than the right one. Though they are sedentary found attached to rocks by means of byssal threads arising from the byssal gland found at the proximal end of foot, they can also move with the help of foot which is a tongue shaped organ capable of contraction and elongation.

Reproductive Biology of *P. fucata*

There is no sexual dimorphism. Hence, male and female can't be distinguished externally. There is a pair of gonads (testis in male and ovary in female) placed over the intestine and hepatopancreas. Duct arising from the gonad opens by means of gonopore located at the anterior end of the gills. Sperm from the testis and ova from ovary are released into the surrounding water where fertilization takes place.

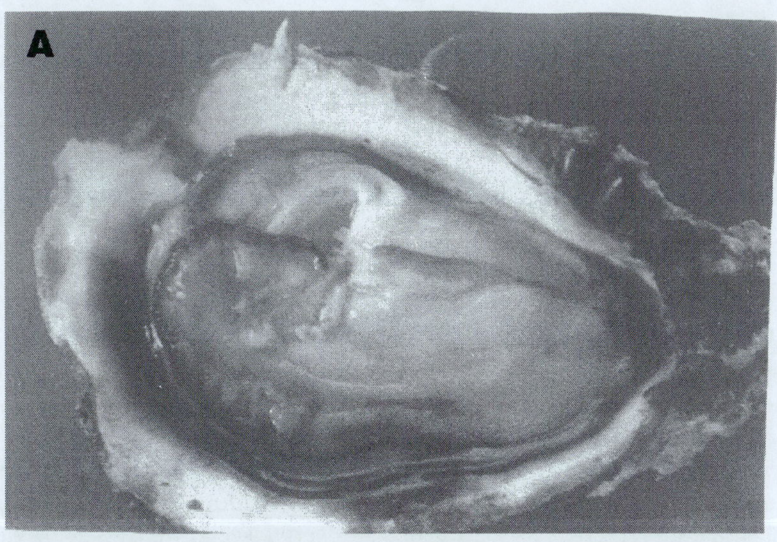

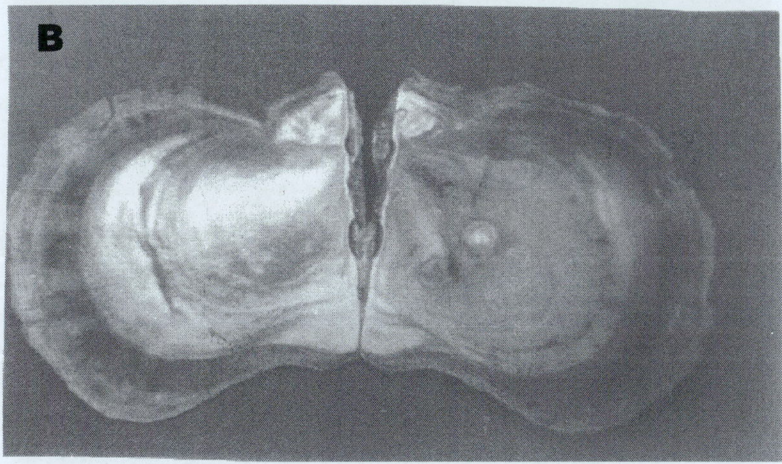

Fig. 9.1: (a) Edible Oyster, *Crassostrea madrasensis* on Half Shell, (b) A marine Pearl Oyster Cultured for Pearl. One Valve is Removed to Show the Visceral Mass Containing Pearl.

(Courtesy, CMFRI annual report, 1994-95)

Life history (Fig. 9.2)

45 minutes after fertilization, segmentation begins. At two celled stage, one micromere and a macromere are produced. The macromere does not devide any further, but micromere undergoes repeated division to form a mass of cells called morula. Each micromere possesses a cilium by means of which it swims. It is attracted towards light and congregates at the surface. Later a blastocoel and a blastopore are developed by the reorientation of cells. This is the blastula stage which is reached five hours after fertilization. Blastula grows to gastrula seven hours after fertilization. Gastrula is double layered and has archenteron. The gastrula develops a flagellum and tuft of cilia at the apical side and postoral tuft of cilia at the rear end. At this stage it is called trochophore larva. This larva swims with flagellum. The ectodermal cell secrete embryonic shell material called prodissoconch- I. The larva looks like letter 'D" and hence it is called D-shape stage or veliger or straight hinge stage. At this stage flagellum and tuft of cilia are lost and velum is developed as the locomotory organ. Veliger stage is reached within 20 hours. it is 50-55 micron in size dosroventrally. Additional shell material called prodissoconclh-II is developed, and veliger reaches *umbostage* At this stage shell valves are equal and mantle folds also develop. Later an eye spot is seen at the base of foot bud and it is called eye spot stage which develops to pediveliger. Pediveliger is a transitional stage from swimming to crawling and the larvae have both velum and foot at this stage. As the foot becomes functional, the velum disappears. The byssus gland becomes active secreting byssus thread. Gill filaments are also developed.

Pediveliger develops to plantigrade having a size of 200 µm. All along the margin except umbo region rapid shell growth takes place. Labial palps, additional gill filaments and byssus threads develop. Plantigrade develops into spat (300 µm). it is a juvenile having left valve larger than the right (Fig. 9.2).

Hatchery Technology

In order to produce the seed of pearl oyster (spat stage) the following are required:

(a) Brood stock development unit
(b) Induced spawning Unit
(c) Larval rearing unit
(d) Algal culture unit

Brood Stock Development Unit

In order to develop the brood stock pearl oyster farming is undertaken. Pollutant free coastal waters where salinity and temperature do not fluctuate widely and having sufficient depth are suitable for pearl oyster farming. Three types of culture techniques such as raft culture, rack culture and onshore tank culture are practised.

Raft culture : Raft culture is undertaken where water depth is 5 meter and more. A raft is prepared with wooden logs and kept floating on the surface of water with 4 buoys of 200 litre capacity. About 100 culture cages can be hung from this raft.

Rack culture : If water depth is less (<5 m) wooden poles are driven to the sea bottom at intervals of 1 meter and horizontal wooden poles are tied to the vertical ones above the sea level. Culture cages are suspended into the sea water from the horizontal ones.

Onshore tanks culture : large concrete tanks of 75 to 150 tons capacity and of 3 meters depth and used for rearing the oyster sea water is pumped into the tank.

Culture cages consist of box cage of 40 cm × 40 cm × 10 cm with synthetic fabric. Frames are made of mild steel rod. Care is to be taken to eradicate fouling organisms like Barnacles, Ascidians, Bryozoans, boring organisms, predator organisms etc from the osyter farm. At the end of 1st year it grows to 41 mm (8. 3 g), at the end of second year it reaches 64 mm (31. 6 g) size and at the end of of 3rd year it grows to 75m m (45 g).

Ripe oysters from the farm or natural ground are kept in FRP tanks containing 50 litres sea water. Adequate aeration, and mixed algal food containing 80% chaetoceros is given at the rate of 4 litre per oyster per day. This is maintained in air conditioned room at 22-25°C.

Induced Spawning Unit

Fully ripe oysters spawn by thermal stimulation or chemical stimulation.

Thermal Stimulation

The spawning tanks (76 × 46 × 46 cm) is filled with freshly filtered sea water. It is provided with an immersion heater which is thermostatically controlled and a thermometer. About 50 to 100 numbers of ripe oysters from conditioning unit are released into the tank. Water temperature is gradually increased to 35°C by switching on the immersion heater. This gradual rise in temperature induces ripe oysters to release the gametes into the medium. Usually, the male releases the sperms followed by the female.

Chemical Stimulation

Spawning tank is provided with sea water and pH of the sea water is raised to 9-9. 5 using either Tris buffer solution or NaOH buffer solution. Oysters are released into the spawning tank. They usually spawn in about 1-2 hours.

After the osyters release the eggs and sperms fully, parents are separated from the tank. Soon after the gametes are released into the water, fertilization takes place and fertilized eggs settle down. The newly fertilized eggs are collected in 30 µm sieve and released into freshly filtered sea water in FRP tanks of 1 ton capacity.

Larval Rearing Unit

About 20 hours after fertilization, the trochophore larva that hatches out developes in to veliger larva. It is a free swimming larva. The veliger larva is to be stocked at the rate of two numbers per ml of water in the FRP tanks. On day one, no feeding is required. The rearing tanks are usually covered with thick black cloth to avoid dust and light. On day two, micro algae, *Isochrysis galbana* @ 5000 cells/larva/day is fed to the larvae. Water is exchanged once in two days using 45 µm sieve. This is continued till 10th day. Between 10th and 12 th day, veliger reaches umbo stage, At this time feeding is doubled 80 µm sieve is used for water exchange. On 15th day, umbo stage reaches eyespot stage. At this time water exchange can be with 100 µm sieve. On day18, when pediveliger stage is reached feeding has to be 15000 cells/larva/day. On day 20, plantigrade stage is reached when water exchange is done using 140 µm sieve. On day 24 plantgrade transforms into spat when isochrysis @ 20, 000 cells/spat/day is given upto 30th day. Water exchange is done using 180 µm sieve.

Spat will settle to the sides and bottom of the rearing tank. Feeding is increased to 30000 cells/spat/ day and aeration is also done. By 45th day spat grows to 1000 µm and feeding dose is increased thereafter to 50000 cells/spat/day upto 60th day. At this time mixed algae cultured out doors containing chaetoceros is mixed with isochrysis is given at the ratio 50 : 50. Chaetoceras is considered good for the spat. It grows to 2-3 mm size and at this stage isochrysis feeding is stopped and only mixed algae is given. By 90th day spat becomes 3-5 mm. It is transferred to farm for further culture.

Algal Culture Unit

Isochrysis is the microalgae required for feeding the larval stage upto 60th day of rearing. Hence, facilities are to be made for its mass culture and also for mixed algal culture as mixed algae is given as feed during the later stages.

Preparation of Stock Culture

Preparation of stock culture to be used as inoculum for mass culture is the first step in algal culture.

Preparation of nutrient solution

A. Potassium nitrate - 100 g
   Sodium orthophosphate - 20gm
   Sodium EDTA - 45 g
   Boric acid - 33. 4 g
   Ferric chloride - 1. 3 g
   Manganese chloride - 0.36 g
   Distilled water - 1000 ml

B. Trace metal solution.
   Zinc chloride - 4. 2 g
   Cobalt chloride - 4 g
   Copper sulphate - 4 g
   Ammonium molybate - 1. 8 g
   Distilled water - 1000 ml

C. Vitamin
   1. Dissolve 200 mg vitamin B_1 (Thiamine) in 100 ml dist. water and refrigerate.

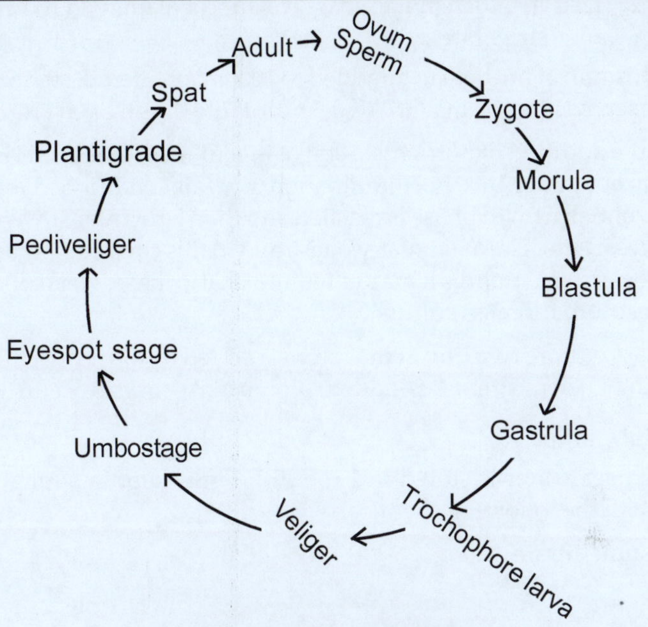

Fig. 9.2: Life Cycle of Pearl Oyster.

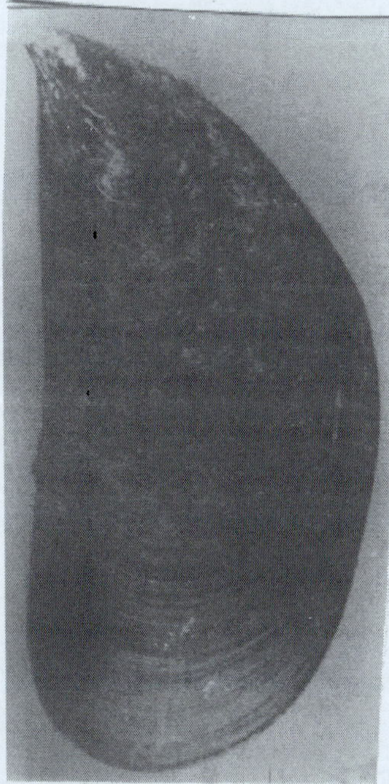

Fig. 9.3: Green Mussel, *Perna viridis*.

2. Dissolve 10 mg Vitamin B_{12} (Cyanocobalamin) in 100 ml distilled water and refrigerate

Composition of nutrient to be added to 1000ml of filtered, sterilized seawater for growth of isochysis is one ml. Of 'A'+0.5 ml. of 'B' +0.1 ml of C (1) and 0.1 ml of C (2)

After the addition of nutrients, 10ml of inoculum of Isochrysis is added and the medium is kept under 1000 lux illumination in an air conditioned room. Within 8-10 days exponential growth phase is reached and then light intensity is reduced to half. After 15 days it enters into stationary phase and in this condition it can be maintained for 2 months. After 2 months it may enter into death phase. Hence, before that this is used as inoculum for mass culture.

For mass culture two litre of inoculum is added to 100 litre of filtered, sterilized sea water to which nutrients are added. Bloom of isochrysis occurs within 5-6 days.

Mixed Algal Culture

Filtered sea water is stored in 1 ton FRP tanks kept in sunlight. The water is fertilized with the following chemicals.

Potassium nitrate	-	13. 2 g
EDTA .		6. 6 g
Sodium orthophosphate	-	6. 6 g
Sodium silicate (in 1000 litre of sea water)	-	6. 6 g

Algae for indoor mass culture is used as inoculum. The mixed algae shall bloom in 3-4 days.

SEED PRODUCTION OF MUSSEL, *Perna viridis and P. indica*

Mussel culture is promoted in the coastal areas of many countries like Thailand, New Zealand and Philippines. During 1999, farmed mussel production in the world was 1451032 mt. worth US $ 524536, 000.In India, Kerala coast is considered as the mussel fishery zone. About 555 ha of green mussel bed with an estimated potential of 15887 t and brown mussel bed with a standing stock of 1586 t reported to exist in Kerala.

There are three species of mussels of the genus Perna such as *Perna viridis* or green mussel (Fig. 9.3), *Perna perna* and *Perna canaliculus* (green-lipped mussel), Along the Indian coasts *P. viridis* and *P. indica* (brown mussel) have been reported. According to Vakily, J. M. (1989) *Perna perna* is synonymous with *P. indica* Being a ciliary - mucoid felter feeder it feeds on phytoplankton, zooplankton and detritus of upto 0.46 mm dia meter and hence is considered a highly suitable candidate for culture in coastal areas as it economically utilizes the primary production available there. It has also a high protein content up to 67g/100g dry weight. Regular seed supply is an important factor in the expansion of this bivalve culture. Natural spat collection could not meet the need of this sector and hence during recent years, hatchery reared spat is made available for mussel culture.

Reproductive Biology

The sexes are separate. Testis is creamy white in colour while the ovary is pink or reddish.

Size at First Maturity

Perna viridis (green mussel) is a faster growing species than *Perna indica* (brown mussel). It has been also found that cultured species grow faster than natural ones and in the open sea it grows faster than in bays or lagoon. *P. viridis* attain sexual maturity at a very small size such as 15.5 to 28. 0 mm length *i.e.* 2-3 months old.

Spawning Season

It varies in different geographical areas. In Kakinada Bay, *P. viridis* spawns during December to July, peak season being January to May. At Madras harbour and Goa, it spawns through out the year. At Calicult in Kerala, the spawning season is from July to November, peak being August-October. At Ratnagiri, Maharashtra it has been shown to spawn twice in an year once from June to early September and another from February to March. At vizingam in Kerala, P. indica has been shown to spawn from May end to September, peak being July-August

Larval Development

Mussels release sperm and eggs freely into the water and fertilization is external. The zygote undergoes cleavage and a morula is developed within 1½ hour. At 6-7 hrs it grows to trochophore larva performing movement by the cilia. This is developed to veliger after 17-20 hours with the development of a D-shaped ciliated velum and it start feeding. Veliger grows to umbo stage aproximately 20 hours after fertilization. It may remain at this stage for a few days (15-20 days) performing locomotion by means of velum. As metamorphosis starts it develops a pedal organ when it is called pediveliger larva. Pediveliger has also got a clearly visible eye spot. Pedal organ is used by the larva to search for a suitable substratum to settle. The larva attaches itself to solid objects by byssal threads. Secretion of adult shell starts. The young one, called plantigrade is the spat. Spat settlement takes place from 20th day onwards and continuous for 5-9 days.

Spat Collection from Natural Spawning Ground

During the spawning season, when there is an onset of spatfall, spat can be collected by providing spat collecting materials in water. In Thailand and Philippine, bamboo poles/wooden stakes are used by the farmers. The spat attaches to those objects easily. Another popular type of spat collectors are ropes made of either coir or from synthetic material such as polypropylene or polyethylene. Ropes made of natural fibre like cover is very successfully used for collecting the mussel spat because of their hairy and crevised nature. There was poorer settlement of spat on smooth ropes. But the disadvantage with coir rope is that it has lesser longevity in water where as synthetic fibre can be used for several years. Several modifications of the rope system has been devised for better efficiency of spat collection.

(1) Poly coco ropes: It consists of a main polyethylene rope, 14 mm in dia, to which coconut coir ropes of 40 mm dia and 30cm long are tied. These are

Fig. 9.4: A Cluster of Green Mussel Grown on Rope.

Fig. 9.5: Venerid Clams of Commercial Importance.

positioned at the middle of each meter length of the main rope (Cheong and Lee, 1984)

(2) In Philippines, a polypropylene or polyethylene rope with sickle shaped pieces of husk stripped from coconut shell inserted at 20cm intervals between the lays of the rope is used for spat collection. Heavy object is tied to the end of the rope to prevent it from floating. Additional bamboo cross piece are also used to give support to the rope.

(3) 'Christmas tree rope' used in New Zealand consist of a three strand, 12 mm diameter polypropylene rope of black colour. One of the 3 layers is covered in fibriliated sacking offcuts. When this rope is immersed in water, surface available for mussel larvae to settle is increased several time.

(4) Tiles suspended from rafts, used tyres suspended in water, or polyethylene netting are used as spat collectors.

(5) It is necessary to season the spat collecting material by immersing in sea water for at least 2 weeks. Settling of young mussel (spat) is successful only if it is seasoned. It is due to the fact that during seasoning other marine organism attach to the substrate prior to spat fall. According to some (Korringa, 1976) presence of barnacle is critical for settlement while other workers (Cheong and Chen, 1980) consider the establishment of a film of primary settlers such as algae and hydroids as necessary pre-requisite for the mussel spat to settle.

Spat Transfer

When there is a heavy spat fall, spat settles on the rope in a clustered pattern and at high densities. During such times, the spat has to be transferred (thining or receeding) to ensure uniform and maximum growth of mussels. This is done by stripping the mussels from the spat collectors and redistributing them to other culture ropes. This is to be done when mussels are not more than 15 to 20 mm long. The spat are detached and placed on cotton netting/mosquito netting and then the same is bound around the culture ropes. Optimum density has to 200-300 per meter for *P. canaliculus* and 250-350/800-1000 for *P. viridis*. By the time cotton netting disintegrates (within 2 weeks) spat will attach themselves to rope.

Hatchery Technique

Mussel hatchery technology has been developed in countries like China, Tahiti and Thailand. In India, CMFRI at Madras and N. I. O Goa have developed technology for the production of spat of green mussels and Vizhinjam centre of CMFRI has developed the spat production technology for *P. indica.*

Brood Stock Rearing

Culture of mussels can be done in various ways as noted below for developing a healthy brood stock.

(a) Stake culture

(b) Rack and bag culture

(c) Raft culture

(d) Longline culture

In stake culture, the spat is grown on poles (teak wood/bamboo) fixed to the bottom of the sea where the water depth is 2-5 meter in the intertidal zone. Either naturally set spat is allowed to grow or spat collected in rope are placed in plastio net tules and spirally wrapped around the pole and allowed to grow till it reachs the marketable size (6-8 months or even upto 12-18 months). In rack and bag culture, mussel seed of 10-15 mm length are loaded in flexible synthetic mesh tubing and suspended from racks in shallow waters. In raft culture, seeded rope are suspended from a raft. Raft is made up of teak wood or bamboo pole placed parallel and across tied with coir rope to make a rigid frame. This frame can be made to float in the water by tying 4 air tight barrels of 200 liter capacity at 4 corner or by floats. Raft is a floating structure where as rack is fixed. The raft is moored at the bottom by anchor. From the raft rope with seed are hung, each rope 0.5 to 1m apart. 10m long rope yields 100 kg of mussels (Fig. 9.4) and the production per raft is 50 tones. The long line system consists of a main line, floats, anchors, marking buoys and connecting rope. About 150 meter long, 16-20 mm dia synthetic rope forms the long line, which is connected with barrel which function as the float and concrete blocks functioning as anchor. Ropes with mussel seed are suspended from the main rope.

Spawning and Fertilization

(a) *Inducing spawning through thermal shok*: A rise in the temperature by 4°C above the ambient temperature induces spawning of green mussels.

(b) Inducing spawning by adding Hydrogenperoxide at a concentration of 100-150 ppm induces spawning.

The mussel is left in the solution for 8 hours, and then placed in clean sea water which induces spawning within 1-4 hours.

(c) *Development and larval rearing*: For development of embryo and larval rearing optimum salinity has been recorded as 25-35 ppt and optimum temperature range as 20-25°C. Below 18°C larvae stopped feeding.

The larvae can be fed on algae Isochrysis and Monochrysis in equal proportions (25000 cells/ml) to feed mussel larvae. After settlement of larvae, the quantity was reduced to 12, 500 cells/ml and *Skeletonema costatum* was added at a rate of 50000 cells/ml. About the time spat is formed *i.e.* 15 days after fertilization, suitable substratum is provided for the larvae to settle. Heavy spat fall occurs from 20th day onward and continue for 5-9 days. Spat may measure 2.7 mm.

SEED PRODUCTION OF CLAMS

Clams alone contributed 2226025 tonnes to the total global aquaculture production in the year 1998. Clam meat is one of the major sea food items having consumer preference in several countries such as Japan, Thailand USA, Australia, Malaysia, Singapore, Kuwait, Belgium, France, Germany, Italy, Netherlands, Spain and Swizerland. Apart from the clam meat, other products such as clam juice, clam strips, clam streaks, clam pickles etc. are also of good demand. Shell of clams is used in the manufacture of cement, lime, calcium carbide etc. Among the cultivable clams belonging to families such as Arcidae (blood clams) veneridae (Meretrix, Paphia, Katylesia) corbiculidae, and Tridacnidae, *Anadara granosa*, *Paphia malabarica* and *Katelysia opima* are the favoured species (Fig. 9.5). *A. granosa* (blood clam) occurs in soft muddy substratum and altains 41 mm length in first year and it is found all along the Indian coast, particularly in Kakinada bay *P. malabarica* grows to 49. 1 mm in one year. *K. opima* attain 26-33. 8 mm in one year and are found in esturaries. Clams of commercial importance found in other countries such as Philippians, Solomon islands, Indonesia, Australia, Fiji etc. are *Tridacna gigas*, *T. derasa*, *T crocea*, *T. maxima*, *T. squamosa* and *Hippopus hippopus*. *T. gigas* is commonly known as giant clams as it is reported to grow to a giant size of 1 meter long and 250 to 300 kg weight (Hviding, E., 1993). Hatcheries for giant clams are established in the above mentioned countries. Other clams of importance are *Venerupis japonica* (Japanese little neck or Manila clam), *Mercenaria mercenaria* (hardclam)) and *Meretrix meretrix* (big clam).

Reproductive Biology

A. granosa

Males attain first maturity at 20 mm length and females at 24 mm length. It is an 'year round spawner' having 2-4 reproductive cycles a year. Fecundity is about 500000 eggs per spawner. Brood stock maybe of 3-4 cm length and 1-1. 5 year old.

P. malabarica

Size at first sexual maturity is 11 to 20 mm. It spawns from September to January in Ashtamudi lake and during October- February in Mulky estuary.

K. opima

Its size at first sexual maturity is 11-20 mm length, main spawning season is October – November in kalbadevi and a minor spawning period during March-April.

The different stages in the development is mostly similar to that of green mussels described earlier, the differences being that D-hinge veliger starts consuming food 1-3 days after fertilization in clam, umbo become clealry visible 7-15 days after fertilization and settlement stage is 21-35 days after the date of fertilization.

Hatchery Technology

Collection and Preparation of Brood Stock

Brood stock (3-4 cm size and 1 –1. 5 years old) collected from muddy substratum is washed thoroughly in freshwater. This makes the animal to keep the shell closed. The washed cockles are then disinfected by bathing them in 10 ppm sodium hypochlorite solution for 15 minutes. This kills all harmful organisms present on the outer surface of shell.

Induced Spawning

Spawning Inducement by Temperature Shock

Sea water is taken into a tank and the same is heated to 34°C –35°C or 4-5°C above ambient temperature. The clams were transferred to this tank and kept at this temperature for two hours. After this, the temperature is decreased gradually to ambient temperature or to lower temperature such as 20°C for two hours. This is heat shock-cooling cycle. If the clam do not spawn with one cycle it may be repeated. This method was very effective for bivalves. 00

Spawning Inducement by Increasing pH and Addition of H_2O_2

pH of seawater was raised to 9.1 by the addition of 10% sodium hydroxide. Then hydrogen peroxide was added to form 150 ppm solution. Clams were placed int his solution for 8 hours. After this keep the clam in a raceway with clean flowing seawater. After 1 to 4 hours, bivalves with ripe gonads spawned.

Inducing Spawning by the Injection of 0.1 N Ammonium Hydroxide into Body Cavity

1 cc of 0.1 N ammonium hydroxide is injected into the body cavity of the clam. Then it was kept in the tank. After 30 minutes it spawned.

Development and Larval Rearing

The developmental stages are similar to that described for mussels. The larvae were stocked @ 10 number per ml in 50 litre fibre glass tanks at ambient salinity *i.e.* 32 ppt. EDTA was added to the tank each day after water exchange so that its final concentration was 5 ppm. Larvae were fed on algae isochrysis sp. from one day after fertilization @5000 cells cm^{-3}. Feeding was done twice daily (morning and evening) – batch feeding or 4-5 times a day –beginning with small amount and continuing throughout the day when the water is cleared of foods – pulse feeding-the quantity was increased to 30000 cells cm^{-3}/day depending on the size of the larvae.

The spat stage is reached 21-35 days after fertilization. Cockles do not need special setting substrate as they set on the bottom of the rearing tank. After setting, growth is very rapid and within a few days, it is easy to screen off cockle spat with 280 μm or large screen to a nursery. They grow better when reared on a mud bottom.

SEED PRODUCTION OF SEA CUCUMBER

Introduction

Holothurians, commonly called as sea slugs or sea cucumbers is a group of economically important echinoderms which are consumed either fresh or boiled or prepared as beache-de-mer (trepang) by people of Japan, Korea, China etc. Body wall (namake), ovaries (Kenoko) intestine (Konowata) and respiratory tress (minowata) are considered delicacies for Japanese. FAO statistics indicate a downward trend in world sea cucumber harvest through natural fisheries and hence attempts have been made to promote the holothurian fisheries through culture of juveniles produced in hatcheries

Holothurians exhibit a world wide distribution. About 1200 species are recorded as belonging to the class Holothuroidea. Along the Indian coast *Holothuria scabra* commonly called 'sand fish' (Fig. 9.6)is the predominant species whereas in Japan and Korea *Stichopus Japonicus* predominates. Some species live on hard substrates like rock, coral reef etc. and vary in their habitats from foreshore to deep water zones.

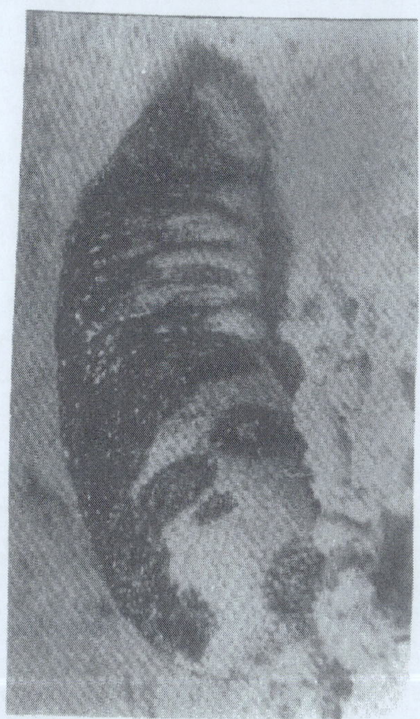

Fig. 9.6: *Holothuria scabra* (sand fish).

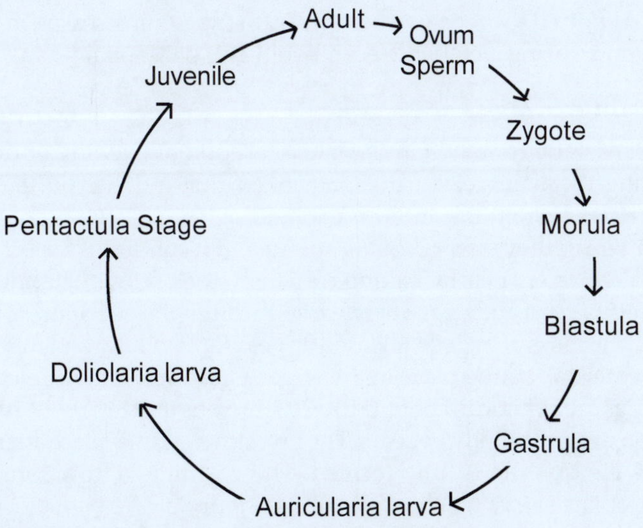

Fig. 9.7: Life Cycle of Sea Cucumber.

The other Holothurians of commercial value are *Holothuria nobilis* (black teat fish), *H. fuscogilia* (white teat fish) *Actinopyga echinites* (deep water red fish), *A. milaris* (black fish) *Thelenota ananas* (prickly red fish), *Holothuria ata* (Lollyfish) *H. fuscopuncata* (elephants trunk fish) etc.

Reproduction

H. scabra attains a length of 400 mm and weight of 500 g and lives on sandy-muddy bottom and becomes sexually mature at 18 months. The size at first maturity is 210 mm. The sexes are separate. There is no distinct sexual dimorphism. The ovary in female and testes in male are in the form of a tuft of tubules attached to the dorsal mesentary through which the gonoduct passes terminating in gonopore situated on the dorsal side near the oral region. The gonadal development is distinguishable into five stages such as immature, resting, growing, mature and post spawning phases. During the immature and resting stage, the ovarian tubules are transparent, short and thin distal end of tubules are club shaped, during growing phase the tubules are having opaque spherical oocytes 20-120 microns in diameter, during mature phase, the tubules are swollen containing ripe oocytes of 150-200 μ, during post spawning phase some tubules are seen as in mature stage, others are shorter showing atresia a few ripe oocytes are in the tubules and signs of resorption/atresia are seen.

In males during stages I and II testes is hardly distuinguishable from ovary but, during stage III, the testis tubules become more branched and whitish, length and diameter also grow, spermatids develop inside the tubules, during stage IV, tubules grow laryer in size, more whitish with swellings and spermatozoa are seen in tubules, during stage V the tubules appear evacuated in appearance with only a few spermatozoa remaining in the tubles. In general ripe tests are milky white whereas the ovaries are tansluscent.

It has been reported to breed twice in a year first spawning season in from March to May and second during October- December in gulf of Mannar

Life history (Fig. 9.7)

Holothurians usually spawn in the late afternoon or evening or during night. During spawning, the males release the spematozoa first and then the female releases the eggs. The males first lift the anterior end and perform swaying movements for sometime after which they start releasing sperm and it continues for 1-2 hours. Ripe females, if any are present nearby, exhibit responsive behaviour. The anterior region of the female gets bulged and eggs are released through the gonopore forcefully in a continuous jet. The mass of eggs so released appear light yellow and mucus like. The fertilization is external, taking place in the water. One adult female releases about 1 million eggs. Eggs are spherical, about 180 to 200 microns in size. About 20-30 minutes after fertilization first polar body appears. The first cleavage take place after 15 minutes and in 3 hours the blastula is fully formed. The gastnula is fully formed after 24 hours. The auricularia larva hatches out after 48 hours

The auricularia larva [Fig. 9.8(a)] is transparent, pelagic and measures 560 microns in size. It is slipper shaped and performs locomotion by the movement of flagella of the ectodermal cells that forms ridges or bands. Along the sides of the larva

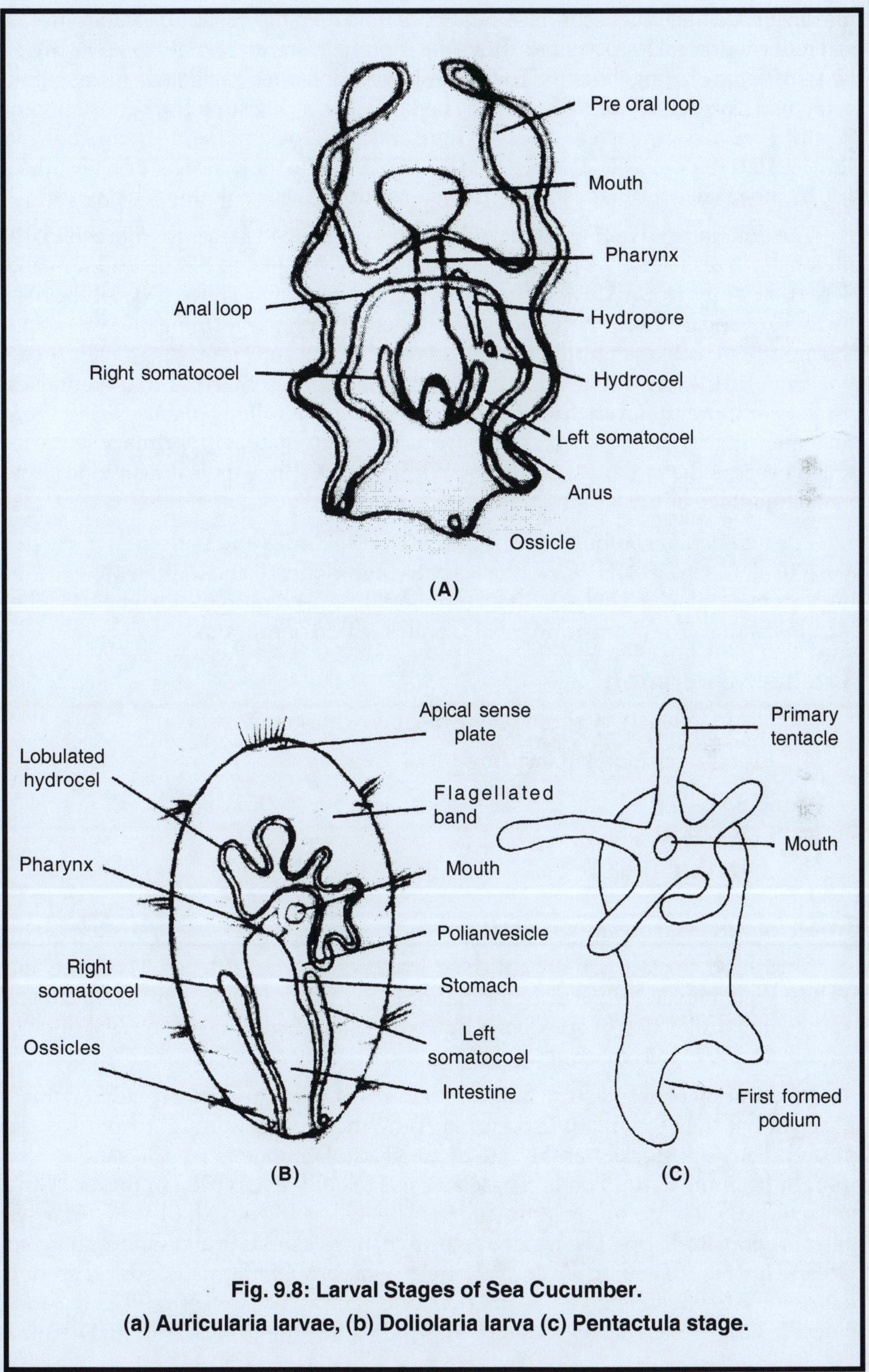

Fig. 9.8: Larval Stages of Sea Cucumber.
(a) Auricularia larvae, (b) Doliolaria larva (c) Pentactula stage.

this flagellated band proceeds in sinous curves and then dips ventrally at the anterior part to form preoral loop located above the mouth. Posteriorly it ciurves forward as the anal loop encircling the anus. These bands were described as ciliated but according to Hyman (1992)they are flagellated. The larva has a digestive tract consisting of mouth, pharynx, stomach and anus. There are three coelomic sacs –hydrocoel and right and left somatocoels. The hydrocoel is connected to the hydropore on the dorsal side by means of a tubular canal. It grows to doliolaria larva within ten days.

The doliolaria larva [Fig. 9.8(b)] which is barrel shaped measures more than 620 microns in size. There is a partial degeneration of flagellated bands leaving short pieces that joins to form 3 to five numbers of flagellated rings. The hydrocoel encircles the fore gut and eventually, the ends meet together and fuse forming the water ring. The mouth is located ventrally which leads to pharynx and stomach. The hydrocoel that encircles the fore gut develop five lobes which represent the primary tentacles. Immediately after this a sixth lobe is developed which is called polian vesicles. From the water ring five more processes grow out that alternate with primary tentacles which become the radial water canals. Within two to three days it developes in to pentactula stage [Fig. 9.8(c)].

The pentactula is tubular and it creeps over the sides and bottom. It measures 600-700 micron in size. It has 5 tentacles at the anterior end, and a single tube foot at the posterior end which helps in its locomotion. They feed on benthic algae and other detriitus matter. They transform into juveniles within one month.

Hatchery Operation

Holothurian hatchery should have the following components :

(a) Brood stock maintenance unit

(b) Spawning unit

(c) Larval rearing unit

(d) Algal culture unit

Brood Stock Maintenance Unit

The brood stock is usually collected from commercial catches. The large and healthy specimens which are not injured or eviscerated during capture are chosen for breeding. Collection of breeders is done during the breeding season *i.e.* March to may or October to December.

Preparation of brood stock tanks. FRP tanks of 1 ton capacity provided with 6″ thick sand at the bottom are used for keeping the breeders brought from natural ground. They are stocked at the rate of 20-30 adults in one tank. The sand is also brought from the natural beds. These animals usually live buried in the sand and hence the sandy bottom is recommended. The tank is filled with filtered, clear sea water of about 32-35 ppt. The water has to be changed everyday and sand is changed once in a fortnight. Feeding is done with fresh algae brought from the sea and ground to a fine paste which is given in the tank once in a week. Excess food may cause waters fouling. In case of any water fouling, the sea cucumber eviscerate and become useless for breeding. As algal paste settles to the bottom, the sea cucumbers ingest the same with sand. If the feeding is not proper the animals get shrunken and the gonads are reabsorbed making the specimen not fit for spawning purpose.

The brood stock rearing tanks are kept in an air conditioned room to maintain a low temperature of 18-25°C.

Spawning Unit

Spawning is carried out in rectangular FRP tanks of about 100 litre capacity. The provision for an immersion heater with thermostat, Jumo thermometer and aerator are provided in the tank for thermal stimulation of spawners. After introducing the spawners into the tank having filtered, clear and clean sea water, the temperature of the water is raised by 3°C –5°C by using the immersion heater. This thermal stimulation, induces the sea cucumbers to spawn. This is the most widely used and most reliable method to induce the holothurians to spawn.

Apart from the thermal stimulation there are three other ways to do the breeding of holothurians. These are (1) natural spawning (2) stripping and (3) stimulation through drying and powerful jet of water

Natural Spawning

The male and female may release the gametes into the surrounding water without any artificial stimulation.

Stripping

This is done mainly on an experimental scale only. The animals are cut open from cloaca to mouth through the dorsal side. The ovary which is transluscent is taken out from the female and the same is slighly dried in a shade. It is then placed in sea water in a petridish and punctured with the scissors to release the eggs into the sea water. In the same way, the testis is taken out and cut into pieces. When the sperm move out in the water it is mixed with the eggs kept in a beaker with seawater. Mild aeration failitates higher rate of fertization.

Stimulation Through Drying and Powerful Jet of Sea Water

The breeders conditioned for more than a weak in the hatchery are utilized for this purpose. First of all water in the brood stock tank is removed and specimens are dried in a shade for about half an hour. After this a powerful jet of water is sprayed on the specimens for a few minutes. Then the animals are put back into the tanks with sea water. After 1-2 hours, the animals move up the tank wall and exhibit swaying movement indicating the release of gametes. First the males release the sperm and after one hour the females release the eggs.

Larval Rearing Unit

After the spawn and eggs are released, the breeders are removed from the tank carefully. The fertilized eggs are removed to the rearing tanks. The auricularia larvae hatch out after 48 hours. the healthy larva occupies the surface layer of water whereas dead or deformed one settle at the lower layer of water column or at the bottom of the tank. Those settled at the bottom are siphoned out. The healthy larvae are collected in a sieve and counted using a plankton counting chamber. Then the larvae are released into the rearing tank containing clean, clean filtered sea water at density of 300 to 700 numbers per litre. The larvae are taken out once in 3 days to clean the tank to avoid infestation of other organisms. On other days, the water level is reduced to more than half by keeping the sieve inside the tank taking care to prevent the escape of larvae along with siphoned out water

The larva is fed on micro algae *Isochrysis galbana,* two times a day. About 20000 to 30000 cells per ml of rearing tank water is sufficient initially. This may be increased ior decreased depending on the stage of larvae. After 4-5 days, the larvae may be fed with mixed culture of phytoplankton mainly having chaetoceros

Environmental Factors Affecting Larval Rearing

Temperature

The ideal temperature is reported to be 27°C-29°C.

Dissolved Oxygen

Aeration is carried out in larval rearing tanks to maintain oxygen at saturation level.

pH

pH of the water may be between 6 and 9.Normal sea water has a pH of 7.5 to 8.6 which is suitable

Salinity

Normal sea water salinity (32 to 35 ppt) is favourable. If salinity falls lower than 12.9ppt. the larvae shall die. Most favourable salinity has been shown to be between 26 to 32. 7ppt.

Ammonia

Ammoniacal nitrogen should not exceed 500 mg per cubic meters. It tolSerates a range of 70-430 mg/m^3.

Under the above conditions of rearing, the auriclaria develops to Doliolaria larva on 10th day. The doliolaria larva transforms into the pentactula larva within 2-3 days. The pentactula is the creeping stage. Late Doliolaria larvae settle on hard surface provided suitable substratum is provided in the tank. Hence, artificial settling bases (settlers) are provided for them to settle. 2 types of settling bases are tried (1) Polythene sheets are taken and kept in a tank kept outdoors having enough sunlight. For 4-5 days, filtered sea water is circulated continuously. Benthic diatoms and other algae settle on these sheets. These are kept suspended in the water having late doliolaria larvae which are about to settle. The larvae settle on them as it gets food and hard surfaces. In another type of settlers used, the polythene sheets are kept in a tank having sea water. In this, filtered algal extract (50µ filter) is added. Usually species of algae sargassum are used to make the extract. Algal extracts shall stick to the sheets. Fresh extract is put daily and the water is also changed daily. After 4-5 days. when the sheet is covered with algal extract it is given as settling base for the larvae. Juveniles that settle to the hard substrate has very weak motility. Hence, algal extract is again given daily twice - morning and evening- which is filtered through 40 microns seive. After one month, 80 micron seive can be used. After one month, large size juveniles of 15-20 mm size are separated and put in tank with very fine sand. There are also fed on algal extract. Optimum density of larvae should be adjusted to 200-500 individuals per square meter.

Algal Culture Unit

The hatchery should have an algal culture unit to provide sufficient quantity of desired species of algae such as *Isochrysis galbana, Dicrateria* sp., and *Dunaliella* sp. In addition mixed culture predominated by chaetoceras also shall be required. For details of algal culture please refer to the section dealing with seed production of pearl oysters.

Chapter 10

Genetic Upgradation of Fish Seed

INTRODUCTION

Increase in productivity from unit area is the primary objective of any aquacultural research. Hitherto scientists have tried to achieve this objective mainly through the manipulation of the environment. However, every technology has its own limitations, after certain level of production it may not be economical to manipulate the environment to increase productivity. Moreover, intensive culture system involves large quantity input of feed and fertilizer. A consequent probable effect shall be imbalance in the pond environment. In this situation, increase of productivity could be possible through introduction of genetically improved varieties of fish. Genetics played a very vital role in enhancing production in agriculture as well as in animal husbandry. The high yielding varieties of different crop species, increasing yield of milk, egg and flesh of different animals are testimony for this. It was possible through the rational application of genetics in these fields, which not only improved the production level but also improved the quality of the product. In aquaculture, particularly in India, the potentials of genetics are not fully exploited. Application of aquaculture genetic technologies has not been done so far in its true sense for enhancement of production. However, after the success of induced breeding of Indian Major carps, genetic improvement work has been initiated starting with simple hybridization, effective selective breeding to the modern genome manipulation and gene transfer techniques (Reddy, 1999)

HYBRIDIZATION

Hybridization is a simple means to improve the genetic status. By combining the haploid genome of two different parent species belonging to either same genus (inter-specific) or two different genera (inter-generic) and some times even between two strains of the same species (intra-specific) hybrids can be produced.

Several workers have reported the occurrence of fish hybrids in nature from time to time. Slastenenko, 1957 has given a list of natural hybrids of the world which suggested that species belonging to the cyprinid group are more prone to interbreed than other groups of fishes and Indian Major Carps belong to the same group. Several causes have been suggested to account for the occurrence of hybrids in nature. Hubbs, 1955 hold the view that hybridization in nature is facilitated when there is scarcity of one species and dominance of an allied species in close proximity. In bundh breeding occurrence of natural hybrids may be due to congestion in the spawning ground. Due to limited space, there is every likelihood of ova of one species getting accidentally fertilized by the sperm of another species.

Hybridization work among Indian Major carps, between Indian Major Carps and Chinese carps and among Chinese carps has been initiated with a view to study the positive or useful traits like smaller head, wider body, more flesh content, pond breeding habit like *C. carpio*, advantageous feeding habits etc. Initially six inter-specific and 13 inter-generic hybrids have been produced for evaluation.

Inter Specific Hybrids

Inter-specific hybrids are obtained by crossing individuals of same genus but different species. Most of the inter-specific hybrids of Indian Major Carps show intermediate characters to their parental species. Some of the important inter-specific hybrids and their characters are discussed below.

Labeo rohita* × *L. calbasu (female rohu with male calbasu) and L. calbasu* × *L. rohita hybrids (female calbasu with male rohu)**

These hybrids show high percentage of fertilization and hatchability. The offspring of both the crosses are highly viable and their growth rate was superior to one of the parent *i.e.* slow growing *L calbasu.* Hybrids of both the crosses are found to be fertile and attained maturity within two years (Reddy, 1999)

***L. fimbriatus* × *L. rohita* hybrid**

The growth of this hybrid was evaluated by Basavaraju *et al.* (1990). In a 182 days rearing experiment the hybrid growth was 34. 75% higher than fimbriatus and 23. 55% less than Rohu. The hybrid also showed higher survival *i.e.* 90% compared to Rohu with 86. 67% and fimbriatus 76. 67%. In terms of percentage of weight gain, specific growth rate and feed utilization the hybrid performed better than both the parent species.

Other inter-specific hybrids like *L. calbasu* × *L. bata, L. gonius* × *L. calbasu, L. rohita* × *L. bata* and vice versa did not exhibit any useful trait. They exhibit poor hatching and slow growth rate.

Inter Generic Hybrids

Within Indian Major Carps (Fig. 10.1)

Hybrids obtained by crossing of individuals from two different genera are termed as inter-generic hybrids. Like inter-specific hybrids inter-generic hybrids also show intermediate characters to their parents. Out of 13 inter-generic hybrids only four crosses *i.e.* Rohu × Catla, Mrigal × Catla, Rohu × Mrigal, Fimbriatus × Catla have been found to possess useful traits in terms of growth (Reddy, 1999). In all the cases

---

* While representing a hybrid, first name indicate 'female' and second name 'male'.

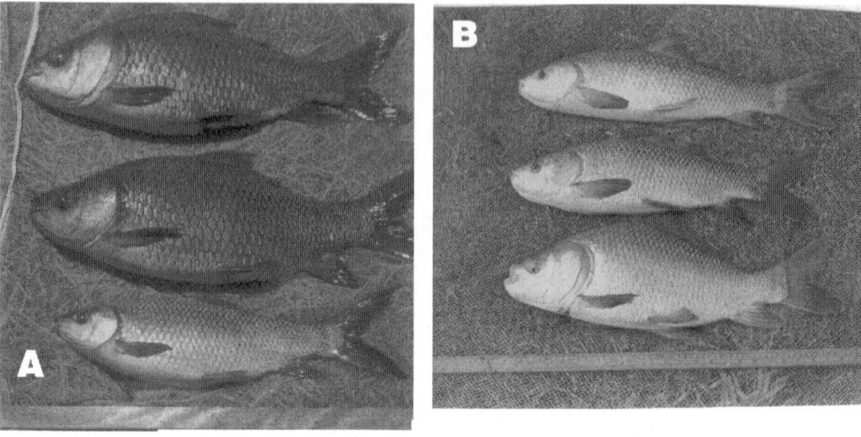

Fig. 10.1: Hybrids of Indian Major Carps. Hybrid is Shown inbetween Parents.
(a) Catla-Rohu (Female Catlax × Male Rohu),
(b) Rohu-Catla (Female Rohu × Male Catla),
(c) Kalbasu-Rohu (Female Kalbasu × Male Rohu).
(Courtesy, CIFA, Kausalyaganga)

hybrids are observed to be better than one of the slow growing parents. Rohu × Catla and vice versa shows positive trait in terms of dressing percentage 58% compared to Rohu (52%) and Catla (48%).

Between Indian and Exotic Carps

Common carp *(C. carpio)* and Indian major carps hybrids show some positive traits for culture practices. All these hybrids are reported to be sterile, as IMC chromosome number is 2n=50 while *C. carpio communis*, chromosome number is 100 and in the hybrids, because of aneuploidy condition, sterility occurs. Sterile common carp has its own advantages in culture ponds as unwanted increase of population due to pond breeding habit can be checked.

However almost all fish hybrids exhibit intermediate heterosis effect *i.e.* superior to one of the parents. Hence any hybridization program should consider this aspect and careful evaluation has to be made with regard to the type of hybrids or progeny that results from a particular hybrid cross. Another fish that has been used extensively for hybrid generation is Tilapia (for details of tilapia hybrids chapter on Tilapia may be referred).

Hybrid Index

To evaluate the tendency of hybrid trait towards the parents, the hybrid index calculation appears to be effective. Formula of Hybrid index is

$$Hi = (Hv - M1) / M2 - M1 \times 100$$

Where,

Hi = Hybrid index

Hv = Hybrid value

M1 = Value of maternal parents

M2 = Value of paternal parent

Impact of Hybridization

Adaptability of any hybrid to the environment is determined by genetic introgression *i.e.* flow of genes from one species gene pool to another species. If introgression is < 0.1% it may help in increasing capability of adaptation against natural selection. A large amount of gene flow may disrupt the adoptive gene complexes, which have evolved overnight to permit a species to effectively use its particular environmental niche, so while releasing any hybrid to the nature, care should be taken to evaluate it properly. Reduction of genetic diversity can also occur because of indiscriminate hybridization.

CYTOGENETICAL STUDIES

The study of chromosome started in the last decade of 19th century, has been widely used to ascertain the taxonomic position of different fish species. However, modern cytogenetics is also helpful to determine genetic relationship between different species for chromosome manipulation and banding pattern studies.

Chromosomes

Chromosomes can be classified to different types on the basis of position of centromere and its arm. The terminal end of chromosome is called telomeres, which confer stability to their ends. The portion of the chromosomes on either side of the centromere is called the chromosome arms. The total number of chromosome arms in a karyotype is called fundamental arm number. Ratio of arm length determines the type of chromosome.

Arm ratio = Length of long arm/Length of short arm

Chromosomes are divided in to 4 types according to position of centromeres and arm ratio. They are as follows:

Sl. No.	Type of Chromosome	Centromere Position	Arm ratio
1	Metacentric (m)	Median or near centre	1. 0-1. 7
2	Submetacentric (sm)	Mid way between centre and end of chromosome	1. 71-3. 0
3.	Sub-telocentric (st)	Subterminal	3. 01-7. 0
4.	Telocentric/Acrocentric (T/A)	Terminal	7. 01 and above

The complete set of chromosomes arranged according to their size is known as karyotype. The karyotype of a normal individual usually consists of two sets of chromosome and derives one set of chromosome from each parent. Karyotype of different species differs in shape, size and number of chromosomes. Within a species male and female may differ due to difference of sex chromosome (one pair). They are known as sex chromosome, while rests are known as autosomes

Chromosome Banding

Through chromosome banding technique it is possible to make precise identification of individual chromosome or even part of chromosome, their structural and molecular organization, chromosome change during evolution, detection of change in karyotype, allopolyploid and anueploid identification and detection of interspecific translocation.

Chromosome banding technique is also useful in clinical, somatic cell genetics or chromosome mapping. "G", "C", "Q" and NOR banding are some of the banding pattern used for identification of different fish species and mapping of chromosomes. By careful analysis of banding sequences, translocations, rearrangement, pericentric inversion, centric fusion, addition or deletion of chromosomes have been detected.

'C' Banding Technique

'C' banding is the most common banding process available for fishes. Most species invariably possess C-bands. By C-banding technique, the centromeric hetrochromatin containing highly repetitive satellite DNA is identified as distinct darkly stained zone on the chromosome. This form of banding demonstrates a specific type of heterochromatin. The procedures employed for the display of C-band are based on the principle of partial extraction of non C-band DNA and retention of C-band DNA on the chromosomes. The C-band DNA is presumably protected during extraction being highly repetitive and intimately bound to certain non histone protein. Generally C-band appears near the centromeres and frequently at the chromosome tips (telocentric). C-band may also be found along chromosome arms and as entirely heterochromatic short arms of acrocentric chromosomes. The function of constitutive hetrochromatin is essentially unknown.

About 3-4 days old chromosome slide are hydrolysed for 15-20 minutes in 0.2 N HCL at 37^{0}C, denatured in a saturated Ba $(OH)_2$ solution at room temperature for 2-7 minutes, then incubated at 65^{0}C for 90 minutes in humidity chambers. Chromosome staining is carried out on air dried slide using 2% Giemsa for 10 minutes.

'G' and 'R' Banding Technique

Dark and light bands can be seen over the chromosomes by treating the chromosomes with dilute trypsin or hot saline citrate and staining with Giemsa. The dark bands along the chromosome contains facultative heterochromatin region, popularly called G-band and the euchromatin region, popularly called as 'R' bands. The G-bands are heterochomatic AT-rich region and R-bands are GC rich region. However, G and R bands are most often not reproducible.

'Q' Banding Technique

When the chromosome slides are treated with 5% quinacrine dihydrochloride (0.2g/40ml distilled water) for 7-15 minutes and rinsed for 5 minutes in running distilled water then the slides are observed under fluorescence microscope and chromosomes exhibit bands of fluorescence. The brightly fluorescing regions or segments are called Q-bands.

NOR Banding Technique

Certain chromosomes are specifically associated with nucleolus formation. The specific association sites of these chromosomes are termed as nucleolus organiser regions (NORs). The NORs represent secondary constriction, which are not recognisable by using staining procedures. Since specific sets of chromosomes are associated with NORs, this technique provides additional parameter for identification of chromosomes (Sahai *et al.*, 1989). Nuclear Organizer Regions (NORs) are identified using silver staining technique. In this process the metaphase spread chromosome slides are placed on a slide warmer and the temperature is fixed at 50^{0}C. A colloidal developer solution (gelatin) is placed over the slide. The silver nitrate solution (0.5g/ml) is added to the slide. A cover slip is placed over the slide to allow the mixing of the developer and silver nitrate solution. The slides are allowed to remain on the slide warmer till the colour turns to light brown or golden brown. Then they are washed under running distilled water after dropping the cover slip. The slides are allowed to air dry before they are stained in 2 % Giemsa. NORs and nucleoli have the highest amount of silver deposition and are stained as dark brown or black. Chromosome arms are light brown or yellow.

Mechanisms of Chromosome Banding

On the basis of banding patterns and other observations there are three major types of chromatin: Centromeric constitutive heterochromatin, corresponding to C-bands; intercalary heterochromatin, generally corresponding to G-bands and euchromatin, corresponding to R-bands. Experiments with acid-treated chromosomes indicate that non-histone rather than histone proteins are primarily involved in banding. The heterochromatic segments are consists of repetitive sequence of DNA composed exclusively AT and GC nuclotides, or of repetitive sequence where the relative proportion of AT: GC nucleotides may be biased in either way. Q banding is due to a difference in base composition of DNA along the chromosome. AT base pair sequences enhance Q-fluorescence where as GC rich base pairs effectively quench it. It is also suggested that Q and G bands are result of dye binding to adjacent sites on folded molecules of native DNA, the appropriate conditions necessary for such binding depends to a great extent on the relationship between DNA and its associated non histone proteins. The major factor in C-banding appears to be the presence of non-histone proteins that bind specifically to centromeric heterochromatin and protect it from extraction by barium hydroxide and salt (Lavania, 1980)

Genotoxic Study

Sister Chromatid Exchange (SCE), chromosome aberrations and micronuclei formation can be used in very sensitive genetic assays for detecting environmental mutagens and genotoxic pollutants even at sub-toxic level (John and Reddy, 1989). Living organisms, particularly fishes in the water are in stresses due to use of insecticides, pesticides etc., which may cause instant mortality or genetic damage to the living ones. In vivo sister chromatid exchange is a short-term test to measure the mutagenicity of water borne chemicals

CHROMOSOME/GENOME MANIPULATION

Genome refers to the chromosome complement of a given individual. Through genome/chromosome manipulation a new genetic status can be attributed to an individual by altering set (s) of chromosome. Chromosome manipulation involve addition of extra set (s) of chromosome to the existing set of chromosome, resulting in triploid or tetraploids individual or replacement with a duplicate set of chromosome from the same individual, the resultant product being either gynogens with maternal inheritance or androgens with paternal inheritance.

Gynogenesis

Gynogenesis is the process of embryonic development with solely maternal genome and without paternal genetic input. It is a specialized form of parthenogenesis wherein embryos develop after activation of egg by the genetically inactivated sperm but there is no genetic contribution from the paternal genome. However, the resulting zygotes are haploid. Restoration of diploidy may occur spontaneously in natural gynogenesis or can be induced through shock treatment (thermal/pressure) in artificial gynogenesis .

Natural Gynogenesis

In some fish species of family Poecilidae *i.e. Poecilia formosa* and Cyprinidae *i.e. Carassius auratus gibelio* gynogenesis has been the method of reproduction. Natural gynogenesis has also been reported among members of Pleuronectidae family.

Induced Gynogenesis

Gynogenesis can be artificially induced by eliminating/denaturing the genetic material (DNA) of the sperm through irradiation either by UV or Gamma rays or chemicals and activating the eggs with irradiated milt. Giving thermal (cold and heat) or hydrostatic pressure shock can restore diploidy.

Methodology to Induce Gynogenesis

Ovaprim/pituitary injection to matured male and female brood fishes is done to obtain gametes as in induced breeding technique. The milt may be stripped (preferably from another species) an hour before the expected ovulation time of the female fish. The advantage of using milt from different species is that it will be easier to distinguish gynogens from hybrids. The eggs, which will be fertilized by unaffected spawn by UV irradiation, will develop in to a hybrid.

After stripping the male, milt may be mixed thoroughly with freshly prepared Hank's solution in a ratio of milt one-part and Hank's solution 4 parts.

Then milt may be divided in to two parts. One half may be kept as control (normal) under refrigeration. The second half may be poured in to small petridish to a 1-2 mm thick column.

Then the petridish may be placed on an ice tray over the mechanical shaker in the UV chamber. The UV lamp and mechanical shaker may be operated at a medium speed (care may be taken so that the milt in the petridish gets properly and continuously mixed by the action of the shaker).

The irradiation of milt process may be continued for 17-20 minutes in case of Indian Major Carps and 35-40 minutes in case of common carp for complete irradiation. After completion of irradiation, milt may be stored in the refrigerator at 4-5^{0}C.

Then the female brood fish may be stripped and eggs may be divided into four groups. After thawing the milt properly, two groups of egg may be mixed for activation of egg in presence of little water. Another group of egg may be mixed with normal milt of same species to know the quality of egg used for the experiment and the fourth group of egg may be fertilized with the donor species milt to know the quality of milt.

After activation, egg may be washed gently and allowed to develop in a mixer of tap and filtered pond water.

One group of egg activated with irradiated milt may be given thermal/pressure shock to restore diploid nature (due to absence of paternal genome in those eggs, only maternal genome (n) is present).

Other group of eggs with irradiated milt activation may be kept as such without any shock treatment. These embryos will give us an idea whether the genetic inactivation of milt is adequate or not. If all the embryos exhibit haploid syndrome with curved or bent caudal region then the irradiation is proper.

Types of Gynogenesis

Restoration of diploidy in a haploid gynogen egg can be achieved in two ways. In the first case it can be achieved by suppressing metaphase-II in the second meiotic division in other words by preventing the extrusion of the second polar body (Fig. 10.2). This type of gynogen induction is known as meiotic gynogenesis.

In the second case gynogenesis can be achieved by blocking the first cleavage (Fig. 10.3) and they are termed as mitotic gynogenesis.

Both meiotic and mitotic gynogenesis was successfully induced in *Catla catla, Labeo rohita, Cirrhinus mrigala* and *L calbasu*. The first report on successful induction of mitotic gynogenesis was reported by Reddy *et al.*, 1993 for inducing catla and Rohu with the help of milt from common carp and mrigal respectively. The yield of mitotic gynogens as reported by Reddy *et al.*, 1993 was 15-30% up to fry stage and 0.66-10% up to fingerling stage with heat shock and 30-4. 3% up to fry and fingerling stage respectively through cold shock treatment. The average survival rate ranged from 1.13% with heat shock to 4.3% with cold shock. Success of induction of gynogenesis is species specific to type of shock treatment. In case of Rohu, yield of gynogens is better in cold shock than heat shock. However, yield of meiotic gynogens in Indian Major Carps is higher than mitotic gynogens (John *et al.*, 1984 and Reddy *et al.*, 1993)

Qualitatively, meiotic gynogens requires about 4-5 generations to achieve complete homozygosity (Nagy *et al.*, 1984). Where as same result can be achieved with one generation with mitotic gynogens, as blocking of mitotic division before first cleavage results in retention of two identical replicated mitotic product (Mair *et al.*, 1986)

Top Crossing of Gynogenetic Line

Once the induction of gynogenesis is over, the gynogen population may be reared separately. Because of their high homozygocity, proper feed and aeration is required for growth and maturity. The resulting gynogenetic offspring may all develop to female where homogamety exists.

One group of gynogens may be treated with Methyltestosterone (MST) for reversal of sex from female to male. MST have to be provided along with feed to the gynogenetic larvae from the day of yolk sac absorption for varying period (2-3 weeks or more) depending on the species. This will result in production of genetic female with physiological male. This can be crossed with gynogenetic female to produce next generation offspring. The highly homozygous gynogens can be top crossed with heterozygous stock to realise heterosis effect.

Androgenesis

Androgenesis is a process which, results in all paternal inheritance. Like gynogenesis, androgenesis also occurs in nature and can be induced. Induction of androgenesis is difficult than gynogenesis.

Natural Androgenesis

Natural or spontaneous androgenesis has been observed in certain hybrid-related individuals with remotely related individuals or those with not so compatible genome as in the cross between *C. carpio* female and grass carp male. Indian carp being highly compatible among themselves, no such instances were ever reported in any of the hybrid crosses.

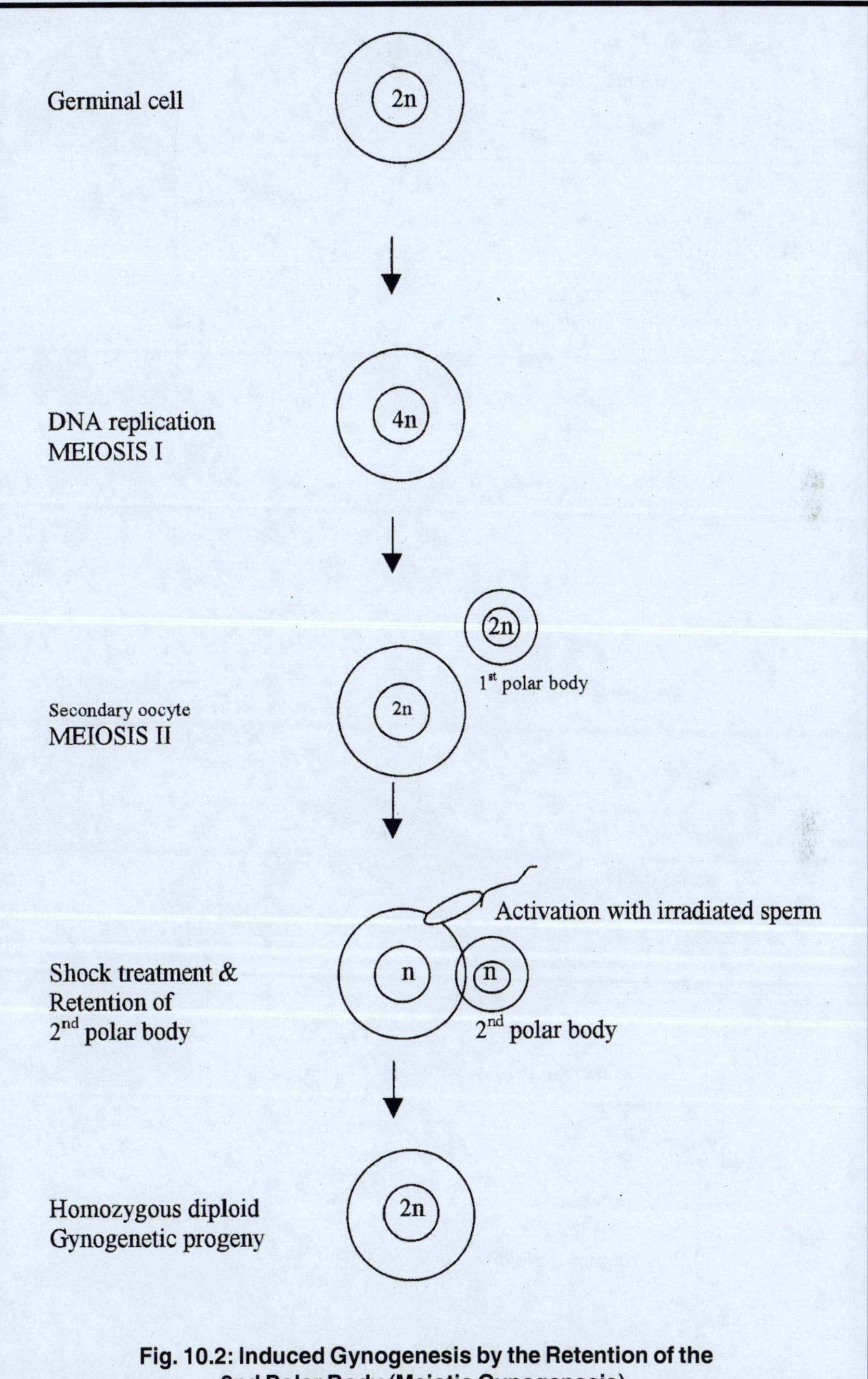

Fig. 10.2: Induced Gynogenesis by the Retention of the 2nd Polar Body (Meiotic Gynogenesis).

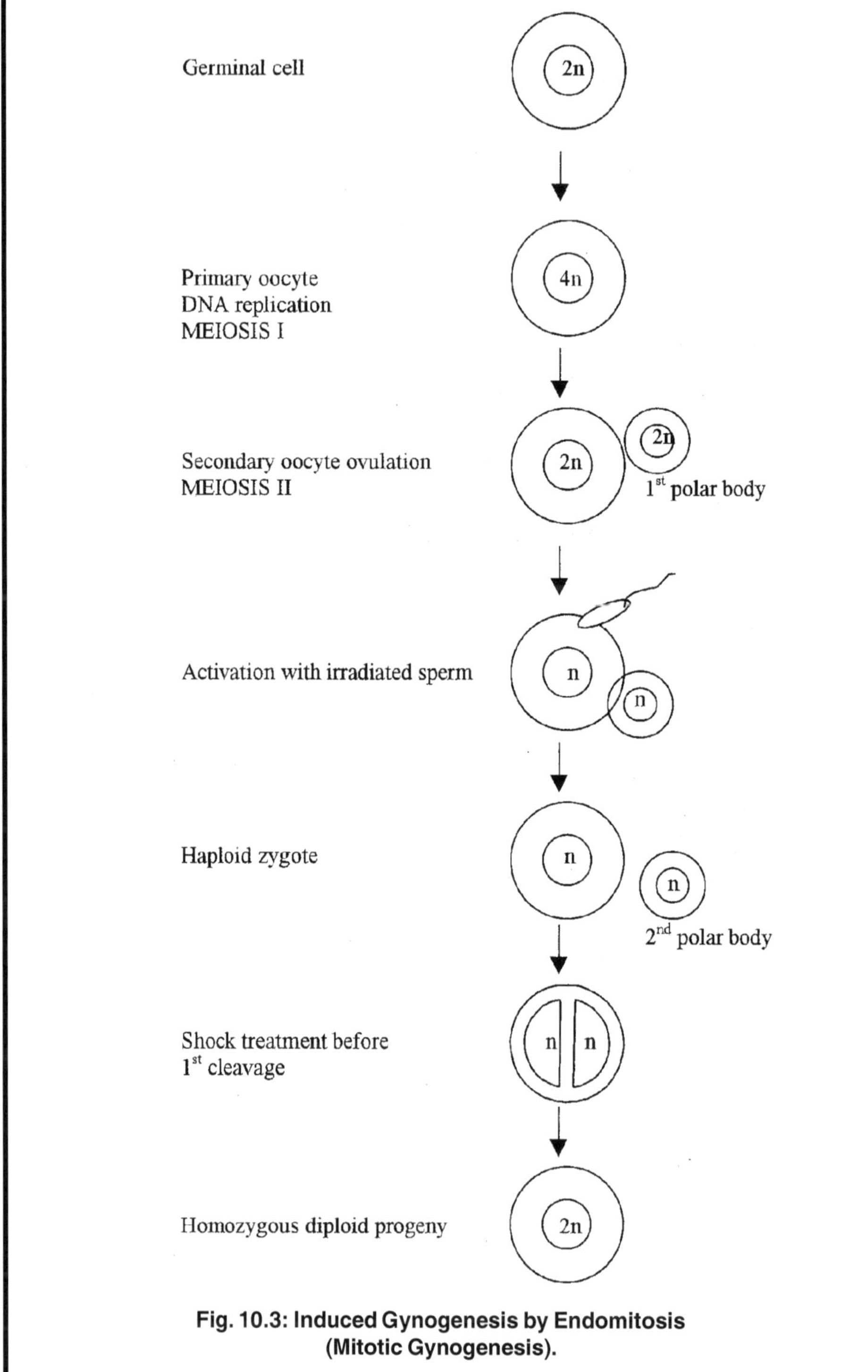

Fig. 10.3: Induced Gynogenesis by Endomitosis (Mitotic Gynogenesis).

Induced Androgenesis

Eliminating maternal genome in a similar way as is done for gynogenesis can induce androgenesis. The process of inducing androgenesis involves the genetic inactivation of egg and its activation with normal sperm of the corresponding species. Shock treatment has to be administered to restore diploidy. But the mechanism of restoring diplody in androgenesis is quite different from that gynogenesis. The process probably involves dispermy or other mechanism.

Attempts of inducing androgenesis in Indian Major Carps have been made earlier but were unsuccessful mainly due to the difficulties in the effective irradiation of the egg. The egg consists of huge mass of yolk that prevent the proper exposure of the DNA to UV rays in the nucleus, which specially in carps is oriented towards one side of the cytoplasm and usually facing downwards covered by cytoplasm and yolk. UV irradiation of egg may need a longer duration of exposure and that may be lethal.

Utility of Gynogens and Androgens

Gynogenesis and androgenesis provide shorter routes and are effective methods to build homozygous inbred line in a faster way. By top crossing these inbred lines with heterozygous stocks, progeny with higher growth rates 30-40% over normal ones can be expected. Production of monosex population is also possible through gynogenesis and androgenesis.

Polyploidy

Polyploidy is a process in which there will be addition of one or more set (s) of chromosomes to the original diploid complement. Like gynogenesis and androgenesis, polyploidy also occurs in nature and can be induced. In nature, polyploidy occurs when very distantly related fish species are crossbred. For example cross between grass carp and big head carp. The hybrids of this cross are reported to be triploid. However, none of the inter-specific or inter-generic hybrid crosses among Indian carps have been reported to produce such triploids. Polyploid is induced in the same way as diploid gynogenesis (Fig. 10.4). However, unlike the latter the former is induced by subjecting the fertilized egg (by normal sperm) to the usual shock treatments. Triplody can be induced by preventing the extrusion of second polar body, while tetraploidy is induced by blocking the first cleavage in the zygote (Fig. 10.4).

Artificial Induction of Polyploidy

Reddy *et al.*, 1987 made the first attempt of inducing polyploidy in Rohu, *Labeo rohita* by using colchicine. They could induce only tetraploids and mosaics. However, Reddy *et al*, 1990 successfully induced triploidy in Rohu, tetraploidy in Rohu and Catla by using thermal shock. Triploidy was induced in Rohu by administering heat shocks to the zygotes, seven minutes after fertilization at 42± 0.5 ^{0}C exposed for a duration of 1-2 minutes. However, the incidence of triploidy was only 12 %. Rohu zygotes exposed to heat shock, prior to first cleavage at 39±0.5 ^{0}C for two minutes yield 70% tetraploids and those exposed to cold shocks of 10-15 ^{0}C for 10 minutes

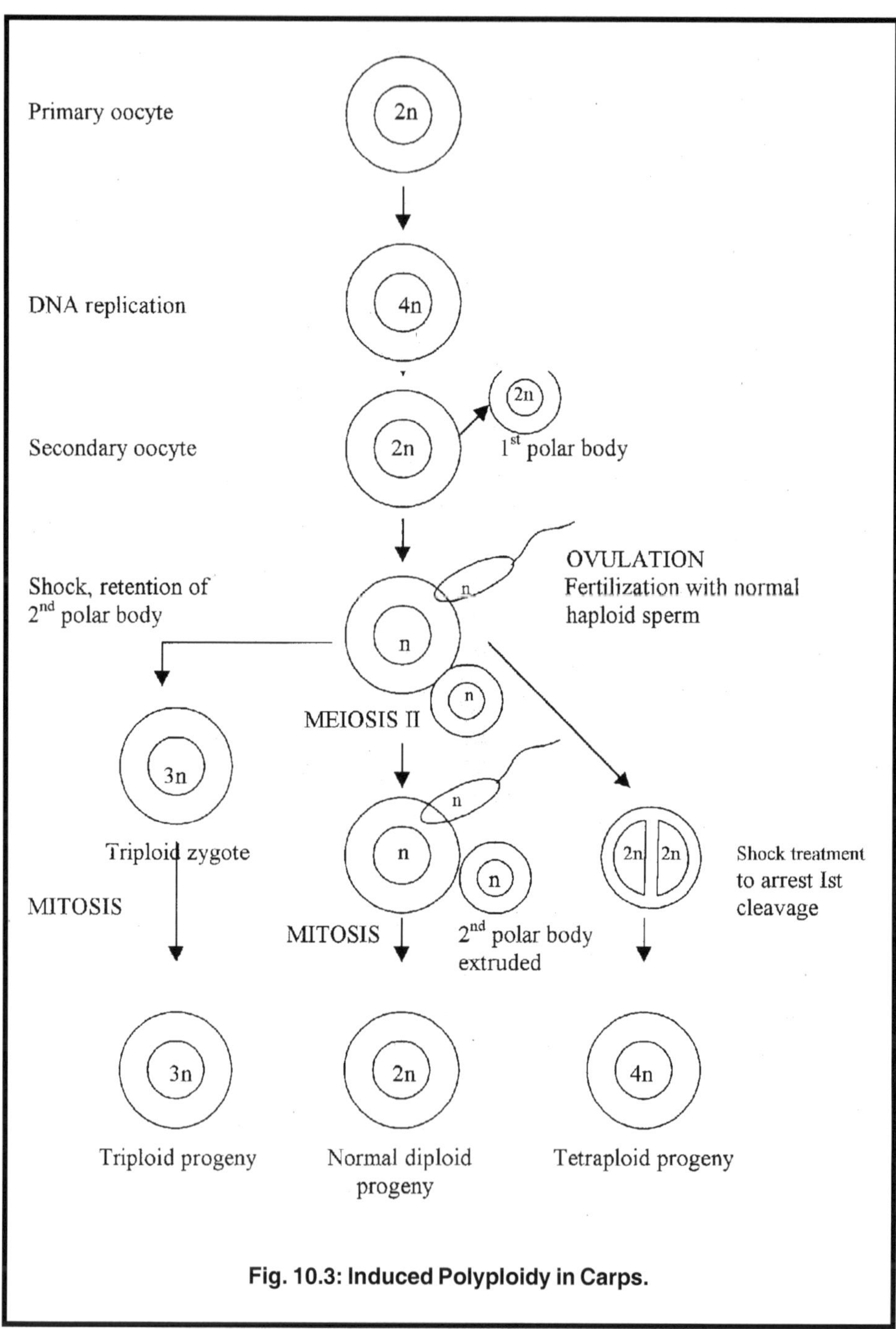

Fig. 10.3: Induced Polyploidy in Carps.

yield only 30-55% tetraploids. In catla, heat shock to the zygote at 40°C for two minutes prior to first cleavage yielded 30-65% tetraploids (Reddy, 1999).

Zhang, 1990 induced tetraploidy in *C. mrigala* for the first time. It was reported that heat shock at 39-40°C for two minutes to the embryo after 22-25 minutes of fertilization could induce tetraploidy in mrigal as well as in Rohu. However, the percentage of tetraploidy ranged from only 10-40% and 60% triploidy could induced by applying heat shock at 40°C for two minutes after 4 minutes of fertilization.

Utility in Aquaculture Practices

Sterility is one of the great advantages in modern fish culture, as energy spent for maturation of gonadal development may possibly be diverted or utilized for increased somatic growth, especially in species like common carp, which have shorter maturity cycle. Triploid common carp have shown significantly higher growth rate than its normal diploid counter part (Reddy *et al.*, 1998). Because of its sterility, overpopulation is also checked as the species has pond- breeding habit.

Sterile triploid grass carps can be safely released in open water systems to check the aquatic weeds without fear of its establishment through reproduction or its influence on the indigenous fauna.

Allopolyploidy

Allopolyploidy is a process in which there will be addition of set (s) of chromosomes to the diploid hybrid genome. Here male and female of different species are used. Allotriploidy can be achieved by preventing the extrusion of second polar body so here female genome contribution is 2n and male contribution is n. Allotetraploidy can be achieved by blocking the first cleavage in the developing hybrid zygote.

Allopolyploids may be possible to attribute some additional traits to the hybrids such as faster growth rate and disease resistance.

SEX DETERMINATION AND MANIPULATION

Sex determination methods are known for very few fishes. Sexuality in fishes has been shown by hermaphroditism, unisexuality, bisexuality etc. However, sexuality is under a low grade of differentiation in majority of fishes. In some species sex chromosomes are morphologically distinct and can be identified but in majority of cases it is not morphologically distinct. Some fishes show primitive type of polygenic sex determination in which case male and female determiners are located in many chromosomes and the outcome of the particular sex depends on the balance of these genes. More advance type of sex determination as male heterogamety (XY) or female heterogamety (WZ)is also observed in some fishes. However, there are nine types of known systems in fish for determination of sex. Out of which sex is controlled by sex chromosome in eight cases while the ninth one is controlled by autosome.

The most common system of sex determination in fish is the XY sex determining system. Here females are homogametic (XX) and males are heterogametic (XY). In this system 'Y' chromosome determines the sex. Offspring that receives 'Y' chromosome become male, while those receive 'X' chromosome become female. Example: Channel catfish (Davis *et al.*, 1990), Rainbow trout (Thorgaard, 1977), Common carp (Nagy *et al.*, 1981), Silver carp (Mirza and Shelton, 1988) and Grass carp (Stanley, 1976).

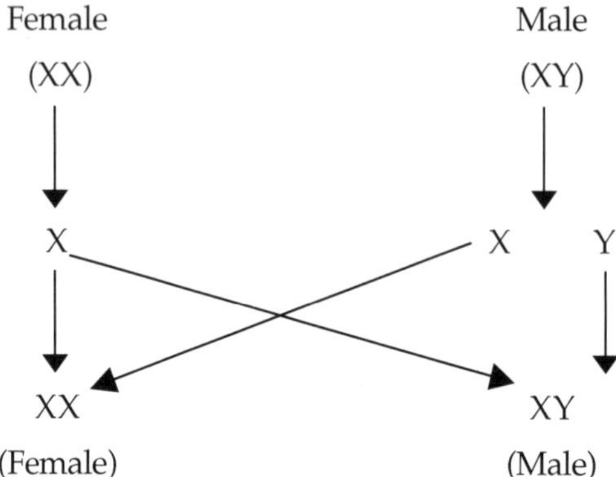

The second system is WZ sex determining system. This is reverse of previous system. Here male is homogametic (ZZ) and female is heterogametic (WZ). Presence of 'W' chromosome determines the sex of the fish *i.e.* the offspring who receives the 'W' chromosome become female. Example: *Tilapia aurea* (Guerrero, 1975), *Tilapia hornorum* (Chen, F.Y., 1969) and Japanese eel (Park and Kang, 1979).

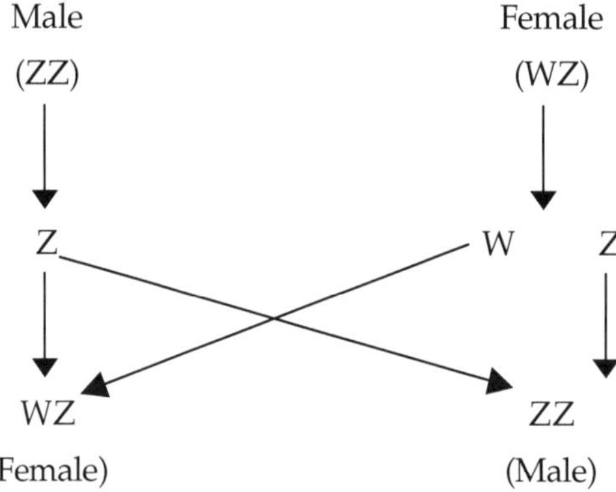

The third, fourth and fifth sex determining system have multiple sex chromosome. In the 3rd system females are $X_1X_1X_2X_2$ and males are X_1X_2Y. Male is heterogametic here. Those who receives 'X_1X_2' becomes female and those receive 'Y' chromosome become males. Example: Caterina pupfish (Uyeno and Miller, 1971), Freshwater gobi (Pezold, 1984)

In the fourth system, Males are ZZ and females are ZW_1W_2. Here female (heterogametic) determines the sex of the fish. Those who receives W_1W_2 are females and those receives 'Z" chromosome become males. Example: Virolito (Filho *et al.*, 1980)

The fifth system is multiple 'Y' chromosome system. In this system males are XY_1Y_2 and females are XX. Male determines the sex of the fishes. Those receive 'Y_1Y_2'

chromosomes become males while those receive X chromosome become female. Example: *Hoplias* species (Bertolo *et al.*, 1983).

The sixth system is the WXY sex determining system. The 'W' chromosome is the modified 'X' chromosome which block the male determining ability of the 'Y' chromosome. Thus XY and YY fishes are males while XX, WX and WY fishes are female. Both male and female can be either homogametic or heterogametic. Example: Platyfish (Gordon, 1946).

The seventh and eight system are those where only one sex chromosome exists *i.e.* XO and ZO system. In XO system, females are XX and male are XO (O means no chromosome). Male determines the sex of the chromosome. Those who receives 'X' chromosomes from male are females and those receives no sex chromosome from male become males. Example: Dollar hatchetfish (Chen, T.R., 1969)

In ZO system, Females sex chromosome is ZO and males are ZZ. Female determines the sex of the offspring. Those receive 'Z' chromosome become male while those receive no sex chromosome become female. Example: Dwarf gourami (Rishi, 1976)

The ninth method of sex determination is not controlled by sex chromosomes. Some species of fish do not have sex chromosomes. In those species, the number of male or female genes located in the autosome determines sex. Example: Swordtail and Blue poecilia (Kosswig, 1964)

Although sex determination is primarily under genetic control, environmental factor such as temperature, salinity, photoperiod and crowding can also determine the sex in some fishes.

Manipulation of Sex

Monosex population in aquaculture practice is gaining importance because of following reasons,

Growth

In certain species of fish male and female shows variable growth. In tilapia, male grows faster than female while in carp females grows faster then males. So higher production can be achieved through monosex population.

Population Control

Population control of common carp can be achieved through monosex culture.

Sexual Maturity

In common carp due to early maturity of fishes, growth efficiency and flesh quality declines, which can be checked through monosex culture.

So sex reversal can be one of the methodology to improve production level.

Fishes are different from other animals because it can fertilized externally and embryological development takes place without the protection of an internal womb or a hard shelled egg. This combination allows aquaculturists to manipulate the sex phenotypically. During early embryology, an embryo is phenotypically neither male nor female. It does not possess ovaries, testes or other characters associated with reproductive systems. Instead, an embryo possesses embryological precursors of ovaries and testes (primodial germ cells) and at this stage the embryo is 'totipotent' because it could develop to either male or female. At a specific time during embryological development, a chemical signal originated from a gene or sets of genes and this signal informs the totipotent tissue to develop. Once this is over, the tissue completes its development and the fish becomes either phenotypic male or female. Once this occurs, it is not possible to alter phenotypic sex

There is a specific time when phenotypic sex can be altered and the timing is species specific. If a fish ingests or absorbs anabolic steroids during this period the steroid can direct the development of the totipotent cells.

The direct alteration of phenotypic sex by administration of hormone is a common approach to alter the sex.

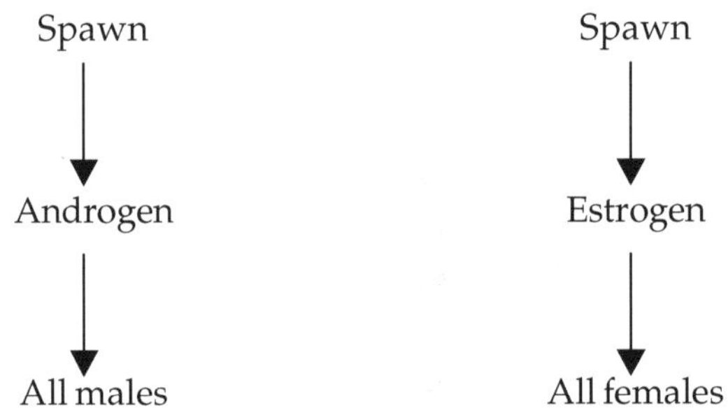

(General method of production of monosex population)

The hormone administration to spawn started from 1st feeding day to 30-60 days. For sex reversal to male, 17 α Methyltestosterone (MST) was applied while in case of sex reversal to female 17 β Estradiol is applied for 40-60 days. The ration of hormone is species specific. The hormone can be supplied to the fishes through dietary supplement, immersion of spawn in the hormone or through injection. The amount of hormone consumed by each fish is negligible and most is eliminated rapidly. So use of these hormones may not create any health hazard to the fishes. But some times it is risky from a market aspect in a health conscious society. Incorporating sex reversal in to the breeding programme can reduce this problem.

In XY sex determination system, monosex population can be produced by creating supermales (male having 'YY' chromosomes instead of 'XY'). Mair et al., 1997 developed a strategy for production of all male population in *O. niloticus*. Which can be presented as follows:

Sex reversal breeding program is gaining importance and it can be applied to carp culture unit also. Over half the rainbow trout farm in UK and 40% of Chinook salmon farm in British Columbia are all female progeny produced from sex reversed breeding program.

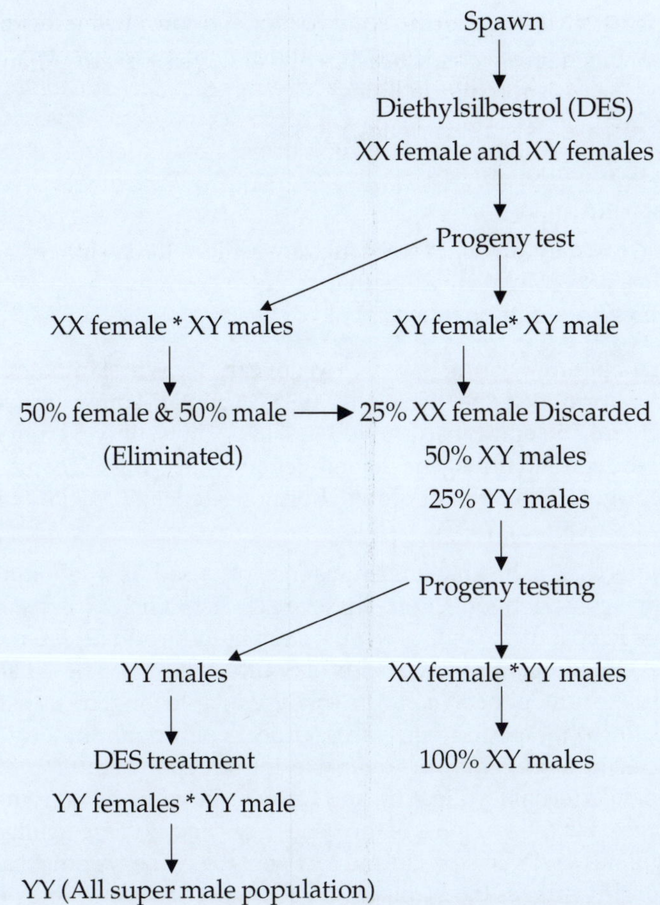

[Method for production of YY (super male) offspring, *after Mair et al.*, 1997]

GENE TRANSFER AND TRANSGENESIS

Early in 1980's revolutionary technique was developed to introduce isolated genes in to animals and investigate their function in vivo. This technique is the microinjection of foreign gene in to the eggs and the production of animal lines carrying the injected gene. Individuals carrying such foreign genes are called transgenic.

Transgenesis is the process of transferring foreign gene into the fertilized egg so that altered phenotypes having desirable characters are generated. The first report of the production of a transgenic animal was that of Palmiter and coworkers in the year 1982 when they successfully generated a transgenic mouse having dramatically faster growth rate by introducing metallothionein-human growth hormone fusion gene into the fertilized egg of mouse and implanting the transgenic embryo into the uterus of the female. Ever since this discovery, several scientists have attempted to produce transgenic fishes for aquaculture with the following objectives:

(a) Production of fish with faster growth rate and better feed conversion efficiency.

(b) Modification of biochemical characteristics of fish with a view to enhances nutritional and/or organoleptic qualities.

(c) Production of disease resistant varieties.

(d) Control of sex differentiation.

(e) Induction of sterility.

(f) To develop cold resistant/drought resistant transgenic of Indian major carps and other fishes of commercial value that can grow in regions exhibiting extremes of climatic conditions such as cold or heat.

Gene transfer or genetic engineering in Indian Major carps was first initiated by injecting human growth hormone gene (Alok *et al.*, 1995). About 434 embryos were microinjected, of which only 50 embryos survived. Up to 90 days, only 5 embryos survived and only 2 of the contain transgene in their genome, carrying 1-2 copies of human growth hormone gene. Isolation of growth hormone gene in *L. rohita and C. catla* done successfully (Mazumdar, 1997).

During the last decade, 2 fishes have been extensively used as a laboratory model for studies in transgenic gene function. These are (1) Zebra fish *(Brachydanio rerio)*, a small oviparous teleost from Ganges river system in India and (2) Japanese medaka *(Oryzias latipes)* that inhabit the rice fields of South East Asia. These fishes are easy to breed and large number of eggs can be easily available for gene transfer. Moreover, by manipulating photoperiod, eggs can be produced throughout the year, as and when required. As the fish eggs are of larger size, introduction of foreign gene into the egg is also easier. External fertilization and faster embryonic development are additional advantages. Further, in most of the fishes eggs and milt are available by stripping which facilitates a scientist to plan and carry out the work systematically. But fish egg has some disadvantages too in this regard foremost of which being poor visibility of zygote nucleus making the transfer of foreign gene into the nucleoplasm impossible. Hence microinjection is done into the ooplasm only. Further, the chorion, the outer egg covering, is very tough and resistant making the microinjection difficult.

There are four steps for production of transgenic fish viz, (i) introduction of foreign gene into the fertilized egg (ii) integration of the same with the original genome of the recipient, (iii) expression of the gene in the transgenic fish and the resultant alterations, if any in the phenotype of the recipient and (iv) transmission of the foreign gene to next generation.

Introduction of Foreign Gene

There are four methods commonly used for introducing foreign gene into the embryo. These are microinjection, electroporation, beakonization and sperm mediated gene transfer.

Microinjection

This is the most commonly used method for introducing transgene into the developing embryo at the one cell state. In this method a solution containing the foreign DNA is injected into the cytoplasm by a microinjection needle. After collection of fertilized egg, dechorionation of eggs has to be performed. This step is aimed at removing or softening the hard egg covering called chorion. For dechorionation, eggs are washed extensively in dechlorinated water at room temperature and then the eggs are rinsed in 10%. Hank's balanced salt solution (HBSS). Eggs are incubated for 5 minutes at room temperature under gentle agitation. The chorion turns slightly brown and paliable under gentle pressure. The eggs are then emptied into 300-ml water. When the eggs settle to the bottom of the beaker, the water is decanted and fresh dechlorinated water is added. This procedure can be repeated at least 5 times. Then the eggs are taken to 10% HBSS again for microinjection. Some workers have also followed treatment of eggs with glutathione to make the chorion soft.

Preparation of Microinjection Solution

It consists of 50ng/ml plasmid DNA in 0.25m KCl and 0.2% phenol red. This solution is iso-osmotic with egg interior. The phenol red is added to visualize the injected fluid inside the egg.

The microinjection is done by means of a microinjection pipette prepared in the lab. Using glass capillary tubes of 1.2 mm (OD) and 0.44 mm (ID) by pulling the same by a horizontal pipette puller. The solution shall be drawn towards the tip by capillary action within 3 minutes. The loaded pipettes are held vertically, tip down in a humidified container prior to use.

The dechorionated embryos at 1- cell stage are placed on a cavity slide in a drop of 10% HBSS at room temperature and placed on the stage of a stereodissecting microscope. Then the eggs are gently shaken till the blastodisc is clearly visible. The injection is done using a course micromanipulator and a picoinjector. The pipette with DNA solution is inserted horizontally into the manipulators arm, the arm is tilted forwards to an angle of 80^0 and pipette carefully plunged into the solution containing egg. Gently the pipette end is tapped at the bottom of the slide, while doing so the tip will be broken producing an opening of approximately 3 - 5 mm. If the opening is narrower it is difficult to inject the solution, so also if it is larger it is likely to damage the eggs. Approximately 250 pl of solution is injected into each egg. Microinjection is done carefully by penetrating the egg membrane with the pipette tip and delivering the solution by a single pulse and gently withdrawing the tip. The injected solution may be visible inside the egg. In fishes such as rainbow trout where dechrionation is difficult microinjection is performed through micropyle. In fishes where micropyle is not distinguishable a small hole is drilled in the chorion for introducing the needle. Injection into germinal disc provides higher frequency of gene retention than other methods such as injection through micropyle. Some workers were also successful in transforming foreign genes through injection into perivitelline space. For all cases linear form of DNA is retained more effectively than circular form.

The Quantity of DNA and Volume of DNA Solution

The volume of DNA solution injected and quantity of DNA varied in different experiments. When the nuclear size of zygote is larger, it will be able to accommodate more quantity of DNA. For example the size of zygote nuclear of medaka is 135mm, in

Gambusia and Guppy it is between 150-200mm, when compared to that of mouse, it is 1000 times larger. Usually the concentration of DNA injected is between 10 and 15mg/ml. According to Stuart *et al.*, 1988 if DNA concentration is more than 15 mg/ml it is toxic to Zebrafish. Size of plasmid used for injection also differed varying from 2.4 to 9.4kb. The number of copies of transgenes introduced into the egg nucleus also varied. In case of mammals where the injection is directly done into the zygote nucleus, only a few hundred copies are sufficient but in medaka if injection is done into the nucleus more than 50000 copies are required for transgenic fish production and if injection is done into the cytoplasm 1 to 2 million copies are required.

Hatching Rate of Injected Eggs

This varied in different fish groups studied so far. In low mortality group such as medaka and salmonids where only 20% mortality is found in both injected and control groups (Ozato *et al.*, 1992; Winkler *et al.*, 1991 Chourrout *et al.*, 1986, Fletcher *et al.*, 1988; Shears *et al.*, 1991; Rokkones *et al.*, 1989) and carps where 60% mortality was found in both injected and control groups (Zhang *et al.*, 1990; Chen and Powers, 1990) there was no additional mortality due to microinjection. In the moderate mortality group such as Zebra fish, there was 40% additional mortality of injected eggs when compared to control. In the high mortality group such as cichlids, catfish and sea bream there was higher rate of mortality, varying from 40 to 90% and these fishes have been found to be too sensitive to microinjection. This sensitivity may be due to several factors such as presence of chorion, diameter of needle used, quantity of injected DNA, quality of egg and time taken for injection. The time taken for injection into one egg varies from 0.02 minutes to 2 minutes by different workers.

Disadvantages of Microinjection Method

In this method only a limited number of eggs can be treated as it is a complicated, time consuming process requiring great deal of skill. As the time for injection is limited to a specific period of development *i.e.* from fertilization to 2-cell stage, only a few eggs can be injected. For example in a tropical fish such as *Heteropneustes fossilis*, the time is limited to a period between 15 minutes and 40 minutes after fertilization (in *H. fossilis* sufficient cytoplasm for injection is seen only 15 minutes after fertilization.) Hence only 25 minutes shall be available for microinjection in this fish (Pandian T. J and A. Marian, 1994) If one takes one minute for injecting into one egg only 25 eggs can be injected.

Electroporation

In this method of transfer of foreign genes, the fertilized eggs are immersed in appropriate buffer containing foreign DNA and appropriate electric pulses are applied which causes alteration in the permeability property of egg membrane and consequent diffusion of foreign genes into its cytoplasm. After the treatment, the treated eggs are rinsed with distilled water and incubated at appropriate temperature until hatching. For successful transfer of foreign gene, several factors such as pulse condition, state of chorion, developmental stage of egg etc. is found important.

Though this method permits mass production of transgenics, it is difficult to assess the actual quantity of foreign gene entering the cell.

Beakonization

Zhao *et al.*, 1993 proposed an alternative method termed Beakonization for introduction of transgene. They have suggested for applying high voltage but low current electric pulses to egg-DNA mixture from electrodes placed apart from mixture.

Sperm Mediated Foreign Gene Transfer

Khoo *et al.*, 1992 demonstrated successful introduction of foreign gene into zebra fish by incubating sperm in foreign DNA solution and utilizing such treated sperm for fertilizing eggs. Adults developed from such eggs and also some of the F_1 and F_2 generation fish also were transgenics. In course of time several workers reported on the successful use of sperm cells as vectors for introducing foreign gene, into egg and there by to offspring through electroporation of sperm cells *i.e.* by applying appropriate electric pulses to sperm - DNA mixture.

Transgenes/gene constructs introduced in fishes so far belong to either prokaryotic or eukaryotic sources. The genes from eukaryotes tried so far are (1) human growth hormone (hGH), (2) bovine growth hormone (bGH) (3) rat growth hormone (rGH) (4) antifreeze protein gene (AFP) (5) rainbow trout growth hormone gene (rtGH) (6) Chinook salmon growth hormme gene (csGH) (7) chum salmon melanophore concentrating gene (8) *Lates calcarifer* growth hormone gene (9) melanin concentrating hormone gene (MCH) (10) firefly luciferase gene (luc) an insect gene. Prokaryotic genes used so far are (1) chloramphenicol acetyl transferase (CAT) (2) b-galactosidase (b-gal) (3) neomyson phospho transferase (neo), (4) Hygromysin. Two types of transgenes have been used such as (1) full length nuclear gene having all coding sequences such as promotors, introns etc and (2) complimentary DNA (cDNA) containing only the gene that codes for specific protein. As promotor or enhancers may be absent in cDNA, foreign promoters are used for gene constructs. Three types of promotors are used (1) promotors from normal cellular genes that are active in most tissues *eg.* metallothionein- I (MT-I) and b-actin (2) promotors active only in specific tissues *eg.* Long terminal repeat (LTR) protein of Rous sarcoma virus (RSV) of avis. (3) Viral promotors active in many tissues especially those infected by virus. It is reported to be more effective. For example, Trout growth hormone along with RSV promoter has expressed in carp, medaka and sea beam. But human growth hormone gene with metallothionein promotor was expressed in loach, cruciancarp but not in Tilapia and cat fish. As it is not advisable to produce transgenics with viral promotors, (as it may cause health hazard), fish promotors such as AFP or bactin are used for making gene constructs.

Integration of Foreign Gene

Several studies have been conducted on the fate of foreign genes introduced into the fertilized eggs and these studies have indicated that while a portion of the trangene may be integrated with the genome of the recipient, a portion may persist in the cytoplasm itself and another portion may get degraded by DNA-ases. Injected DNA may be amplified after fertilization, by replication but only a small proportion of replicated DNA is retained after gastrula stage due to degradation activity of DNA-ase. Salmonids can tolerate large amount of transgene in the egg *i.e.* up to 200 pg

DNA in a single egg (Fletcher and Davies, 1991) where as zebra fish can tolerate only up to 50 pg DNA. These studies have indicated that quantity of transgenes in the transgenics decreased with time and injection dose. The reason for this is attributed to the extrachromosomal persistence of injected transgene, which causes unequal distribution in daughter cells leading to mosaic tissues, organs, and individuals. This mosaicism has been observed by almost all workers. Large differences exist in the copy number of injected gene persisting in different tissues. It has been suggested that delay in genomic integration prior to second cell division in zygote may be leading to mosaicism.

Expression of Transgene

It has been reported that introduction of trangenes of fish origin facilitates a better and higher percentage of expression, though frequency of genome integration was more for genes of mammalian origin than for fish genes.

Expression of a foreign gene in a transgenic animal can be detected by estimating the corresponding mRNA through Northern blot analysis or by estimating the corresponding protein by immuno blotting or radio immuno assay. Phenotypic expression such as growth rate can be recorded by noting the required parameters such as length, weight etc. Some times marker genes such as luciferase or melanin concentrating hormone has been used to identify the transgenic fish. Level of expression of a gene also depends on the numbers of copies of desired gene in the tissue or organ. For example, Fletcher *et al.*, 1988 injected flounder AFP gene into salmon was liberated into the blood stream and it measured 5 ng AFP/ml serum. In order to protect the fish in temperature below freezing point it should have higher AFP level. Flounder has 40 AFP gene copies, expression of many genes together elevates AFP level in serum which is required for freeze resistance. Another factor to be considered in this connection is whether the gene is homologous or heterologous. Heterologous genes are not ideal as it may produce underirable effects such as tissue specific expression that are harmful and may cause side effects due to impairment of other functions. For example, several studies have indicated undesirable effects such as less fecundity (zebrafish) lack of enhaced growth (as reported in rabbit, pig and sheep with human growth hormone gene) and infertility as reported in female mice expressing high levels of hGH. To prevent such ill effects introduction of homologous gene has been recommended. By this time homologous growth hormone gene from chum salmon, cohosalmon, chinook salmon, rainbow trout and grass carp have been sequenced.

TRANSGENE TRANSMISSION

Mode of transmission of transgenes from parents to off springs has been studied by only a few workers in selected fishes. In this regard, for fishes like salmon and carp information available is very meager because of the prolonged generation time. A few studies have been conducted on zerbra fish and these studies have indicated that transgenic characters are not transmitted as per the expected Mendelian ratio because of the mosaicism described earlier in transgenic fish. Khoo *et al.*, 1992 utilized sperm cells as vectors for introducing foreign DNA into zebra fish and such transgenics were crossed with non transgenic controls. Four male founder parents gave transmission rate of 4/4, 2/5, 0/8 and 0/1 where as 3 female founder parents gave transmission rate of 1/3, 0/5, and 0/1. Surprisingly some of the F_1 negative progenies produced by the founder parents when crossed with untreated fish produced positive F_2 progenies. This indicates that original founder parent was mosaic. As the presence of transgene in this fish was determined by southern analysis of pectoral fin, absence of transgene from pectoral fin does not imply its absence from other tissues such as germ cell etc. Further, it indicated that transgenes were not integrated with genome but remained extrachromosomally.

Though there is immense scope for revolutionizing aquaculture through the production of transgenic fish seed exhibiting faster growth rate better feed conversion efficiency, disease resistance, cold resistance, drought resistance etc. there is need for exercising restraint and control in adopting this technology without intensive researches and field trials and in allowing such genetically modified fishes to enter into natural ecosystem and cross breed with the natural fish fauna. Some of the workers have even suggested to limit the foreign gene transfer into triploid eggs only as the fish developed from such eggs are likely to be sterile.

SELECTIVE BREEDING

Role of selective breeding in increasing production level is well established in Agriculture and animal husbandry. Today the high yielding crops and land animals are totally depending on genetically improved domesticated breeds. This has not been true for aquaculture. Proper exploitation and utilization of genetic potential is lacking in aquaculture. Less than 2% of the total output of the aquaculture production is coming from improved breeding progamme. Aquaculture species are thus genetically much closer to their wild counter part than the land animals and plant species. During the last few years it has been well documented that high selection response can be obtained in fish as well as in shell fish for economic important traits like growth, disease resistance, flesh quality etc

Before success of induced breeding through hypophysation in late fifties rivers were the main source for seed collection of Indian major carps. With the introduction of induced breeding technique, hatcheries have been able to produce enough quantity of carp seed. The hatcheries in India hardly follow any genetic norms to produce carp seed. A limited number of brood fishes are used repeatedly for successive generations. As a result the quality of carp seed is showing negative effect of inbreeding and genetic drift *i.e.* slow growth rate, disease proneness etc.

Inbreeding

Inbreeding occurs due to mating of closely related individuals. Genetically inbreeding leads to homozygosity. Almost all individuals carry deleterious recessive genes, which are hidden in heterozygous state. Related individuals are likely to share common genes and probability of pairing of deleterious recessive genes gets enhanced with the increase of closeness between the parents. Due to which inbreeding depression occurs to the population with decrease growth efficiency, disease resistance and survival.

Genetic drift

Genetic drift is the random change in the gene frequency created by sampling error. The sampling error may occur naturally or may be man made. Through natural calamity like flood, earthquake etc. a portion of the population may be isolated. Inaccurate sample collection also leads to sampling error creating genetic drift.

Change in gene frequency as a result of genetic drift occur during production of new generation *i.e.* in spawning season or during acquisition of population from other sources. When a population is sampled there is a chance that the population does not accurately reflect the population. The smaller the population, the greater the likelyhood that inaccuracies in the sample will occur. So effective population size/number has inverse relationship with genetic drift. Genetic drift can be expressed as:

$$\sigma^2\Delta q = \frac{pq}{2N_e}$$

Where,

$\sigma^2\Delta q$ = Variance in the change of the gene frequency

P & q = Frequency of "p & q" alleles for a given gene

N_e = Effective population size/number

If the frequency of an allele changes from 0.5 to 0.4 as a result of genetic drift, the genetic effects on the population may not be visibly great. The major damage that is caused by genetic drift occurs when the frequency of an allele goes to 0.0.Then the allele is lost and no longer exists in the population. Rare alleles are lost more easily than common ones. Loss of alleles through genetic drift increase homozygosity in a population and it reduces genetic variance which is the raw material for genetic improvement programme *i.e.* selection.

The loss of genetic variance can produce irreversible damage to a population's gene pool. The loss can also prevent future genetic improvement programmes and also viability in the population that are stocked in open water system *i.e.* lake and reservoir.

Effective Population Size

Effective population size is one of the most important concepts in the management of a population in that it gives an indication about the genetic stability of the population. It depends upon several factors such as total number of breeding individuals, sex ratio, mating system and variance of family size. The effective population size can be calculated by following formula,

$$N_e = 4N_f X N_m / N_f + N_m$$

Where,

N_e = Effective population size

N_f = Number of female brood fishes used for seed production

N_m = Number of male brood fishes used for seed production

Effective population size is inversely related to inbreeding

$\Delta F = 1/2N_e$

So $\Delta F = 1/8N_f + 1/8N_m$

Where,

ΔF = Rate of inbreeding per generation

N_e = Effective population size

N_f = Number of female brood fishes used for seed production

N_m = Number of male brood fishes used for seed production

So to improve the genetic status of any population, hatchery managers should decrease inbreeding rate and increase effective population size.

Selective breeding plays vital role to improve genetic status of fish in a positive direction which has already been demonstrated in case of Salmon at Norway, Tilapia at Philippines and Rohu at CIFA, India.

Objectives of Selective Breeding

Objectives of a selective breeding programme is to change the average performance of the targeted trait *i.e.* growth rate, disease resistance, better flesh quality, feed conversion efficiency etc. of the population in a favorable direction.

Selection Process

Selection is an age-old process in nature. Fittest organism survives and other eliminated. Selection also can be achieved artificially. In this process best individuals are selected as parents so that parents pass on their superior genes to their progeny and better progeny can be obtained. Selective breeding based on principle of quantitative genetics. Which indicated that that phenotype of an individual, which can be measured or scored, could be partitioned in to two components. One attributable to the influence of genotype *i.e.* the particular assemblage of genes possessed by the individual and other one attributes to the influence of environment *i.e.* all non-genetic components.

$$\text{So, } P = G + E$$

Where,

P = Phenotype of an individual

G = Genotype

E = Environmental (Non-genetic) component

Quantitative phenotype exhibit continuous variation, the only way to study them is to analyze the variance that exists in a population. The phenotypic variance (V_P) that is observed for a quantitative trait is the sum of the genetic variance (V_G) and environmental variance (V_E) and the interaction that exists between the genetic and environmental variance (V_{G+E})

$$\text{So, } V_P = V_G + V_E + V_{G+E}$$

Genetic variance is the component of interest in selective breeding program. V_G is further subdivided in to three componentss *i.e.* additive genetic variance (V_A), dominance genetic variance (V_D) and the epistatic genetic variance (V_I).

$$V_G = V_A + V_D + V_I$$

V_A, V_D and V_I differ from each other according to their mode of inheritance. Dominance genetic variance is the variance that is due to the interaction of the alleles at each locus. Because of this, V_D can not be inherited; it is created a new in each generation. Since it is the interaction between alleles at each locus so V_D is a function of the diploid state as alleles occur in pairs. During meiosis homologous chromosomes and allelic pairs are separated during reduction division and the chromosome complement reduced to half. So V_D is not transmitted to next generation.

Epistatic genetic variance (V_I) is due to interaction of alleles between two or more loci. Epistatic interaction occurs across loci so it is also not transmitted from parents to offspring. This also created a new in each generation.

V_A or additive genetic variance is the additive effect of genes. It is the sum of the effects of each allele that is responsible for phenotype. It does not depend on specific interaction or combination of alleles so it is not disrupted due to meiosis. Additive genetic variance transmitted from parents to offspring. It is transmitted in a reliable and predictable manner. V_A is also called the variance of breeding value.

Heritability

Heritability describes genetic component that is not disrupted by meiosis. The proportion amount of phenotypic variance (V_P) that is controlled by V_A is called heritability (h^2).

$$h^2 = V_{A/}V_P$$

Once you know the heritability response of selection can be predicted as

$$R = S * h^2$$

Where,

R = Response to selection,

S = Selection differential (difference between offspring and parent generation)

h^2 = Heritability of the trait.

Selection Methods

Several selection methods are available for obtaining additive genetic improvement. The methods differ with respect to which type of relatives that provide information used for the selection decisions. The objective of all the methods is to maximize the probability of correct ranking of animals with respect to their breeding value, an estimate of each individual ability for producing high/low performing offspring. The breeding value of an individual cannot be estimated on basis of the phenotypic value of the traits). In fish such records are usually obtained from the individual itself (Individual selection), full and half sibs (family selection) or from all three sources of information (combined selection).

Individual Selection

When breeding candidates are selected according to their own phenotypic performance only then the method of selection is known as Individual selection. Individual selection method is most frequently applied method among fishes. A prerequisite for using individual selection is that the trait (s) selected for can be measured on the breeding individual itself while being alive. It also requires substantial attention in order to avoid inbreeding. In individual selection, individual marking is not required so it does not involve expensive method of tagging. From spawn stage small definite quantity of seed from each group can be mixed up and can be reared communally which reduce the cost of separate rearing to greater extend. Individual selection is effective when the heritability is high. However, individual selection has its own limitations such as traits like carcass quality for fat content and flesh colour can not be measured as the data collected from dead fishes and stocking of large fullsib families for testing may easily result in a high representation of selected breeders from a few families which leads to negative effect of inbreeding. The problem can be minimized by pooling a restricted number of individuals from each family after fertilization.

Family Selection

When heritability is low or trait like carcass quality to be measured then individual selection is not possible for selection then family selection is undertaken. So information about its family (halfsib, fullsib) is taken in to consideration for breeding value estimation and ranking of individuals. This type of selection method may give higher selection response for traits of low heritability. It can also reduce environmental effect common to fullsib and halfsibs. For this large number of families required to be produced.

Within Family Selection

In this type of breeding program, each family is considered as a temporary sub-population and selection occurs simultaneously and independently within each family. Individual with in the family is selected or rejected basing on the relationship to its family mean. It is advantageous to adopt when there is a large component of environmental variance common to members of a family. This type of selection will eliminate large non-genetic components.

Combined Selection

In combined selection method, information from fullsib, halfsib and individuals are taken in to consideration for breeding value estimation. This method combines all the information in a optimal way and it represent the general solution for obtaining the maximum rate of genetic gain therefore generally considered to be the best selection method. Selective breeding of Rohu at CIFA, Kausalyaganga based on combined selection method.

Genetic Gain

Major objective of any selective breeding programme is to move the value of trait (s) of selection in the desirable direction. The change in population mean from one generation to the next generation is termed as genetic gain. It is the difference of mean phenotypic value between the offspring of selected parents and the whole of the parental generation before selection *i.e.* offspring from mean of the parents population. The measure of selection applied is the average superiority of the selected parents which is called the selection differential. It is the mean phenotypic value of the individuals selected as parents expressed as a deviation from the population mean. The genetic gain or selection response varies from generation to generation. In case of Rohu selective breeding at CIFA after three generations of selection average response was observed to be 17%.

Dissemination of Genetic Gain

Utilization of full potential of genetic improvement requires effective system for dissemination to the ultimate target group *i.e.* fish farmer. Before releasing the improved seed to the users, field testing at different agro-climatic zone is required along with its impact on environment. For dissemination work the research station where the genetic improvement programme has initiated may be the nucleus of the dissemination programme. Different state governments and leading private hatchery owners, fish seed development corporations may act as multiplier unit. They will receive the improved seed from the nucleus unit and can multiply and supply seed to the fish farmers.

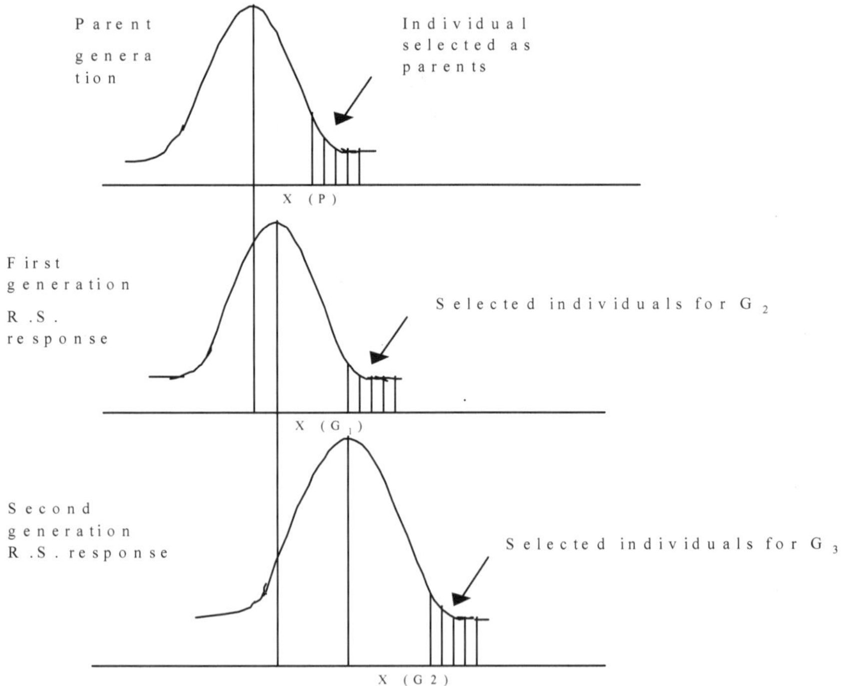

Fig. 1 : Genetic gain obtained by two generations of selection

Annexure-1: Chromosome Numbers and its Formula of Some of the Economically Important Fishes

Sl. No.	*Species*	*No of Chromosome*	*Chromosome Formula*
1	*Catla catla*	50	8m+16sm+14st+8t+4T
2.	*Labeo rohita*	50	10m+14sm+6st+20T
3	*Cirrhinus mrigala*	50	6m+8sm+14st+16t+6T
4.	*Labeo calbasu*	50	10m+10sm+14st+16T
5.	*Ctenopharyngodon idella*	48	14m+20sm+8st+2t+4T
6.	*Hypophthalmichthys molitrix*	48	20m+12sm+6st+8t+2T
7.	*Labeo bata*	50	6m+18sm+16st+6t+4T
8.	*Cirrhinus reba*	48	18m+2sm+6st+4T
9.	*Cyprinus carpio*	100	24m+24sm+52T
10.	*Amblypharyngodon mola*	50	12m+20sm+8st+10T
11.	*Puntius japonicus*	50	6m+14sm+8st+6t+16T
12.	*Punctius stigma*	48	4m+2sm+42T
13.	*Punctius ticto*	58	14m+22sm+6st+8t
14.	*Tor putitora*	100	10m+24sm+14st+52A
15.	*Tor tor*	100	24m+24sm+6st+46A
16.	*Rita rita*	54	14m+34sm+6st
17.	*Notopterus chitala*	42	42T
18.	*Notopterus notopterus*	42	42T
19.	*Mystus aor*	52	20m+14sm+10st+8T
20.	*Mystus gulio*	58	30m+12sm+2st+14t
21.	*Mystus vittatus*	54	20m+24sm+10st
22.	*Pangasius pangasius*	58	14m+6sm+18st+6t+18T
23.	*Wallago attu*	86	12m+6sm+2st+66A
24.	*Claris batrachus*	50	18m+20sm+8st+4t
25.	*Heteropneustes fossilis*	56	18m+10sm+12st+16t
26.	*Amphipnous cuchia*	42	4m+12t+26T
27.	*Chanda nama*	48	2sm+46T
28.	*Chanda ranga*	44	4m+40T
29.	*Nandus nandus*	48	4m+36sm+6st+2t
30.	*Anabas testudineus*	46	4st+2t+22T
31.	*Channa gachua*	78	12m+12sm+4st+14t+36T
32.	*Channa marulius*	44	4m+40T
33.	*Channa punctatus*	32	14m+12sm+6st
34.	*Channa striatus*	40	8m+2sm+16st+14T
35	*Chela bacaila*	50	10m+12sm+10st+4t+14T

(*Source*: FISH CHROMOSOME ATLAS by NBFGR, Lucknow, India)

Chapter 11

Culture of Livefeed Organisms for Larviculture

Introduction

The newly hatched larvae of fin fish/shell fish, soon after yolk absorption, start feeding on live diet. Hence production of adequate quantity of live diet to meet the feed requirement of growing larvae form the key component in the successful operation of hatchery of fin fish/shell fish. Usually, in fishes with large size eggs, absorption of yolk sac takes longer time and hence by the time it starts feeding, it may have a well developed alimentary system and large mouth. For example, salmon hatchling possesses a large yolk sac which provides required nutrition for the first three weeks of its development. Once the yolk is fully consumed and exogenous feeding starts, it develops sufficiently large mouth to feed on even formulated feeds. But in species with smaller size eggs, yolk reserve in an egg lasts only for 1-2 days. When the larva starts feeding, its mouth opening measures less than 0.1 mm and has very primitive digestive system. In case of larvae such as that of shrimps, size of mouth is not the only problem, as it passes through different stages changing from herbivorous filter feeding habit to carnivorous predatory habit.

Larval live diet can be categorised into 3 types such as:

(a) Different species of microalgae ranging in size from 2 to 20 mm.

(b) Zooplankters such as rotifers, cladocerans and copepods of size range 50 to 200 mm

(c) Brine shrimp (*Artemia*) larvae of 200-500 mm.

While selecting a live diet for culturing larvae of a given species, it is necessary to know the larval preference for a particular diet which shall be depending on the mouth size and digestibility of the larva and acceptability of the food item. For example, the larvae of tiger shrimp prefer diatoms such as *Chaetoceras*, *Skeletonema* where as the larvae of molluscans such as Pearl oyster, mussel and clams prefer phytoflagellates such as *Isochrysis galbana*, *Tetraselmis* (*Platymonas*), *Pavlova*, *Dicrateria* and *Chromulina* etc. Though attempts are made to rear the larvae on plankton collected from wild source, it has its limitations with regard to its availability, purity, acceptability, digestibility and with regard to meeting the energy and nutritional requirements of the larvae. Hence, in order to upscale aquaculture as an industry, standardisation of live diet culture on a commercial scale is absolutely necessary.

Live Feed Enrichment Technology

Another progress in live feed culture for larval rearing is the live feed enrichment technology with a view to enhance the nutritional quality of the feed. It has been found that live feed such as artemia nauplii and rotifer contain low quantity of essential amino acids (EAA) and essential fatty acids (EFA). So also only half of the methionine and tryptophan requirement of the fish larvae is met by feeding with artemia or rotifers. They lack DHA (Docosahexaenoic acid) and AA (Arachidonic acid). Hence various enrichment techniques of live feed are used recently with a view to increase the survival rate, improve growth, enhance metamorphosis, reduce deformities of skeleton, enhance pigmentation and stress resistance of fish larvae fed on live feed. (Harel, M, and place, A. R. 1998)

CULTURE OF MICROALGAE

The common species of microalgae cultured for larval rearing are as noted below:

Class 1: Bacillariophyceae

Chaetoceras gracilis, C. densum, C. danicum, C. calcitrans, C. simplex, C. curvisetus, Skeletonema costatum, Thalassiosera pseudomonas, T. fluviatilis, Phacodactylum tricornutum, Ditylum brightwelli, Scenedesmus sp.

Class 2: Haptophyceae

Isochrysis galbana, Dicrateria inornata, Cricosphaera carterae, Coccolithus huxley.

Class 3: Chrysophyceae

Monochrysis sp.

Class 4: Prasinophyceae

Pyraminimonas grossii, Tetraselmis suecica, T. chuii, Micromonas pusilla.

Class 5: Chlorophyceae

Dunaliella tertiolecta, Chlorella autotrophica, C. ellipsoidea, C. pyrinodosa, Chlorococcum sp. ; *Nannochloris atomus, Chlamydomonas coccoides, Brachiomonas submarina.*

Class 6: Chryptophyceae

Chroomonas salina

Class 7: Cyanophyceae

Spirulina maxima, S. platensis, S. fusiformes

Tetraselmis suesica, Dunalieiia tertiolecta, Dunaliella salina, Chlorella, Nannochloris etc. are commonly called unicellullar green algae. *Isochrysis galbana, Monochrysis lutheri, Pseudoisochrysis etc, are called* unicellullar brown algae. *Phaeodactylum tricornutum* is a unicellullar diatom where as *Skeletonema costatum* and *Chaetoceros calcitrans* are diatom chains.

Out of the above listed algal species, those commonly used in shrimp hatcheries are diatoms like *Chaetoceras, Skeletonema*, and *Cyclotella* and unicellular brown algae like *Isochrysis*, and unicellular green algae like *Tetraselmis* or *Platymonas* (Fig. 11.1).

Various steps in the culture of microalgae are as noted below.

(a) Isolation of pure algal strains from water

(b) Preparation of culture media

(c) Maintenance of stock culture

(d) Flask culture

(e) Mass culture

(f) Harvest of culture.

Isolation of Pure Algal Strains

Various methods such as pipetting method, centrifuge method, method by taking advantage of phototactic movement, agar plating method, and serial dilution culture technique are utilized for isolating microalgae of desired species from water. By pipetting methods, organisms are viewed under a microscope and desired species are pipetted out using a micropipette and the same are transferred to culture tubes having suitable culture medium. Centrifuge method is done by centrifuging at different speeds and collecting the sedimented organism for further culture. Phototactic method is used for isolating flagellates which exhibit phototaxis. This can be done by placing water sample in a glass container in a dark chamber having a hole on one side. By keeping a candle near the hole, the flagellats agregate near the light and the same are pipetted out for culture. By agar plating method, the desired species, taken by a platinum needle is inoculated in a agar plate and the same are incubated at 25^0c for 7-8 days under 1000 lux light. By this time, the algal species multiply in the agar plate and the same is taken out by platinum loop into culture tube containing suitable culture media for culture. Serial dilution techinque is the most prevalent method. In this method, the sea water is filtered through 10 micron sieve and the filtrate is inoculated in 5 series 15ml of culture tubes at different concentration and the same is kept under light (1000 lux) and uniform temperature of 25^0c. After 15 days, growth of microalgae can be noted in culture tubes. Then it is subcultured in 50ml flasks, and then in 500ml, and later one litre flasks. Once the culture is fully purified, it can be transferred to culture flasks as stock culture. The filtration of water and enrichment of medium has to be done two days prior to inoculation.

Preparation of Culture Medium

Different species of microalgae require different concentrations and types of nutrients for growth. The three catagories of nutrients used are inorganic macronutrients such as $NaNO_3$, NH_4Cl, NaH_2PO_4, Na_2 glycerophosphate, Na_2Si0_3 $9H_20$.The inorganic micronutrients are FeEDTA, Na_2EDTA, $FeCl_36H_20$, $CuSO_4.5H_20$,

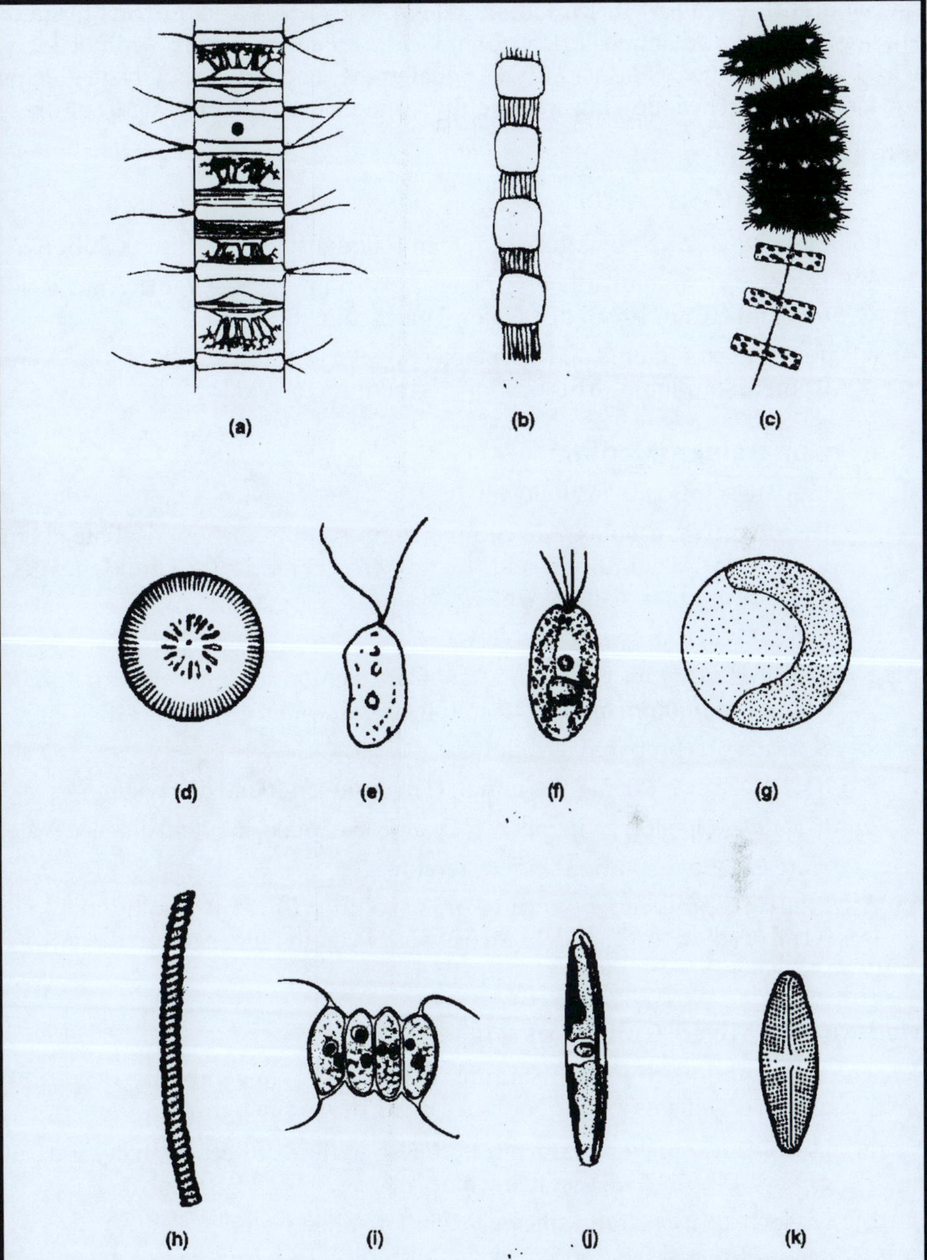

Fig. 11.1: Common Species of Micro Algae Used as Feed for Fin Fish / Shell Fish Larvae.

(a) *Chaetoceras* sp., (b) *Skeletonema* sp., (c) *Thalassiosera* sp., (d) *Cyclotella* sp., (e) *Isochrysis* sp., (f) *Platymonas* sp., (g) *Chlorella* sp., (h) *Spirulina* sp., (i) *Scenedesmus* sp., (j) *Nitzchia* sp., (k) *Navicula* sp.

Zn $SO_4.5H_20$, $C_0Cl_2.6H_20$, $MnCl_2.4H_20$, $Na_2M_00_4$ $2H_20$ and boron. Organic micronutrients are Thiamine HCl, Nicotinic acid, Ca Pantothenate, b-Aminobenzoic acid, Biotin, i-Inositol, Folic acid, Cyanocobalamine, Thymine, tris, Glycyl glycine, and liver extract. The following are the culture media used for microalgal culture.

Miquel's Medium

This can be prepared as follows:

(a) Dissolve 20.2gm potassium nitrate in 100ml dist water. This is solution 'A'.

(b) Dissolve 4 g sodium orthophosphate, 2 g calcium chloride, 2gm ferric chloride and 2ml HCl in 100ml dist water. This is soln. 'B'.

(c) Take 1 liter of filtered and sterilized seawater and add 0.55ml of soln. A and 0.5ml of solution'B'. This is Miquel's medium for algal culture.

Conway or Walne's Medium

(a) Prepare solution 'A' as follows:

Dissolve 100gm potassium nitrate, 20gm sodium orthophosphate, 45gm EDTA (Na) 33.6gm Boric acid, 1.3gm Ferric chloride, 0.36gm Manganese chloride in 1 liter distilled water.

(b) Solution 'B' is prepared as follows

Dissolve 2.1 g Zinc chloride, 2.0gm Cobalt chloride, 2gm copper sulphate, 0.9gm Ammonium molybdate in 100 ml. dist water.

(c) Solution C is prepared as follows

(C, 1). Disslove, 200mg Vitamin B_1 (Thiamine) in 100ml dist water

(C. 2). Dissolve 10mg vitamin B_{12} (Cyanocobalamine) in 100ml distilled water

Store the above solutions in refrigerator

Prepare the medium by adding 1ml of solution 'A', 1 ml of solution 'B' and 0.1ml of solution $'C'_1$ and 0.1 ml of solution $'C'_2$ in 1 liter of filtered, sterilised sea water.

Medium for Mixed Culture of Algae

The following medium is recommended for growing mixture of various phytoplankton organisms in outdoor tanks using direct sun light.

(a) Dissolve 1.32gm Potassium nitrate, 0.66 g sodium orthophosphate, and 0.66 g (Na) EDTA in 25ml distilled water.

(b) Dissolve 0.66 g sodium silicate in 25ml distilled water.

(d) Obtain 100 liter of fresh seawater and filter through 0.33 mm mesh organdy net to remove zooplankton. Add soln. 1 and 2 into this seawater. Pour this into fiberglass tanks or 3-4 basins and keep in open sunlight. Within 24 hours diatom and nannoplankton grow in this. Under higher temperature (28 to 35^{0}C) and bright sunlight (20×10^3 to 120×10^3 lux), blooming of Chaetoceros occurs (75 % Chaetoceras+25 % others such as Thallassiossira, Skeletonema, Navicula, Nitzchia). If water temperature is between 26^0 and

28^0 C Thalassiossira will dominate. At temperature below 26^0 C Skeletonema will dominate.

Guillard and Ryther's Modified 'f' Medium

Guillard & Ryther's modified f or f/2 or f/4 media are recommended for culture of algae, out of which f/2 is considered more suitable. This medium can be prepared from a working stock solution of different categories such as (i) nitrate -phosphate, (ii) silicate, (iii) trace metal (iv) vitamins and (v) $FeCl_3$, (vi) EDTA. Working stock solution of vitamins and trace metals are prepared from a primary stock solution.

Preparation of Primary Stock Solution of Tracemetals and Vitamins

Trace Metal Solution (in Distilled Water)

Four primary stock solution of trace metal (A, B, C, D) are prepared

Trace metal solution A

$CuSO_4.5H_2O$	1.96g
$ZnSO_4.7H_2O$	4.4 g
Distilled Water	100 ml

Trace metal solution B

$Na_2MoO_4.2H_2O$	1.26g
Or	
$(NH_4)_6Mo_7O_{24}4H_2O$	6. 43g
Distilled Water	100 ml.

Trace metal solution C

$MnCl_2.4H_2O$	36g
Distilled Water	100 ml

Trace metal solution D

$CoCl_2.6H_2O$	2 g
Distilled Water	100 ml

Primary Stock Solution of vitamin B_{12} (cyanocobalamin)

B_{12}	0.1 g
Distilled Water	1 lit.

Primary Stock Solution of Biotin

Biotin	0.1 g
Distilled Water	1 lit.

Preparation of Working Stock Solutions

N-P solution

$NaNO_3$	42.074 g

$NaH_2PO_4.H_2O$	5 g
Distl. Water	1 lit.

Silicate solution

$Na_2SiO_3.9H_2O$	16.5g
Distilled Water	1 lit

Trace metals

1 ml of primary stock trace metal solution A + 1 ml of B + 1 ml of C+ 1 ml of D in 1 lit of distl. water.

Vitamins

Vitamins B_1	0.2 g
Vitamins B_{12} (Primary stock solution)	10 ml
Biotin (Primary stock solution)	10 ml
Distilled Water	1 lit.

$FeCl_3$

$FeCl_3.6H_2O$	1.45 g
Distilled Water	1 lit.

Na EDTA

Na_2EDTA	10 g
Distilled Water	1 lit.

Preparation of f, f/2, f/4 Guillard Medium from Working Stock Solution

f-type Guillad medium will have 2 ml of N-P. solution, 2 ml of silicate solution, 1 ml of vitamin solution, 2 ml of $FeCl_3$, 1 ml of EDTA and 1 ml of trace metal solution in 1 lit sea water.

f/2 Guillard medium shall have 1 ml of N-P solution, 1 ml of silicate solution, 0.5 ml of vitamin solution, 1 ml of $FeCl_3$, 0.5 ml of EDTA, 0.5 ml of trace metal solution in 1lit sea water.

F/4 Guillard medium shall have 0.5 ml of N-P solution, 0.5 ml of silicate solution, 0.25 ml of vitamin solution. 0.5 ml of $FeCl_3$, 0.25 ml of EDTA & 0.25 ml of trace metal solution in 1 lit sea water.

Different species of algae grow in different salinities and different concentrations of nutrients. For example:

(a) *For growth of Chaetoceras*, F/2 guillard and Ryther's medium is recommended.

(b) *For growth of Thallassiosera* salinity has to be 28-30 ppt, and silicate should be double the amount of others such as N. P. solution,

(c) *For Skeletonema*, salinity has to be reduced to 25 ppt, and use the same media as that of Thalassiosera.

(d) *For culture of Isochrysis*, 10ml of inoculum is added to Coneways medium in 1 lit of filtered, sterilised seawater and keep the same under 1000 lux illumination in an air conditioned room. Within 8-10 days exponential growth phase is reached and then light intensity is reduced to half. After 15 days it enters in to stationary phase and in this conditon it can be maintained for 2 months. Before entering into death phase it is used as inoculum for mass culture. For mass culture 2 lit of inoculum is added to 100 lit filtered sterilised sea water to which nutrients are added. Bloom of Isochrysis occurs with 5-6 days.

(e) *For culture of Chlorella* procedure is noted below:

Preparations of Stock Solutions

The stock solution noted below are prepared is 1000 ml dist water separately.

(a) *Potassium nitrate* 202 g in 1000 ml dist water.

(b) *Sodium dihydrogen* phosphate. 310.5 g in 1000 ml dist water.

(c) *Sodium monohydrogen* phophate 89 g in 1000 ml dist water.

(d) *Magnesium sulphate* 246.5 g in 1000 ml dist water.

(e) *Trace metal solution.* Dissolve 61 mg boric acid, 169 mg magnous sulphate, 287 mg zinc sulphate, 2.5 mg copper sulphate, and 12.36 mg ammonium molybdate in 1 lit dist water.

(f) *Fe-EDTA complex*. Dissolve 6.9 g ferrous sulphate, 9.3 g disodium salt of EDTA in 800 ml distilled water. Boil and cool it and make up to 1000 ml by adding dist. water.

To prepare 1 liter of medium for chlorella, and one ml each of solution 3, 4, 5 and 6 and five ml of soln '1' and two ml of solution '2' to 1 lit distilled water. This medium can be inoculated with chlorella at a density of 70×10^6 cells/ml.

Culture Media of Fresh Water Micro Algae

Two types of culture media are usually recommended for culture of fresh water algae *viz.* Chu 10 medium and Zarrouk's medium.

Chu-10 medium

Preparation of stock solution of 'A_5' or A_5 solution

$MnCl_2.4H_2O$	1.81 g
$MoO_3.4H_2O$	0.0177 g
$ZnSO_4.4H_2O$	0.222 g
$CuSO_4.5H_2O$	0.079 g
H_3BO_3	2.86 g
$CoCl_2$	0.008 g
Distl. Water	1000 ml
$Ca(NO_3)_2.4H_2O$	0.08 g
$K_2HPO_4.2H_2O$	0.02 g

$MgSO_4.7H_2O$	0.05 g
Na_2CO_3	0.04 g
$Na_2SiO_3.9H_2O$	0.05 g
$FeCl_3$	0.0016 g

Dissolved in 1 lit of distl. Water and 1 ml of A_5 solution.

Micro algae such as chlorella, chlorococcum and diatons grow in this medium after the medium is inoculated usually with 10 ml of inoculum.

For growth of Chlorella under pond conditions on a commercial scale, application of N. P. K. fertilizer (15:15:15) such as 'suphala' '@ 250 g in 200 litres of fresh water is recommended.

Zarrouk's Medium

B_6 solution

NH_4NO_3	22.9 g
$NiSO_4$	47.8 g
Na_2WO_4	17.9 g
$TiSO_4$	4.0 g
$Co(NO_3)_2.6H_2O$	4.4 g
Distl. Water	1000 ml
$NaHCO_3$	18 g
$KH_2PO_4.2H_2O$	0.5 g
Na_2CO_3	4.03 g
$NaNO_3$	2.5 g
NaCl	1.0 g
$MgSO_4.7H_2O$	0.2 g
$FeSO_4$	0.01 g
K_2SO_4	1.0 g
$CaCl_2$	0.1 g

Dissolve in in lit of distl. water and add 1 ml of A_5 solution and 1 ml of B_6 solution. This medium is meant for culture of Spirulina. The solution is to be inoculated with 10 ml of water containing spirulina.

Growth Pattern of Algae

When a medium containing nutrients is inoculated with a given species of algae and the same is exposed to suitable environmental conditions such as light, temperature and aeration algal cells increase following a definite pattern as shown below (Fig. 11.2).

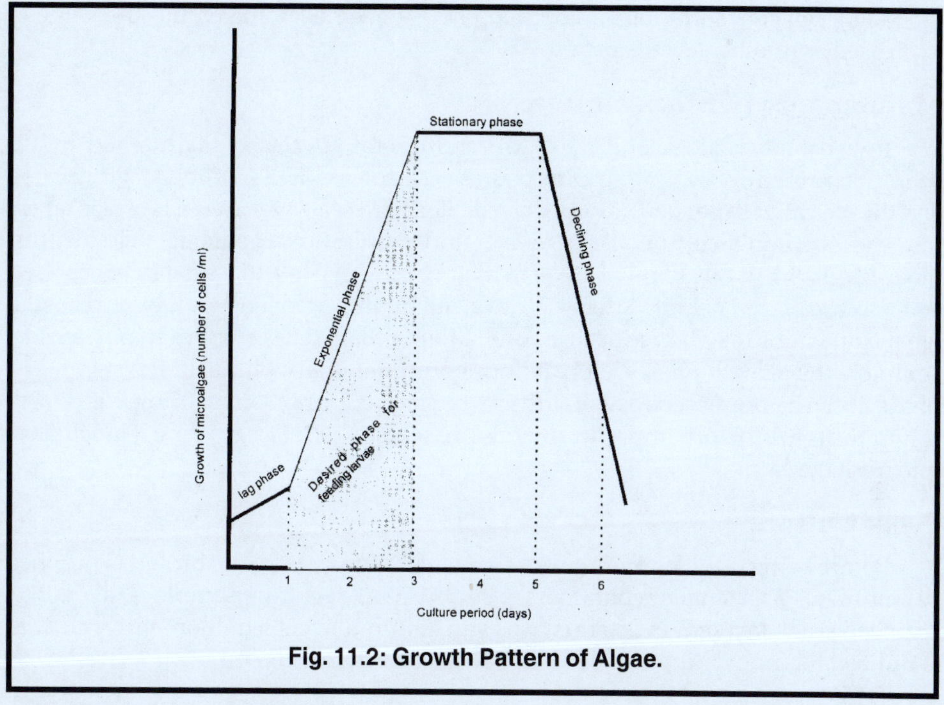

Fig. 11.2: Growth Pattern of Algae.

Lag Phase (Induction Phase)

When a given algae is inoculated into a fresh culture medium, it takes a few hours to acclimatise to the new medium. During this time there will be no cell division though cells may increase in size and this phase is called lag phase or induction phase.

Exponential Phase (Log Phase)

Once the algal cells are acclimatised to the new surrounding, it will start multiplication by division and undergoes rapid growth. As a result of this, it reaches maximum concentration. This phase of growth is called exponential phase.

Declining Phase (Transitional Phase)

Once the algal cells reached the maximum concentration, growth and multiplication is slowed down, this phase of stunted growth is called declining phase.

Stationary Phase

After the decline in growth, algal cell concentration will be stationary for a few days, when reproduction is balanced by death. This phase is a prolonged one in case of flagellates.

Death Phase

After a period of stationary phase, cells start dying and death rate exceeds growth. During this phase, the culture will not be useful for reculturing or for feeding.

Algal cell concentrations is counted using a haemocytometer and the value is expressed as number of cells per ml.

Maintenance of Stock Culture

In a hatchery, stock culture of different algal species is maintained in air conditioned room, observing strict measures of cleanliness and sterility. All glasswares and the sea water to be sued are autoclaved. Sterilised sea water after it is cool is taken into the Haufkin's culture flasks (50ml) and required nutrients are added. It is inoculated with about 10 ml of algae in the growing phase and 2 tube lights each of 1000 lux are placed near it. After 8-10 days, maximum exponential phase is reached when tube light may be reduced to one. Most of flagellates require two weeks to complete the growth phase before entering into stationary phase. In the stationary phase, it can be preserved for 2 months in the stock culture room. During this phase, colour turns to brown. Sample during this phase also shall start growing if inoculated into fresh medium.

Flask Culture

Before being used for mass culture, the stock culture is taken through a number of culture steps in smaller containers. Usually, 20ml of culture sample is taken into enriched sea water kept in 250 ml flasks and cultured to get 5 million cells/ml. This is utilized for inoculating smaller flasks to get 3 million cells/ml.

Mass Culture

Mass culture can be done either indoors or out doors. 200 liter FRP cylinders containing enriched sea water is inoculated and maintaining optimum environmental conditions such as light, temperature and aeration, algae can be made to grow and reach the exponential phase At this phase it is given to the larval rearing tanks. The mass culture can be done outdoor also by inocultating enriched sea water taken is fibre glass tanks of 100 lit capacity and utilizing sunlight.

Harvest and Feeding

Algal water from indoor and outdoors can be supplied directly into the larval rearing tanks. Harvesting is to be done during the exponential phase as during the declining or stationary phase, metabolites will be very high and the cells may not be in healthy condition.

The volume of the feed is determined after measuring the cell density with haemocytometer.

Diatom cells can also be preserved in a frozen condition. For this purpose, first the diatom cells are concentrated using a dairy cream separator. The concentrate is removed to a polythene container and preserved at -4^0C to -20^0C in a freezer. This can be used whenever required.

Infrastructure and Water Sterilization for Algal Culture

Usually, an algalculture section of a hachery will have three units:

(a) Unit for maintaining pure strains and small culture vessles,

(b) Unit for duplicating small cultures, and

(c) Unit for mass culture.

Small cultures are done in containers ranging from 20 ml test tubes to 18 litre carbuoys. The vessels should be made of materials that can withstand sterilization process (boro silicate glass, Poly carbonate, P. E. T). These vessels are kept in glass shelves lighted by fluorescent tubes. The section should also have facilities for supply of pretreated water, such as sand filters and sterilizers and also facilities to prepare and store nutrients.

In marine hatcheries sea water is the medium to prepare the algal culture. The water should be made free of pathogens, pollutants and other organisms. This is done by settling, filtration and sterilization.

Settling and Filtration

Suspended solids and contaminant organisms are removed by settling and sand filtration. Water is pumped into a sedimentation tank where it is allowed to stand for some time which allows the heavier particles to settle at the bottom.

Settled water is filtered through a sand or bag filter which retain particles as small as 50 mm 10 mm or 5 mm. This water is then sterilized as shown below.

Sterilization

U. V. Light Sterilization

U. V light with a wave length of 265 nm has a strong germicidal effect as it damages the DNA of the microorganisms. This is produced by special high or low pressure mercury vapour lamps. Its efficacy depends on its power, transparency to UV, type and quantity of micro organisms to be destroyed, degree of purification required, water flow (contact time) and temperature. The water is allowed to flow through sealed chambers where it is irradiated by one or more lamps placed inside quartz tubes. Thickness of water film inside the chamber should be such as to allow maximum sterilization effect. Intensity of at least 40mj/cm^2 removes 99% of microorganisms.

Chlorine Sterilization

Active chlorine is a strong oxidising agent. It is available as sodium hypochlorite (NaOCl) or as bleaching powder ($CaOCl_2$). Commercial grade Naocl contains 5 – 15 percent active chlorine, where as $CaOCl_2$ contains 60 – 70%.

5-10 ppm active chlorine is used to sterilize sea water. Contact time between water and chlorine should be atleast one hour. Residual chlorine is neutralised with sodium thiosulphate. This technique is used for larger vessels and for culture equipments

Use of an Autoclave or Wet Vapour Sterilization

This is applicable to small volumes. Water is sterilized in culture vessels (5- 6 ltrs) made of pyrex glas. (120^0 C and 2 atmospheric pressure.) Time range from 10 minutes (100ml flask) to 20 minutes (200ml flask) to 30 minutes for 5 – 6 ltrs vessels.

Dry Vapour Sterilization

Instead of autoclave, oven is used for the purpose. Water in vessel is heated to 160 – 170^{0}C for 2 – 3 hours

CULTURE OF ROTIFERS

Introduction

Rotifers (Lat. r*ota* means *wheel, fero* means to *bear* or wheel bearers), commonly called wheel animalcules, form the major component in the zooplankton population on which the fry of fish feeds soon after yolk sac absorption. Of the 1500 species of rotifers known at present (Hyman, L. H. 1992), majority are free living whereas some are sessile while some are enclosed in a tube or envelop. They are mostly found in fresh water while some are marine (order seisonacea) and some in brackish water. Because of its better nutritive value and digestibility, in certain countries like Japan rotifers like *Brachionus plicatilis* (Fig. 11.3) is cultured on a mass scale for feeding marine fish larvae.

Biology of Rotifers

They are microscopic multicellular animals of size ranging from 0.04 mm to 2 mm average being 0.5 mm. Its body has a fixed number of about 1000 cells and the

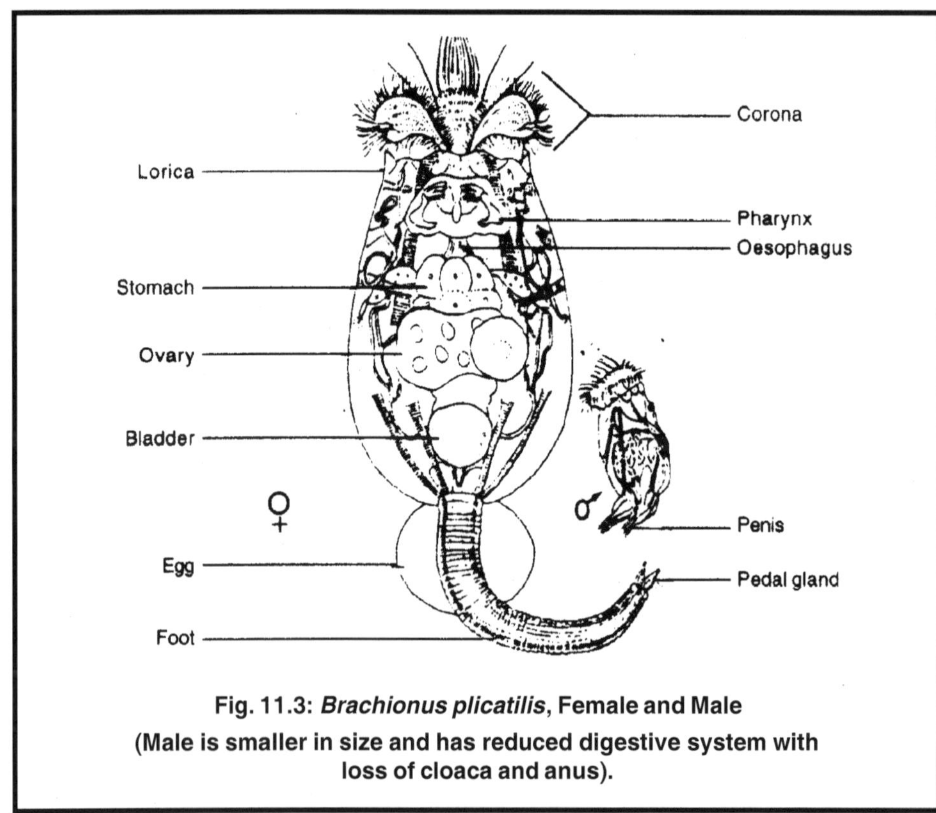

Fig. 11.3: *Brachionus plicatilis*, Female and Male (Male is smaller in size and has reduced digestive system with loss of cloaca and anus).

growth is not by cell division but by growth of cytoplasm (Moretti, A. *et al.*, 1999) Their shape varies from elongated, sacciform or spherical. Typically elongated body is divisible into an anterior end provided with ciliary apparatus, (corona), elongated trunk and a terminal region, tail or foot. Body wall consists of cuticle, syncitial epidermis and sub epidermal muscles. The cuticle which is secreted by epidermis is not chitinous but made up of scleroproteins. In many species the trunk cuticle is thickened to form a lorica which may cover the entire trunk or in some anterior and/or posterier part may be left free. It has alimentary canal starting from mouth, pharynx, oesophagus, stomach and intenstine which opens by anus. The ring of cilia present on the corona produces a whirling water movement which facilitates injestion of small food particles like algae and detritus. This also helps in locomotion. In female a cloaca is present as the oviduct opens into the last part of alimentary canal but in male cloaca is not found as the sperm duct opens separately by genital pore. Rotifers posses a characteristic pharynx (mastax) having 7 numbers of hard cuticular pieces or trophi in the inner wall which are named as unpaired 'fulcrum', paired 'rami' and paired 'unci'. The excretory system consists of protonephridia which opens posteriorly into urinary bladder. They are bisexual and exhibits distinct sexual dimorphism. Males are usually smaller than the female. In ploima group, males are only ½ to 1/8th of the female and also exhibits reduced digestive system with loss of cloaca and anus. They also differ in the form of lorica, corona and lack of urinary bladder. In rotifers of the sub-order Floscularіceae and Collothecaceae males are 1/10 or less than female and also has no distinct corona-corona being replaced by simple ciliated anterior end. In those of order Bdelloidea males are entirely absent, reproduction being only by parthenogenesis. Female reproductive system consists of a syncitical ovary and a syncitical vitellarium bound in a membrane and opening into cloaca by a tubular duct. Male has a testis which leads into sperm duct which receives a pair or more prostate glands and opens out by genital ore. Gonads may be paired in some while in others it is unpaired. There are two categories of females such as amictic and mictic. There are no anatomic distinctions between the two but amictic female lay 'amictic eggs' *i.e.* eggs that give of only one polar body and hence are diploid, where as 'mictic' females produce eggs that give off 2 polar bodies and hence are haploid. If the eggs of mictic female undergo parthenogenetic development; it develops to male, if the egg are fertilized, it secretes a shell around it which can remain dormant for several days or months. So sexual phase is only meant for producing dormant eggs to tide over unfavourable conditions whereas asexual phase (parthenogenetic phase) of amictic female is the common way of propagation. During sexual phase, in some, male introduces the sperms into the cloaca of female during copulation, whereas in a few others, sperms are injected through the body wall into the pseudocoel of female. In these males, sperms are of 2 types, 'typical' with rounded or oval head and a tail and 'atypical' sperms having rod shaped bodies. The rod shaped one helps in penetratiing the cuticle during insemination. Parthenogenetic reproduction appears to be the rule when there is right type of food, optimum water temperature, light penetration and pH. In a mass culture, therefore, environmental conditions and feeding are to be maintained at the optimum required levels so that only asexual reproduction shall take place as sexual reproduction leads to production of resting eggs. The favoured food for *Brachionus* are micro algae such as chlorella, bacteria and yeast. So primary requirement for culturing rotifer is produciton of its feed.

Growth Pattern of Rotifers

The life span of a rotifer depends on temperature. At 25^{0}C it is estimated as seven days. At this temperature the larva becomes adult within 0.5 – 1.5 days. Then the female starts laying eggs once in every four hours. One female can lay about 20 eggs during its lifetime. Rotifers like *B. plicatilis* exists in two morphotypes *viz.* 'S' – or small type 'L' or large type. In 'L' type the average size is 239 mm where as in s type the average type is 160 mm. They also differ in weights, and in resistance to temperature. 'S' type has higher temperature resistance. In case of saltwater rotifers, 'S' type grows better in 18 – 20 ppt where as 'L' type grows at 30 ppt. 'S' type rotifiers are suitable for larvae having small mouth opening. And hence usually 'S' type is preferred for culture. It reaches a density of 500 – 700 per ml in flask culture and 1000 and more in tank culture. 'L' type reaches a density of 150 – 250 per ml in flask and 400 in tank culture system.

Optimum conditions for growth of rotifier is temperature 25 – 30^{0}C pH, 7.5 – 8.5 salinity 20-30ppt. NH_3 less than 1mg per litre and high dissolved oxygen and moderate turbulence. Turbulence is meant to keep the rotifers and food particles in suspension, and this is usually done by aeration.

Growth and multiplication of rotifers in culture conditions can be divided into four phases *viz.* (i) Lag phase or induction phase, (ii) Log phase or exponential phase, (iii) Transitional or declining phase (iv) Decline phase. During log phase they reproduce very fast and exponential growth is seen. Then the growth rate slows down, egg bearing rotifers become rearer. During decline phase only old rotifers without eggs are found. To start new culture rotifer from mid log phase is selected (where at least 20% of population shall be with eggs). Under good conditions harvesting density is reached within 4-5 days.

Culture Systems for Rotifers

The culture system for rotifers like Brachionus should have 3 units. (i) unit for culturing the food organism for Brachionus, (ii) unit for brachionus stock culture, (iii) mass culture unit.

Culture of Food Organisms for *Brachionus*

Micro algae-chlorella is the most commonly used feed for *Brachionus*. (For details of the procedure for chlorella culture please see the section dealing with the culture of micro-algae).

Stock Culture of Rotifers

(a) First step in the preparation of a stock culture of Brachionus is to collect the rotifer from water bodies. About 50 liters of pond water can be filtered using a filter of mesh size 50-100 mm pores. Examine this under a stereoscopic microscope and pick up the rotifer using a fine dropper and place the same in a cavity block containing 3.5 ml of distilled/tap water. pH of this water should be adjusted to that of pond water.

(b) Next step is to take chlorella medium (one million cells/ml) into cavity blocks.

(c) Introduce one rotifer into the cavity block containing chlorella medium and cover it with the glass plate and keep in diffused light.

(d) Replace the medium with fresh chlorella @ 1 million cells/ml at every 12 hour interval transferring adult live rotifer with eggs. Gradually increase the volume to 25 ml in a 50 ml capacity beaker and change the culture daily once every time increasing the volume.

(e) At this stage, it may not be possible to transfer individual rotifers and hence use 50-70 mm size mesh to separate rotifers and continue this process till the density reaches 50 numbers of rotifers per ml and culture volume is 500 ml. At this stage chlorella density may be increased to 3-4 million cells/ml.

Mass Culture

Mass culture methods are of 3 types.

(a) Batch culture,

(b) Semi continuous culture

(c) Feed back culture.

In batch culture, after harvest, entire batch is discarded as in this system culture solution may contain other zooplankton. Then a fresh batch is initiated. Every time, a fresh batch is initiated, filtered sea water is enriched with fertilizers such as urea (5 g/ton water) ammonium sulphate (100 g per ton water) and super phosphate (20 g/ton water) and the same in inoculated with chlorella. The medium is aerated and then inoculated with rotifers once the suitable density of algae is reached. Rotifers grow and multiply rapidly and once the density reaches 100-300 individuals per ml, it is harvested. In semi -continuous culture, a particluar quantity of water harvested is replaced by fresh quantity of water with or without algae. In Feed back culture, faecal and other particulate matters settled at the bottom of rotifer culture tanks are allowed to undergo bacterial decomposition which is used as fertilizer for algae cells.

Rajmani, M. et al. (1999) recommended fertilizing the medium with Neem oil cake (250g/tone), urea @ 10 g/ton and superphosphate 5 g/ton for mass production of rotifiers (*B. plicatilis*) @277 numbers/ml within 9-12 days of culture which was directly related to the maximum production of chlorela achieved with the fertilizer.

Further, recent researches (Hirayma, K and I. Maruyama, 1991) have revealed the significance of vitamin B_{12} as a limiting factor for mass production of rotifer, *B. plicatilis*. They have reported that either chlorella or Baker's yeast alone cannot support sustained growth of *B. plicatilis* unless it is fortified with vitamn B_{12}. The dosage is 100 ml of stock solution of vitamin B_{12} (refer culture of microalgae for preparation of stock solution) per m^3 or 1000 litres of rotifer culture medium in tank. Vitamin is added along with inoculum. For small vessel culture dosage of vitamin B_{12} can be 1ml per litre.

Besides *B. plicatilis* which is euryhaline, other fresh water species like *B. rubens* are suited for mass culture in fresh water and they are fed with Scenedesmus grown in fertilized out door ponds. It is likely that in course of time, Brachionus may be

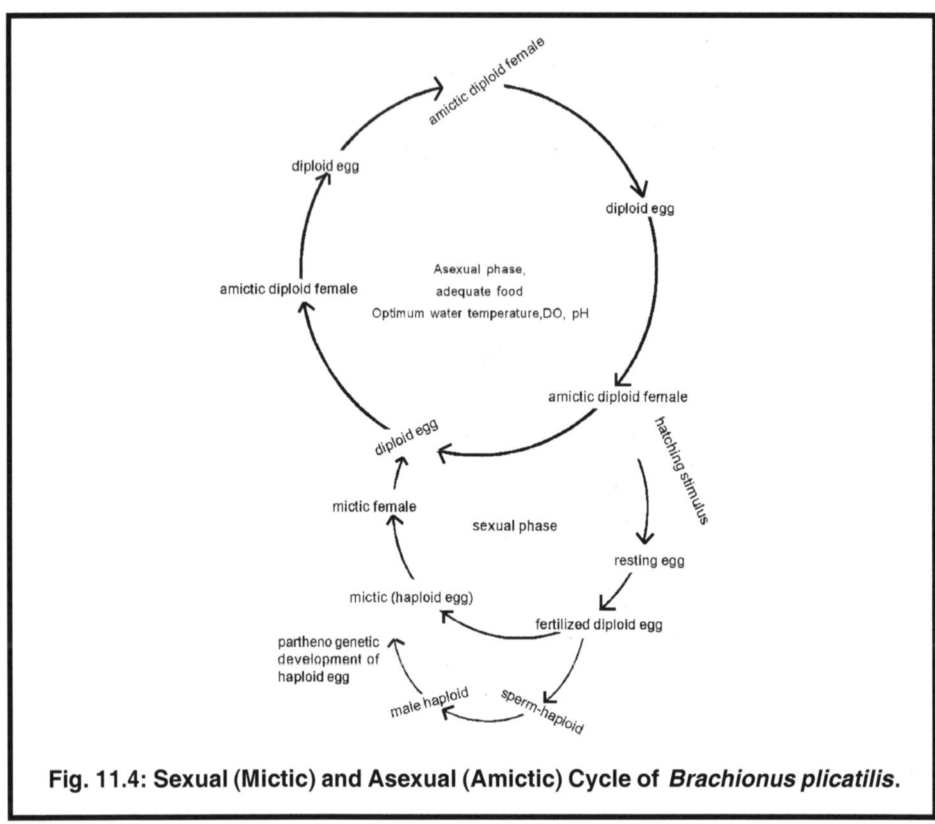

Fig. 11.4: Sexual (Mictic) and Asexual (Amictic) Cycle of *Brachionus plicatilis*.

made available on a commercial sale for hatchery use as a substitute for artemia nauplii.

Enrichment of Rotifers

This is a technique to upscale the nutritional value of rotifers by feeding them with microalgae such as *Chlorella, Nannochloris, Isochrysis* which are rich in PUFA or with artificial diet like Selco which are rich in EPA, DHA and vitamins just before using them as feed for fish larvae. Rotifers can be transferred to a separate tank @ 500 per mland fed with microalgae such as Isochrysis (@ 5 million cells per/ml) or *chlorella* and *Nannochloris* @ 12 million cells per ml. For 8 hours. Total PUFA to be given is 7 mg per g. body weight. For enrichment with artificial diet the food is added to the rotifer culture medium for atleast 6-12 hours taking care to maintain dissolved oxygen level atleast at 4 ppm. The nutrients eaten by rotifer shall decrease rapidly in its body, if rotifers are not fed to the fish larvae at the appropriate time. It is found that 60 % of the consumed PUFA is lost within six hours at 18^0 cand if algae is fed atleast 40 % is lost. To prevent this loss, enriched rotifers @ 2500 to 3000 numbers per ml can be stored at low temperatures (5to 10 ^{0}c) for upto 14 hours taking care to maintain DO atleast 4 ppm.

CULTURE OF *DAPHNIA*

Biology

Of the 420 species of cladocerans, *Daphnia* (Fig. 11.5) is the most commonly cultured species for feeding the young stages of finfish/shell fish. They are microscopic measuring about 200 mm in size. It has a large head with rostrum and a thorax and abdomen covered by a carapace which is in the form of a bivalved shell. The abdomen which is bent down ward is free of appendages and it is in constant motion sweeping out foreign particles which may have made their way into feet. Among the cephalic appendages, second maxilla is absent, mandibles are not conspicuous, antennules are small but second antennae are large and biramous and the same are locomotory. Paired eyes are fused into one. Five pairs of leaf like feet on thorax creates water current which carries food (detritus/microlage) along with it. There is a brood pouch located between abdomen and carapace of the female. At a time, there may be 20-30 eggs, in the brood pouch. Its reproduction is reported to be fast – a single female of *Daphnia pulex* can produce 13, 000, 000, 000 offsprings within a period of sixty days (Davis Charles, C, 1955). Mostly, the development is parthenogenetic. It resorts to sexual reproduction only during adverse environmental condition. Parthenogenetic development results in producing parthenogenetic females. But during adverse conditions, the eggs develop into males. After the appearance of male, the female lays eggs that are considerably larger than parthenogenetic eggs. These eggs are fertilized by male and soon such fertilized eggs develops a case around it called ephippium.

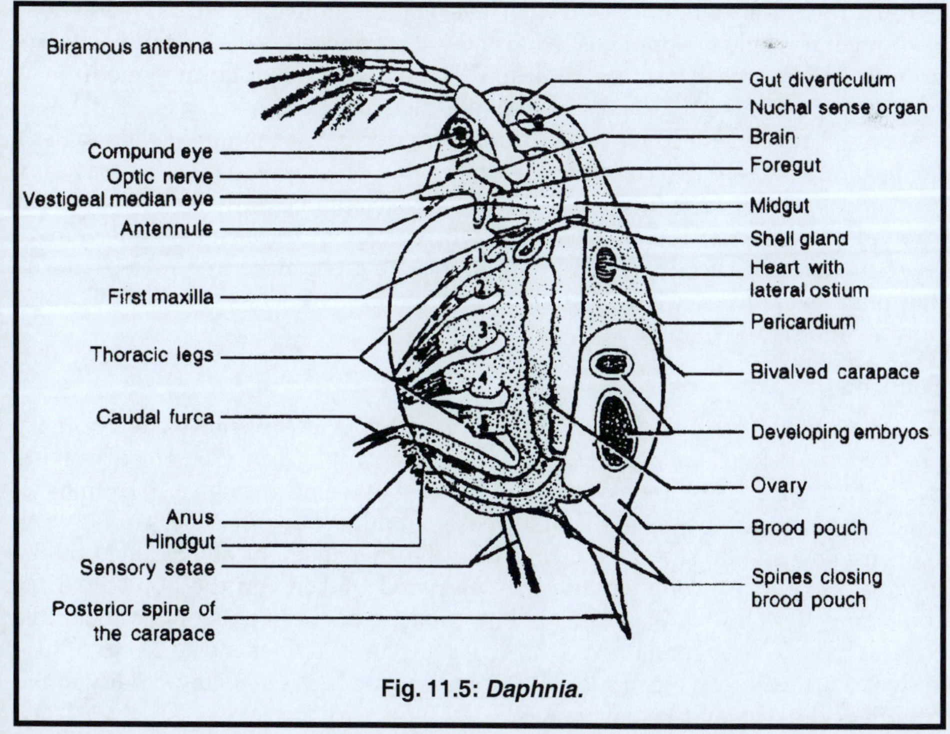

Fig. 11.5: *Daphnia*.

Such eggs can remain dormant for a prolonged period of time and then hatches out with the advent of favourable conditions.

Culture Technique

The initial step in the culture of Daphnia is isolation of *Daphnia* from water samples collected from natural water bodies. Zooplanktons are collected using 500 mm mesh size dipnet and kept in buckets containing water from collection site. *Daphnia* can be sorted out from the sample by taking the same in a petridish and observing under a stereoscopic microscope and picking up *Daphnia* by means of pipettes and placing the same in fresh water

Once a stock is obtained for inoculating the culture medium, cement tanks are prepared for culture purpose. The size of the tank shall depend on the amount of *Daphnia* to be produced. It has been estimated that in order to produce 1 ton *Daphnia* per week, culture tank should have 2500 tonnes capacity. The culture tank is prepared by adding first sun dried soil in the cement tank up to 5 cm height and lime powder @ 0.2 Kg/ha. Then water is introduced up to 15 cm height. On fourth day, poultry manure is added @ 0.4 Kg/ton of water. On the 12th day water level is raised up to 50 cm and poultry manure is added @ 1kg/ton of water. On 15th day inoculate the water with Daphina @ 10 numbers per ml.

Maintenance of Water Quality

In order to obtain the maximum rate of growth, survival and reproduction of *Daphnia*, water quality parameters such as oxygen, water hardness, ammonia and pH are to be closely monitored. Oxygen level in the culture medium has to be above 3.5 ppm. If it is below 1 ppm it is lethal and if it is between 2 and 3.5 ppm, feeding, growth and reproduction are retarded. Hence aeration of culture medium is recommended. Though, *Daphnia* can tolerate wide range of ammonia levels, it has to be 0.2 mg/l, for better feeding, growtfh and reproduction. If ammonia level reaches 1 mg per litre it causes heavy mortality. In order to level reaches 1 mg/l, it causes heavy mortality. In order to reduce ammonia level water can be partially exchanged at intervals. Hard water is found better for the culture of *Daphnia* as higher rate of reproduction and shorter sexual maturation time are noticed in hard water. The optimum level of water hardness shall be 250 mg/l ($CaCO_3$ equivalent). For this purpose lime is added in the culture tank.

Feeding

Manuring of the culture medium is done with a view to enhance the primary production and feed availability. An extract of the cowdung is prepared by dissolving fresh cowdung @ 10 g/litre in water and filtering the same through a 100 mm mesh to get uniform suspension. In the first week it is added @ 10 litre/day, second week @20 litre/day and third week @ 30 litre/day. Further anaerobically treated poultry manure @4 g/l is recommended. It is prepared by keeping the poultry slurry containing 10 g/litre of water in 100 litre capacity earthern pots under anaerobic conditions for 2-3 days and filtered through a fine meshed filter and added to culture system. Further, green algae such as chlorella are added @40-50 mg/l daily to the culture medium as supplementary feed

Harvesting

Before harvesting, the *Daphnia* population, it is necessary to ascertain the asymptote phase of growth in the population of *Daphnia*. This can be found out by plotting the graph of population density of *Daphnia* against time. Biomass of *Daphnia* is calculated by periodic sampling at 5-7 days interval and counting the number in a plankton counting chamber. Harvesting can be done during asymptote phase. There are 3 methods of harvesting such as non selective harvest, selective harvest and alternate harvest. In the non selective harvest, harvesting of all sizes of *Daphnia* is done. In selective method only individuals of medium size are harvested. This is done by using mesh of appropriate size. In this method, as mature and neonate individuals are released into culture tanks a continuous yield is possible. In the alternate harvest larger and smaller individuals are harvested.

As the harvested plankton may contain other undesirable materials such as feed residue, faecal matter, debris etc. it has to be cleaned by siphoning out the unwanted materials. This is done by the simple technique. Take the harvest sample in a tub and create a circular movement of water using hand. Then leave the hand one minute. During this time *Daphnia* will move from the centre where as other material remain at the centre using a pipette, undesirable materials from the centre are removed. This process can be repeated using fresh water.

Presence of other planktonic organisms such as *Brachionus* and *Conochilus* are harmful for the growth of *Daphnia* in the culture medium. Brachionus being smaller than *Daphnia* it can be eliminated by filtering through appropriate sieve. Conochilus is removed by adding cowdung to culture medium. Apart from this, organisms such as *Calanus* and *Anisops* harm *Daphnia* culture by predating on *Daphnia*.

THE BRINE SHRIMP OR ARTEMIA

Biology

Artemia (Fig. 11.6) is a branchiopod crustacean (Anostraca) found in salt pans or coastal/inland salt lake where salinity range between 100 ppt to 200-250 ppt. It exists in two forms (1) bisexual forms where in both males and females are present and (2) parthenogenetic forms where in only females are present. In bisexual population it attains 10 mm length but in parthenogenetic strains it grows up to twenty millimeter in length. Of the several species of Artemia, *A. salina* is the most common. Sixty different strains of brine shrimps have been identified (Sorgeloos, P., 1976). They differ in the size of the nauplii and also in HUFA contents. It has an elongated body with well developed antennule, eleven pairs of oar like functional thoracic appendages (phyllopods)and two compound eyes. Phyllopods are used for feeding, locomotion and respiration.

They are nonselective filter feeders feeding mainly on algae, detritus and bacteria. During larval stage, setae on the antenna serve in filtering food particles from surrounding water. But in juveniles and adults, phyllopods assist in feeding. When the phyllopods move forward (forestroke), the space between the legs widen as a result, water is drawn into this space from the area below the midline of the body. During back stroke this water is forced out through the space between the legs but the

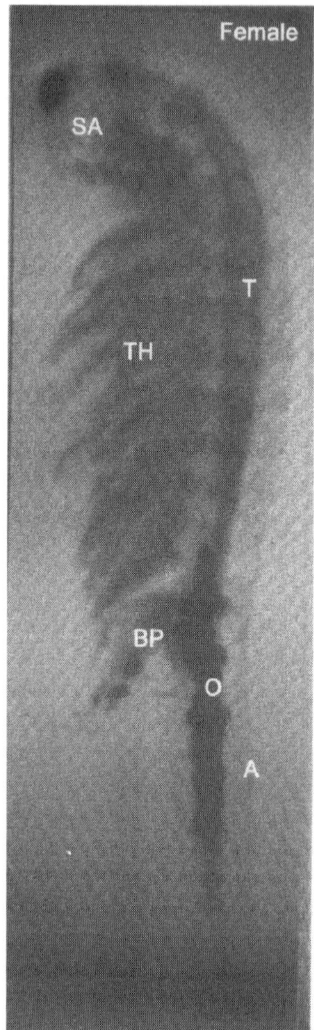

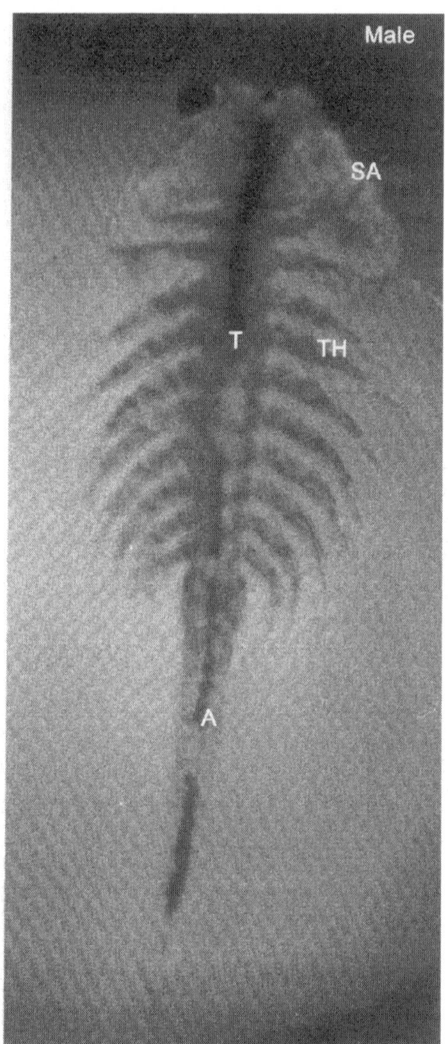

Fig. 11.6: Adult *Artemia* (Male and Female).

Note the second antenna modifed into large hooked 'graspers' or 'claspers' used for holding the female during copulation in male where as in female second antenna is less developed and sensorial in function. Further in male, at the posterior part of the trunk a paired penis is seen, where as in female an unpaired brood pouch or uterus is visible. In both sexes eleven pairs of thoracic appendages (thoracopods) perform locomotory and respiratory functions and also assists in filter feeding. Each thoracopod has three functional part – telopodite acting as filter, oar like endopodite are locomotory and exopodite function as gills.

T: Trunk, TH: Thoracopods, A: Abdomen, O: Ovary, BP: Brood pouch, SA: Second antenna.

food particles (filtered due to the setae persent in the legs) are concentrated in a food groove between the base of the legs. Glands along the groove secrete adhesive material that helps in clumping the minute particles into minute balls. As the food groove leads to the mouth, the food particles so trapped are transferred to the mouth.

Beneficial Effect of Artemia on Salt

As stated above, Artemia are nonselective filter feeders feeding on organic detritus microscopic algae and bacteria. In the evaporation tanks of salt pans. algal blooms occur. If it is allowed to grow and undergo death and decomposition, decomposition products will prevent early precipitation of gypsum which will contaminate the sodium chloride in crystalliser tanks. Further, organic impurities also contaminate the salt. Artemia that are growing in the evaporation tanks, controls the algal bloom by feeding on it. Thus it has a purifying effect on the salt produced.

Its adaptation to high salinity with the most efficient osmoregulatory system is an efficient eccological defense mechanism against predators. At high salinity *i.e.* above 100 ppt, predators of artemia are not found. It can synthesise haemoglobin to cope with low oxygen levels. Further, as stated below it can produce dormant cysts to survive in adverse environmental conditions *i.e.* beyond 150 ppt.

Sexual Dimorphism

Marked sexual dimorphism is exhibited by bisexual strains. Males have second antenna modified to grasping organs (claspers) and pair of penes are also conspicuous at the posterior part of trunk. In female, second antenna is not modified to grasping organ, but a brood pouch can be seen posterior to 11th pair of appendages. Eggs develop in two tubular ovaries located in the abdomen. Ripe eggs, after ovulation passess through the oviduct in to the unpaired uterus or brood pouch.

Mating and Fertilization

The male grasps the female with its hooked antenna in between the uterus and last pair of appendages. In this riding position they swim around for a long time (for many hours), during which time abdomen of the male is bent forward, penis is inserted into the uterus and eggs are fertilized by the sperms ejected by the male.

Reproduction

In normal conditions Artemia is ovoviviparous *i.e.* the fertilized eggs undergo development inside the uterus of the female and the nauplii are set free by the mother in to the water. But in extreme adverse conditions such as high salinity (above 150ppt) and low oxygen, shell glands located in the uterus secrete a shell around the embryo, when it is in the gastrula stage. The embryo enters into a state of dormancy or diapause. These are called cysts and the same are released by the females (oviparous reproduction). One adult artemia reproduce at the rate of 300 cysts/nauplii in every four days and it can live for several days (upto 50/60 days).

Artemia Cysts

Each cyst is 200-400 micron in diameter biconcave having a 3 layered shell. The thickest layer is chorion which is made of lipoprotein impregnated with chitin and

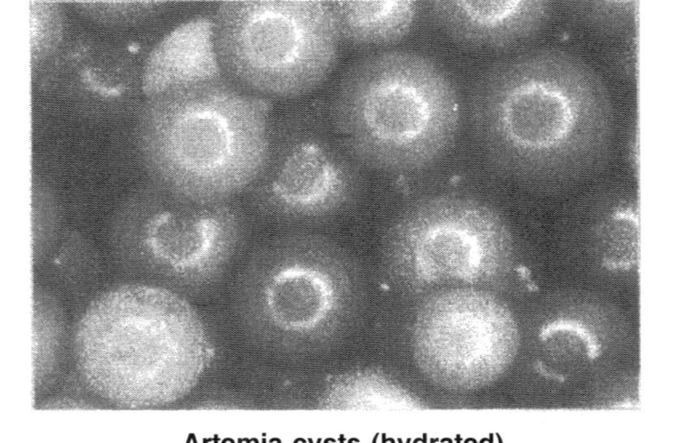

Artemia cysts (hydrated)

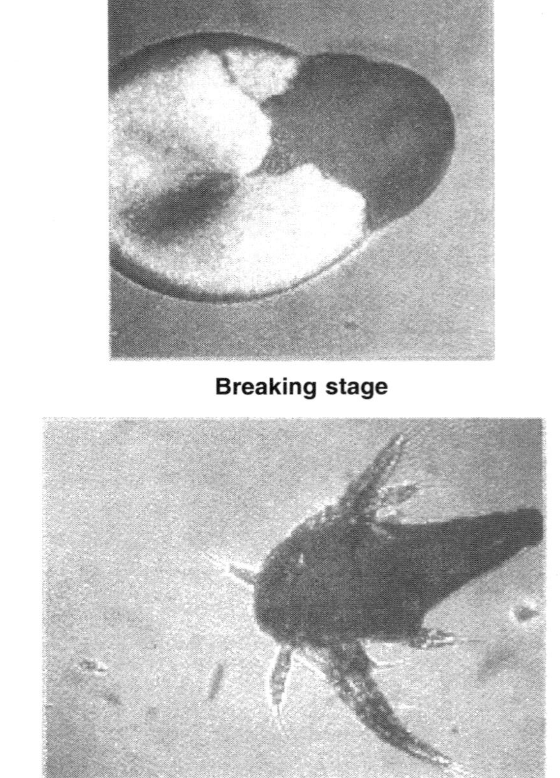

Breaking stage

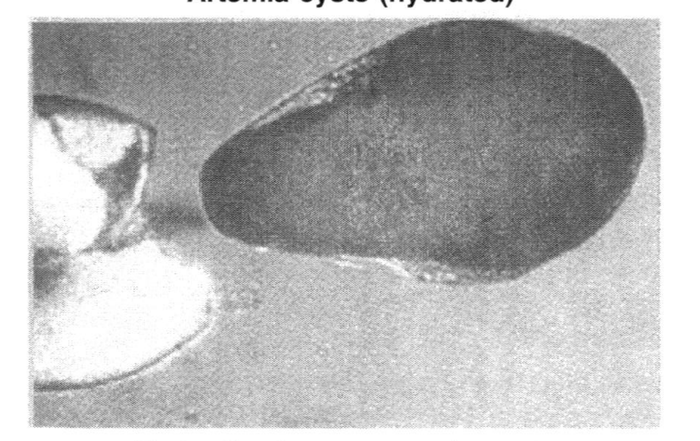

Umbrella stage or parachute stage
Embryo inside hatching membrane leaving the cyst shell

Newly hatched *Artemia* nauplius by the rupture of hatching membrane

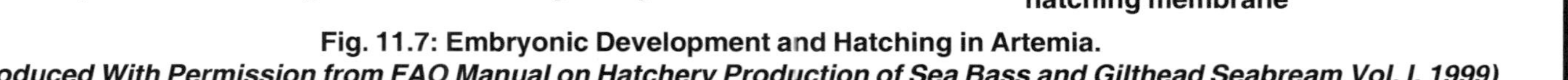

Fig. 11.7: Embryonic Development and Hatching in Artemia.
(Reproduced With Permission from FAO Manual on Hatchery Production of Sea Bass and Gilthead Seabream Vol. I, 1999)

haematin. This layer protects the embryo from mechanical stress or radiation. Outer cuticular membrane which prevents penetration of molecules larger than CO_2 and inner cuticular membrane which is highly elastic are the other two membranes. During hatching, the inner cuticular membrane develops into hatching membrane.

Hatching of Cysts

Artemia cyst production on a commercial scale is taken up at different places by multinationals and the same are available in sealed can for hatchery use. When these stored cysts are soaked in sea water, embryonic development is resumed and a nauplius larva hatches out within 24-36 hours. To achieve optimum level of hatching, the following are to be strictly observed.

Salinity

It is to be maintained between 15-35 ppt. One g cyst can be soaked in 1 litre (not more than 2gm/l)) for better hatching efficiency.

Aeration

Vigorous aeration @ 10-20 litre air per minute is required for the developing cysts so that oxygen level should be nearer to saturation.

Temperature

Ideal temperature can be from 25°C to 32°C.

pH of the Medium

Desirable pH is between 8 to 8.5.This is accomplished by adding one g sodium bicarbonate in one litre of water.

Illumination

Illumination has to be not less than 1000 lux (2000 lux at water surface preferred)

Under the above conditions, the cyst swells and assumes a spherical structure within 1-2 hours. After 15—20 hours of hydration, the cyst shall burst, prenauplius surrounded by hatching membrane becomes visible. The embryo comes out of the shell and hangs underneath the empty shell. This is the umbrella stage. Soon after hatching membrane breaks, and free swimming nauplii larva emerges.

The newly hatched nauplius is 400-500 µ in length, having a brownish organe, colour. It has 3 pairs of appendages corresponding to the antennule, artenna and mandible. There is an ocellus or nauplius eyes between antennules. Mouth and anus are absent

The nauplii do not feed for about 12 hrs. After 12 hours, it starts feeding on algal cells, bacteria and detritus of 1-40 µ size. After 15 moults it becomes adult.

After hatching, the hatching percentage, hatching efficiency, hatching output and hatching rate are worked out. The number of nauplii hatched out per 100 cysts is called hatching percentage. Number of nauplii hatched out per gram of cyst is called hatching efficiency, best being 3 lakhs/g. Naupliar biomass (mg dry weight) produced per gram of cyst is referred to as hatching output best being 600 mg/gram of cyst. The rate at which the nauplii hatch out is called hatching rate. Best products start to hatch 15 hours after incubation in natural sea water at 25°C and attain 90% hatchability within another 5 hours.

Decapsulation of Cysts

Artemia cyst shells are contaminated with bacteria, spores of fungi and other micro organisms. While feeding artemia nauplii to fish larvae if empty shells or unhatched cysts or residues from hatching medium are introduced in the larval tank along with the nauplii the fish larvae shall get infected and shall suffer from diseases and cause heavy mortality. Hence some workers recommended decapsulation of cysts before incubation.

The following procedure is prescribed by Sorgeloos *et al.* (1986) for decapsulation of *Artemia* cysts:

(a) Hydration of cysts for 1 hr by placing them in freshwater or salt water (100 g cysts in 1 lit water)

(b) Collect the cyst on 125 micron mesh sieve and place in hypochlorite solution prepared as follows.

71 g	Bleaching power	Ca $(OCl)_2$
64 g	Na_2CO_3	
1400 ml	Sea water	

Before adding the cyst the solution in cooled to 15°C to 20°C by placing in ice water. Add the hydrated cysts and stir for 5-15 minutes. Never allow the temperature to rise above 40°C by cooling with ice.

(c) When outer parts of the cyst shell is completely dissolved, remove the cysts, rinse with water on a screen of 120μm mesh size untill chlorine smell disappears.

(d) Dip the cysts in 0.1 N HCl or 0.1% Na S_2O_3 solution for less than one minute to totally remove the hypochlorite. Then rinse with water

(e) Incubate the cysts for hatching

Decapsulation also reduces the problem of separating the empty cyst shell during naupliar harvest.

A freshly hatched nauplius larva is a rich reserve of nutrients which is utilized by it for growth till the feeding starts.

After the first molting *i.e.* usually 8 hours after hatching, the Instar II larvae start ingesting any particulate matter nonselectively. Comparing to the instar I stage, instar II larvae have 22 to 39% less energy content and 16 to 34% organic dry weight. They contain lower amounts of nutrients such as free amino acids and they are also less digestible and less acceptable as a prey by the larvae as they are larger and swim faster.

Molting of the larvae to instar II can be avoided by preserving them at lower temperature *i.e.* 10°C. About 5 million larvae/litre can be stored for a period of up to

24 hours. Slight aeration is provided to prevent the accumulation and suffocation of larvae at the bottom of the tank.

Bioencapsulation or Enrichment of Artemia Nauplii

Studies of Leger *et al.* (1989) has indicated that nutritional value of *Artemia* nauplii is determined by the concentration of essential fatty acid 20 : 5n-3 eicosapentaenoic acid (EPA). Certain strain of Artemia such as Great Salt lake (USA) strain has been shown to be dificient in EPA. Hence, a method called bioencapsulation or enrichment or boosting Artemia nauplii (Cure *et al.*, 1992) is practised taking advantage of the primitive feeding habit of instar II. Emulsified concentrates (enrichment emulsion) is added at 300 ppm concentration in the rearing medium having vigorous aeration. The nauplii being non-selective feeders, injest the HUFAS after which they are stored below 10°C. HUFAs are not metabolised by nauplii. In the same way, growth promoters and antibiotics can also be delivered to the larvae through bioencapsulation of the same in the *Artemia* nauplii.

Artemia cysts are hatched out in cylindro-conical tanks of 400 to 5000 litre capacity. It will have facility for continuous aeration and fluorescent lights above. The tank is made up of black fibre gass sheet except at the conical base where it is transparent for allowing the passage of light. It will have a central pipe which is fitted to a drain pipe which can be regulated with a ball valve

When the nauplii are to be harvested aeration is stopped, and the central pipe is removed. Bottom light is switched on so that nauplii get accumulated at the transparent zone. Hatched cysts (empty shells)) float on the surface where as unhatched cysts sink to the bottom. Then the drain is opened slowly which will allow the unhatched cysts to come out first. Then the *Artemia* nauplii are collected into *Artemia* harvesting buckets with 100 µ screen. When all *Artemia* nauplii are collected discard the rest of the tank, which contain only hatched out cysts. Then the *Artemia* nauplii are thoroughly washed with sea water, and stored for feeding the shrimp larvae.

Chapter 12

Cryo Preservation of Fish Gametes

Introduction

Cryopreservation is the technique by which living cells/ tissues/ organs/ embryos/larvae etc. are preserved viable for an indefenite period by storing the same at very low temperature, usually at –196^{0}C. Application of this technique on semen storage for artificial insemination of live stock has made dramatic improvements in animal husbandry practices. Cryopreserved semen of improved strains of cattle are used for selective breeding and hybridization. Cryopreservation of ova/embryo and its subsequent use for impregnation of females has been practiced in livestock as well as in humans to some extent in several parts of the world. In fishes too several workers have attempted to cryopreserve the spermatozoa and ova from various species and to use the same for seedproduction. Though cryopreservation of the spermatozoa of several cultured, vulnerable and endangered species have been done sccessfully (table-1) cryopreservation of female gametes (ova) of fish could not be done successfully so far. Large size, complex structure, existence of several membranes with varaiable permeability are considered as the main obstacles in teleost egg preservation. Many workers have also noticed inconsistency with regard to success in fertilizing ova using cryopreserved spermatozoa. In several cases, though fertilization was successful, there was heavy mortality of embryos during development. Koldras and Bieniarz (1987) observed 98% mortality of common carp embryos developed from eggs fertilized by cryopreserved spermatozoa after 20 hours of normal development. One of the factors for this inconsistency has been reported to be the milt quality. For successful application of cryopreservation protocol with a view to ensure optimum rate of fertilisaton and healthy hatchlings a knowledge on the milt composition and sperm quality is essential.

Table 12.1: Summary of the Results with Cryopreserved Sperm from Different Fishes (modified from Piironen, J. 1992)

Species	*Freezing Technique*	*Cryo-protectant*	*Fertilization Rate (%) Frozen Sperm (max.)*	*Fresh Sperm*	*Authors*
Salmo gairdneri	pellet	DMSO	46. 5 (89. 5)		Borchard 1978
or	ampoule	DMSO	59 (89)1	96-97	Ott & Horton 1971a
Oncorhynchus	straw &	DMSO &	(>90)	a)	Erdahl & Graham 1978
mykiss	pellet	EG			
	Pellet	DMSO	81. 9	87. 9	Stein & Bayrle 1978
	Pellet	DMSO	(80.3)*	58-96	Stoss et al., 1978
	Pellet	DMSO	70.5 (99. 3)	88-98	Bayrle 1980
	Straw	DMSO	75-89	92-96	Erdahl & Graham 1980
	Straw	DMSO	(41. 0)	96. 5	Kurokura & Hirano 1980
	Pellet	DMSO	>80	88. 8	Legendre & Billard 1980
	Pellet	DMSO	69-80	97. 1-95. 5	Stoss & Holtz 1981a
	Pellet	DMSO	75-87	96. 8	Stoss & Holtz 1981b
	Pellet	DMSO	77. 5	91. 5	Stoss & Holtz 1983a
	Pellet	DMSO	50.0-90.3*		Stoss & Holtz 1983b
	Straw	DMSO	51-72	a)	Stein 1984
	Straw	DMSO	67. 3*		Baynes & Scott 1987
	Straw	DMSO	5. 9-37. 9	14. 7-95. 8	Scheerer & Thorgaard 1989
	Pellet	DMSO	49. 1-71. 9	96. 9	Schmidt-Baulain & Holtz 1989
	straw	DMSO	23. 3	a)	Cloud et al., 1990
	straw	DMSO	49. 3* (99. 9)	57. 2-89. 2	Wheeler & Thorgaard 1990
	pellet	DMSO	52-77	79-84	Holtz et al., 1991
	pellet	DMSO	26. 3-47. 0	68. 7-80.5	Schmidt-Baulain & Holtz 1991
	pellet	DMSO	72. 1-95. 5	66	Piironen, 1992
		DMSO+Gly	51. 2-79. 4*		
Oncorhynchus	ampoule	DMSO	(38)	98	Ott & Horton 1971b
tschawytscha	pellet	DMSO	44. 0-77. 5*		Stoss et al., 1981
	Ampoule	DMSO	(54. 3)		Ott 1975
Oncorhynchus	ampoule	DMSO	(79)		Ott & Horton 1971b
kisutch	ampoule	DMSO	(77. 7)		Ott 1975
Oncorhynchus	pellet	DMSO	62. 0-85. 1*		Stoss et al., 1981
gorbuscha	ampoule	DMSO	(83. 2)		Ott 1975
Oncorhynhus	pellet	DMSO	46. 8-89. 5*		Stoss et al., 1981
nerka	ampoule	DMSO	(63. 2)		Ott 1975
Oncorhynchus keta	ampoule	DMSO	(29. 3)		Ott 1975
Salmo salar	ampoule	DMSO	80.0	82. 7	Mounib1978
	pellet	DMSO	36. 0-91. 3*	73. 9-98. 3	Stoss & Refstie1983
	straw	DMSO	65-91	95-96	Alderson & MacNeil 1984
			23-52[b]	89-90[b]	

contd....

Table 12.1–contd....

Species	*Freezing Technique*	*Cryo-protectant*	*Fertilization Rate (%) Frozen Sperm (max.)*	*Fresh Sperm*	*Authors*
Salmo salar	pellet	DMSO	46. 3-54. 3	53. 2	Piironen, 1992
Salmo trutta	pellet	DMSO	41. 2-77. 0	90.2	Stein & Bayrle 1978
	pellet	DMSO	77. 0	82. 0	Stein 1979
		Gly	46. 5		
		EG	5. 3		
	Straw	DMSO	69. 4-71. 8		
	Pellet	DMSO	52. 6 (96. 5)	99	Bayrle 1980
	Pellet	EG	15-20	90-96	Erdahl & Graham 1980
		DMSO	5-10		
	Straw	DMSO	40-60, 98[c]		
		Gly	65[c]		
		EG	60-70		
	Pellet	DMSO	36. 8-54. 8*	73. 9-98. 3	Stoss & Refstie 1983
	straw	DMSO	98	90	Erdahl et al., 1984
		Gly	22		
		EG	87		
	Pellet	DMSO	3		
		Gly	0		
		EG	5		
	Straw	DMSO	43-88	a)	Stein 1984
	Pellet	DMSO	67. 0-78. 8	88. 9-94. 9	Piironen1992
Salvelinus	pellet	DMSO	27. 4	a)	Stein & Bayrle 1978
fontinalis	pellet	DMSO	49. 7 (73. 9)	99	Bayrle 1980
	straw	DMSO	59	77	Erdahl et al., 1984
		Gly	28		
		EG	61		
Salvelinus	pellet	Gly	41. 5-58. 8	75. 0	Piironen, 1992
alpinus		DMSO	35. 6-43. 6		
Hucho hucho	pellet	DMSO	22. 5	a)	Stein & Bayrle 1980
	Pellet	DMSO	18. 5 (31. 1)	a)	Bayrle 1980
Thymallus	pellet	DMSO	46. 8	95	Stein & Bayrle 1980
thymallus	pellet	DMSO	61. 0 (84. 7)	91-95	Bayrle 1980
	straw	DMSO	5-80	90	Lahnsteiner et al., 1991
Coregonus	pellet	DMSO &	39-88*	43-91	Piironen & Hyvarinen 1983
muksun	pellet	Gly	85. 9	85. 8	Piironen 1987
Clupea	vial	glycerol	80-85		Blaxter 1956
harengus har.	(testis)				
Clupea sp.	Straw	DMSO	32. 0	95. 6	Rosenthal *et al.* 1978

contd....

Table 12.1–contd....

Species	Freezing Technique	Cryo-protectant	Fertilization Rate (%) Frozen Sperm (max.)	Fresh Sperm	Authors
Gadus morhua	vial	glycerol	40-89	70-96	Mounib *et al.* 1968
	Vial	DMSO	59	63	Mounib 1978
Pleuronectes platessa	ampoule	glycerol	(39. 0)	41. 0	Pulline 1972
Cyprinus carpio	vial	glycerol	**		Sneed & Clemens 1956
	straw	PG	**		Kossman 1973
	ampoule	DMSO	47 (49. 5)		Moczarski 1977
	pellet	DMSO	0[#]		Stein & Bayrle 1978
	ampoule	DMSO	**		Withler 1982
	straw	DMSO	68. 8 (83. 2)	83. 0	Kurokura *et al.* 1984
	vial	DMSO	8-12. 3		Linhart *et al.* 1988
	straw	DMSO	30-40		Cognie *et al.* 1989
Mugil cephalus	straw	DMSO	2. 7		Chao *et al.* 1975
		Glycerol	2. 47		
	Straw	DMSO	28. 64	45. 17	Chao 1982
Ictalurus punctatus	ampoule	DMSO	**		Guest *et al.* 1976
Esox lucius	pellet	DMSO	25		Stein & Bayrle 1978
	Pellet	DMSO	1-68	50-84	deMontalambert *et al.* 1982
	ampoule	DMSO	75-79. 9[d]	78. 4-80.3	Koldras & Moczarski 1983
Tinca tinca	vial	EG	**		Moczarski & Koldras 1982
Thynnus thynnus	straw	DMSO glycerol	**		Doi *et al.* 1982
Chanos chanos	straw	DMSO	31. 9-67. 5	25. 5	Hara *et al.* 1982
Brachydanio rerio	ampoule	MeOH	103. 0[*]		Harvey *et al.* 1982
Ctenopharyngodon idella	vial	DMSO	32-57	99	Durbin *et al.* 1982
	ampoule	DMSO	**		Withler 1982
Puntius gonionotus	ampoule	DMSO	5. 0	95	Withler 1982
Aristichthys idella	ampoule	DMSO	**		Withler 1982
Pangasius sutchi	ampoule	DMSO	1. 0	90	Withler 1982
Labeo rohita	ampoule	DMSO	58. 0	95	Withler 1982
L. rohita	Straw	DMSO/DMSO+ glycerol	56	80-90	Gupta & Rath, 1991, 1993 Gupta *et al.*, 1995
Epinephalus tauvina	ampoule	DMSO	**		Withler & Lim 1982
Morone saxatilis	vial	DMSO	(87. 7)		Kerby 1983

contd....

Table 12.1–contd....

Species	Freezing Technique	Cryo-protectant	Fertilization Rate (%) Frozen Sperm (max.)	Fresh Sperm	Authors
Prochilodus scrofa	vial	DMSO	**		Coser *et al.* 1984
Salminus maxillosus	vial	DMSO	**		Coser *et al.* 1984
Hippoglossus hippoglossus	pellet	DMSO	71-87	79-85	Bolla *et al.* 1987
	straw		52-84		
Oreochromis Spp.	straw	MeOH	72. 7-93. 4	85. 7-90	Chao et al., 1987
Sarotherodon	straw	MeOH	64. 3	57. 5	Harvey 1983
O. mossambicus	straw	MeOH	45. 7		Rana & McAndrews 1989
O. aureus	cryotube, Straw	MeOH	38. 7-45. 2	77. 3-83. 0	Rana & McAndrew 1989
O. niloticus	cryotube	MeOH	38. 7-93. 4		Rana & McAndrew 1989
	Straw		86. 2-98. 6		
	Straw	MeOH	47. 5-77. 6	94. 0	Rana et al., 1990
Lates calcarifer	vial	DMSO	**		Leung 1987
		Glycerol	**		
Epinephalus malabaricus	straw	DMSO	84. 6-95. 7	91. 9-93. 8	Chao *et al.* 1991
		Glycerol	95. 7	93. 8	
Micropogonias undulatus	straw	DMSO			Gwo *et al.* 1991
Silurus glanis	aluminium disc	Glyce-rol	10.8-48	14. 1-70.8	Linhart *et al.* 1991

DMSO – dimethylsuphoxide, PG – propylene glycol, EG – ethylenglycol, MeOH – methanol-based extender, ** only sperm motility record, Gly – Glycerol based extender,
#: sperm motile, but infertile
a): no control or no information
b): fertility trial of 3000-5000 eggs.
c): frozen to – 79°C
d): none of the embryos hatched
*: expressed as percentage of control.

MILT COMPOSITION AND SPERM QUALITY IN TELEOSTS

Milt is the seminal fluid/seminal plasma containing mature spermatozoa. In spermiation phase, the spermatozoa produced in the seminiferous tubules through spermatogenesis get accumulated in the sperm ducts of the male from where they move towards the vasdeferens mixed with seminal plasma. Seminal plasma is the source of nutrition for the spermatozoa. It also inhibits sperm motility there by preserving their capacity for fertilization. Milt composition and sperm quality of salmonids have been studied by several workers. Motility of spermatozoa of salmonids last only for a brief duration (only upto 2 minutes) after they are activated. Once the motility is lost, they lose their fertilising ability too. But if the spermatozoa are activated in physiological solution, their motility and fertility period can be reinitiated and extended.

Seminal plasma which is secreted by the secretory cells lining the sperm ducts (Lahnsteiner *et al.*, 1993) contains both organic and inorganic components. Organic components consists of glucose, fructose, citric acid, lactic acid, lipids such as phospholipids, cholesterol and its esters, free fatty acids, triglycerides, glycerol etc. The organic constituents of seminal fluid of carps consists of 1.7 nM pyruvate, 243 nM lactate, 15 nM malate, 36 nM citrate, and 2.8 nM a-ketoglutarate (Linhart *et al.*, 1991). The inorganic constituents are sodium, potassium, calcium, magnesium, chloride, phosphates, bicarbonates etc. The ionic composition of seminal fluid of cyprinids is 94-107 mm of Na+, 39-78 mm of K^+, 0.02-1. 2 mm of Mg^{++} and 0.3 –12. 5 mm of Ca^{++}. In addition to these there are enzymes, hormones and vitamines in the seminal plasma. The common enzymes are proteolytic enzymes, phosphatases, glycosidase, cholinesterase and lactic dehydrogenase. The enzymes recorded from carp seminal fluid are LDH, MDH, acetateand butyrate esterases alanyl and leucine aminopeptidase. Hormones include estrogens, androgens, adrenalin, noradrenalin, prostaglandins etc. Common vitamins are ascorbic acid, inositol, riboflavin etc. The exact role of the above mentioned constituents of seminal plasma is not very clear. The monosaccharides and lipids are considered essential for the nourishment of mature spermatozoa. The spermatozoa requires energy for basic cell metabolism while in sperm duct and also for motility after it is ejected into the water (in external fertilization) or into the reproductive system of female (in case of internal fertilization of live bearers). In fish with external fertilization, endogenous energy sources such as preaccumulated ATP, glycogen/glycolipids or glucose present in the middle piece are utilized for sperm motility while in water (Lahnsteiner *et al*, 1991). Presence of K^+ in the seminal plasma has been shown to inhibit flagellar movement/ motility of the sperm. Calcium and magnesium ions are found to inhibit the inactivation effect of potassium on sperm cells. But in the seminal plasma, calcium and magnesium are found bound to citric acid. Hence, the presence of citric acid in seminal plasma is crucial to keep the sperm cells inactive in the sperm duct (Piironen, J and Hyvarinen, H. 1983).

A few parameters have been developed to assess the quality of sperms for cryopreservation. These are miltvolume, milt pH, spermatocrit value, sperm density, osmolality of seminal plasma and prefreezing motility of spermatozoa. The spermatocrit value refers to the packed volume of spermatozoa and other solids in a unit volume of milt. In other words it refers to the proportion of the total solid and seminal plasma. In order to find out the spermatocrit value, haematocrit capillaries are filled with milt and centrifuged for 10-12 minutes in a haematocrit centrifuge at a specific r. p. m. The sediment in the capillary indicates the percentage of packed sperm cells in the milt sample. For the milt of carp, spermatocrit value of 70 and above is considered good. The sperm density refers to the number of sperms per unit volume *i.e.* per cumm. or ml. In case of indian major craps it may vary from 2×10^7 to 3.5×10^7 per cu. mm. Alteration in sperm density is found in the same fish at different seasons of an year - it is found less during pre monsoon and post monsoon periods. Sperm density varies depending on species. In case of *Mugil cephalus* sperm density is 53000 million/ml, in *Tilapia zillii* it is 770 million/ml, *O. mossambicus* has 27400 million per ml, and rainbow trout has 9000 to 26000 million per ml. Sperm motility: The fish spermatozoa are not motile in testis or seminal plasma. When it is activated it performs movement and the same can be seen under a microscope. If percentage of spermatozoa exhibiting movement is about 70% or more it is considered good. Volume of milt refers to the quantity of milt a fish may yield at a given time. Non-induced male of major carps may yield 0.5 to 1 ml of milt per kg body weight. After hormone therapy it increases to 6-10 ml per kg body weight. The age of spermatozoa:The aging of spermatozoa in the sperm duct has been shown to change the quality of the sperm. Hence for cryopreservation, milt from the middle of spawning season has been shown to be good.

PRINCIPLES OF CRYOPRESERVATION

There are three essential steps in cryopreservation *viz.* (1) Freezing (2) storing (3) thawing. If cryopreservation is done without observing necessary precautions, the sperm may die due to cryoinjuries. The injuries to the cell may be due to formation of ice crystals during freezing and thawing or due to osmotic changes.

Basic principles of cryopreservation are represented in Figure 12.1.Rate of freezing and rate of thawing are critical in preserving the vitality and viability of the spermatozoa. Moderate freezing is advisable. However the optimum freezing rate differs in different species. If the cooling rate is too low, the cells get sufficient time to loose water so that it will maintain the osmotic equilibrium with the extra cellular solution. In such cases, due to dehydration the cells die. If the cooling rate is very high, the time for water to diffuse out of the cell to the extra cellular ice crystal is very less and so little or no water escapes from the cell. In such cases, the cells equilibrate by intracellular freezing either by homogeneous hetrogeneous nucleation. The formation of large intracellular crystal is fatal for the cells. When the freezing rate is moderate or optimum only part of the water leaves the cell, and a part of the water vitrifies or form small ice crystals which are tolerable if thawing is fast enough to avoid recrystallisation. In such cases, after thawing, the sperm remain viable for fertilization. The use of cryoprotectants increase the rate of vitrification* and optimise the post-thaw survival.

Cryoprotectants

Cryoprotectants are chemicals that minimise cryoinjuries to the cell due to ice formation or suppress ice formation. Cryoprotectants are of 2 categories *viz.* (1) permeating cryoprotectants and (2) nonpermeating cryoprotectants.

Permeating Cryoprotectants

Cryoprotectants that are permeable to cell membrane are called permeating cryoprotectant. These chemicals function by (1) reducing the rate of diffusion of water

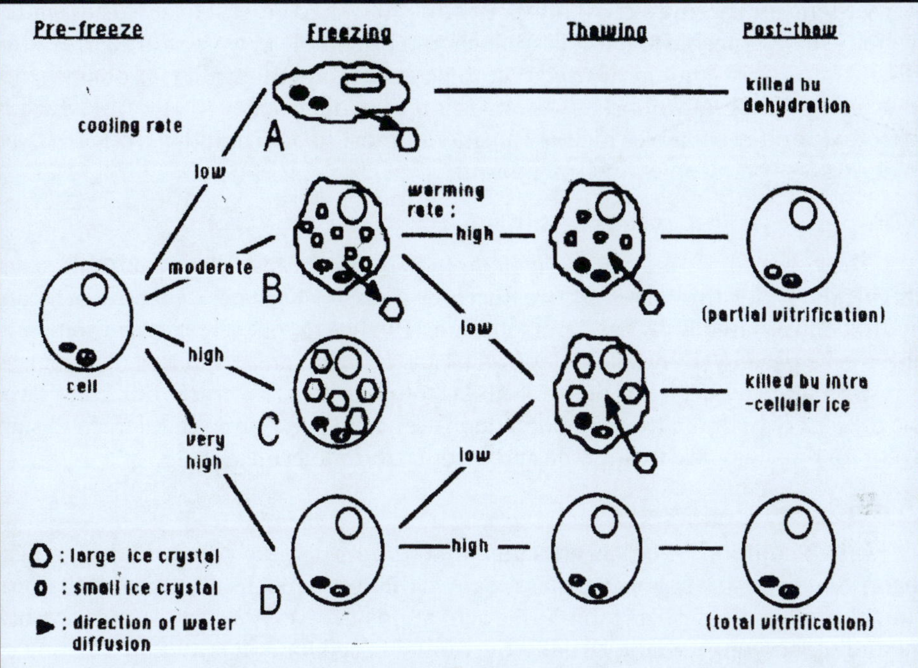

Fig. 12.1: Digrammatic Illustration of a Cryopreservation Model.

(A) When the cooling rate is low, osmotic equilibrium is maintained. Freezable water leaves the cell. Cell death is caused by excessive dehydration (some organisms survive desiccation), (B) Partial vitrification. When the cooling rate is moderate (optimum), osmotic equilibrium is not maintained. Only part of the freezable water leaves the cell. The remaining water vitrifies or forms small ice crystals which are tolerable if thawing is fast enough to avoid recrystallization. The use of cryoprotectants will increase the rate of vitrification, thus optimizing post-thaw survival, (C) When freezing is rapid, little or no freezable water leaves the cell. Large intracellular ice crystals form, when thawing is slow, devitrification and recrystallization occur to form large ice, which is lethal, (D) Total vitrification. When the cooling rate is very high, water vitrifies without forming ice. Such a cooling rate is unfeasible. Vitrification solutions suppress ice formation but allow glass formation.

(Reproduced with permission from Jamieson, B.G.M., 1991)

from cell to extracellular ice crystal, (2) reducing the cell volume change/salt concentration colligatively (3) lowering the homogeneous nucleation# temperature and (4) reducing the rate of ice crystal growth. The common permeating cryoprotectant are (i) Dimethyl sulfoxide (DMSO)m (2) Glycerol, (3) Methanol, (4) 1, 2 propanediol or propylene glycol.

Criteria for chosing a cryoprotectant are (1) it should be least toxic to the cells (2) it should be permeable to the cells (3) it should be soluble in water during freezing. DMSO is the most commonly used cryprotectant as its permeability is not markedly affected by temperature and its toxicity is inter- mediate between glycerol which is least toxic and methanol which is considered most toxic. Though glycerol is least toxic, it is less permeable to cell membrane.

Non-permeating Cryoprotectants

These chemicals are not permeable to cell membrane. The commonly used chemicals under this category are sucrose, glucose, polymers such as dextran, hydroxyethylstarch, polyvinyl pyrrolidone, (PVP), eggyolk serum, skim milk and anti freeze protein. They act by depressing the freezing point and raising the glass transformation temperature of extracellular solution. Thus the cryoprotectants plays the role of a dehydrant, freezing point depressant and extra/intra cellular stabilising agent for the biological membrane and cytoplasmic molecules.

Extenders

It is a solution of balanced salts. Sometimes, organic compounds are also incorporated into it. This solution is meant to dilute the milt as undiluted milt is not suitable for freezing. It also inhibits the activation of spermatozoa and functions as a medium for cryoprotectant. Usually, the duluent used for milt is the combination of extender and cryoprotectant.

Several extenders used by different workers for fish spermatozoa are listed in table 2. For Indian major carps, 3 categories of extenders have been found suitable (Glupta, S. D. and Rath, S.C., 1997) which are as noted below:

	Extender A	*Extender B*	*Extender C*
NaCl	400 mg	600 mg	750 mg
$CaCl_2$	23 mg	23 mg	20 mg
KCl	38 mg	38 mg	20 mg
$NaHCO_3$	100 mg	100 mg	20 mg
NaH_2PO_4	41 mg	41 mg	-
$MgSO_4$	23 mg	23 mg	-
Dist. Water	100 ml	100 ml	100 ml

Three categories of diluents are prepared by mixing cryoprotectants such as DMSO/glycerol with extenders as shown below:

I	II	III
Extender 'A' – 90 ml	Extender 'B' – 90 ml	Extender 'C' – 65 ml
DMSO – 5 ml	DMSO – 8 ml	DMSO – 15 ml
Glycerol – 10 ml	Glycerol – 10 ml	

After the diluent is prepared by mixing the cryoprotetant with extender, it is to be kept in a refrigerator as it exhibits exothermic reaction. Any one of the above diluents can be mixed with milt at a proportion of 4:1 (diluent : milt).

Table 12.2: Composition of Extenders and Cryoprotectants Used in Cryopreservation Experiments with Different Species of Fishes (Piironen, J. 1992).

Species	*Composition of Extender*	*Authors*
Atlantic salmon (*Salmo salar*)	100 mm $KHCO_3$, 125 mM Saccharose (42.7 g/l), 6.5 mm reduced Glutathione (2.0 g/l), 12.5% DMSO	Mounib 1979
	0.3M glucose (59.45 g/l), 10% DMSO	Stoss & Refstie 1983
	5.4% glucose, 9% DMSO, 10% yolk	Alderson & MacNeil 1984
Brown Trout, sea trout (*Salmo trutta*)	128.3 mm NaCl (7.5 g/l), 23. 8 mm $NaHCO_3$ (2. 0 g/l), 3.7 mm Na_2HPO_4 (0.53g/l), 0.4mM $MgSO_4 \times 7H_2O$ (0.23 g/l), 5.1 mm KCl (0.38 G/l), 2.3 mm $CaCl_2 \times 2\ H_2O$ (0.46 g/l), 5.6 mM Glucose (1 g/l), 66. 6 mm glycine (5 g/l), 10% DMSO, 20% yolk (=V2)	Stein & Bayrle 1978
	128. 3 mm NaCl (7.5 g/l), 23. 8 mm $NaHCO_3$ (2.0 g/l), 5.1 mm KCl (0.38 g/l), 5.6 mM glucose (1 g/l), 10% DMSO, 20% yolk (=V2e)	
	0.3 M glucose (59.45 g/l), 10% DMSO	Stoss & Refstie 1983, Piironen, 1992.
Rainbow trout (*O. mykiss*)	101 mm NaCl, 23 mm KCl, 5.4 mm $CaCl_2$, mm $MgSO_4$, 200 mm Triscitrate (pH 7.25), 0.4% bovine serum albumin, 0.75% Promine D, 10% DMSO.	Stoss & Holtz 1981a
	Extenders V2 and V2e	Stein & Bayrle 1978
	Distilled water, 9% DMSO, 20% yolk	Bayrle 1980
	5.84 g/l NaCl, 2.56 g/l KCl, 0.27 g/l Na_2HPO_4, 0.22 g/l $MgCl_2 \times 6\ H_2O$, 0.1025 g/l $CaCl_2 \times 2H_2O$, 0.1 g/l citric acid, 10 g/l glucose, 10 ml KOH (12.7 g/l), 20 ml Bicine (53 g/l), pH 7.8, 7% DMSO	Erdahl & Graham 1980
	slightly modified extender by Mounib (1978) [10% DMSO, 10% yolk]	Legendre & Billard1980 Baynes & Scott 1987, Piironen, 1992
	extender by Alderson & MacNeil (1984)	Wheeler & Thorgaard 1990
	0.6 M saccharose, 10% DMSO	Holtz *et al.,* 1991.
Brook trout (*Salvelinus fontinalis*)	4 g/l NaCl, 8g $Na_3C_6H_5O_7 \times 2\ H_2O$, 20 g/l glucose [=Alsever's solution, Hodgins & Ridgeway 1964], 9% DMSO	Bayrle 1980
	Extender by Erdahl & Graham 1980	Erdahl *et al.,* 1984.
Grayling (*Thymallus thymallus*)	extender V2	Stein & Bayrle 1978
		Bayrle 1980
Whitefish (*Coregonus* Sp.)	0.3 M glucose, 20% glycerol	Piironen & Hyvannen 1983 Piironen 1987

CRYOPRESERVATION PROTOCOL

It can be described in 5 phases: (1) Pre freezing, (2) Freezing, (3) Storage, (4) thawing, (5) post thawing and insemination. (Fig. 12.2)

Pre Freezing Phase

Collection of milt from the milter

During the breeding season, milt is collected from the milter for cryopreservation. An easy oozing milter is selected as the donor fish. The males are induced by carp pituitay extract (5 mg/Kg) or any synthetic compound such as ovaprim or ovatide (0.2 ml per kg.) After a gap of 3-4 hrs, the milt is collected by hand stripping (Fig. 12.3) into an ice cold sterilised tube, taking care to avoid contamination with urine, mucus, faecal matter etc. The collected milt samples are kept in a refrigerator. Before processing, the quality of the milt is assessed as mentioned earlier. Usually, it is advisable to mix milt from several individuals so that even if one fish has aged spermatozoa, it will be made good by the milt of other fish. This will ensure the quality of post thawed milt and optimise fertilization rate. Then the milt is diluted by the diluent (mixture of extender and cryoprotectant.)

Equilibration of milt-diluent mixture

The milt-diluent mixture is kept at low temperature for equilibration. The equilibration time shall vary depending on the type of cryoprotectant used. In case of slow penetrating chemicals such as glycerol, longer equilibration time may be required. But it should never be more than one hour because the sperms usually become mobile upon mixing with diluents and so shorter equilibration time shall minimise exhaustion of spermatozoa. In case of Indian major craps equilibration period can be 45 minutes. The low temperature reduces the toxicity of the cryoprotectant on cell, as permeability is reduced at low temperature for most chemical like DMSO. For glycerol, it may be done at hiqher temperature.

Freezing Phase

The equilibrated milt-diluent mixture is frozen at very low temperatures. The freezing should be correct to prevent cryoinjuries to the cells. There are 3 ways of freezing *viz.* (a) pellet method, (b) straw method (c) vial method (Fig. 12.2).

Pellet method

Pellet method is preferred by many workers and it is extensively used for salmonid sperms. Size of the pellet may vary from 50 to 200 ml for which the freezing rate can be 20-35°C per minute. In this method, the sperm - diluent mixture is formed into pellets in small hole drilled/melted into solid dry ice (-79^{0}C). The pellets so obtained are removed to liquid nitrogen for storage. For pellet freezing no equilibration time is required and immediately after mixing sperm and diluent it can be used for pellet making.

Straw method

Straws are available in different volumes (0.25 to 4 ml). This method is more popular during recent years as it has been found to yield better post thaw quality of

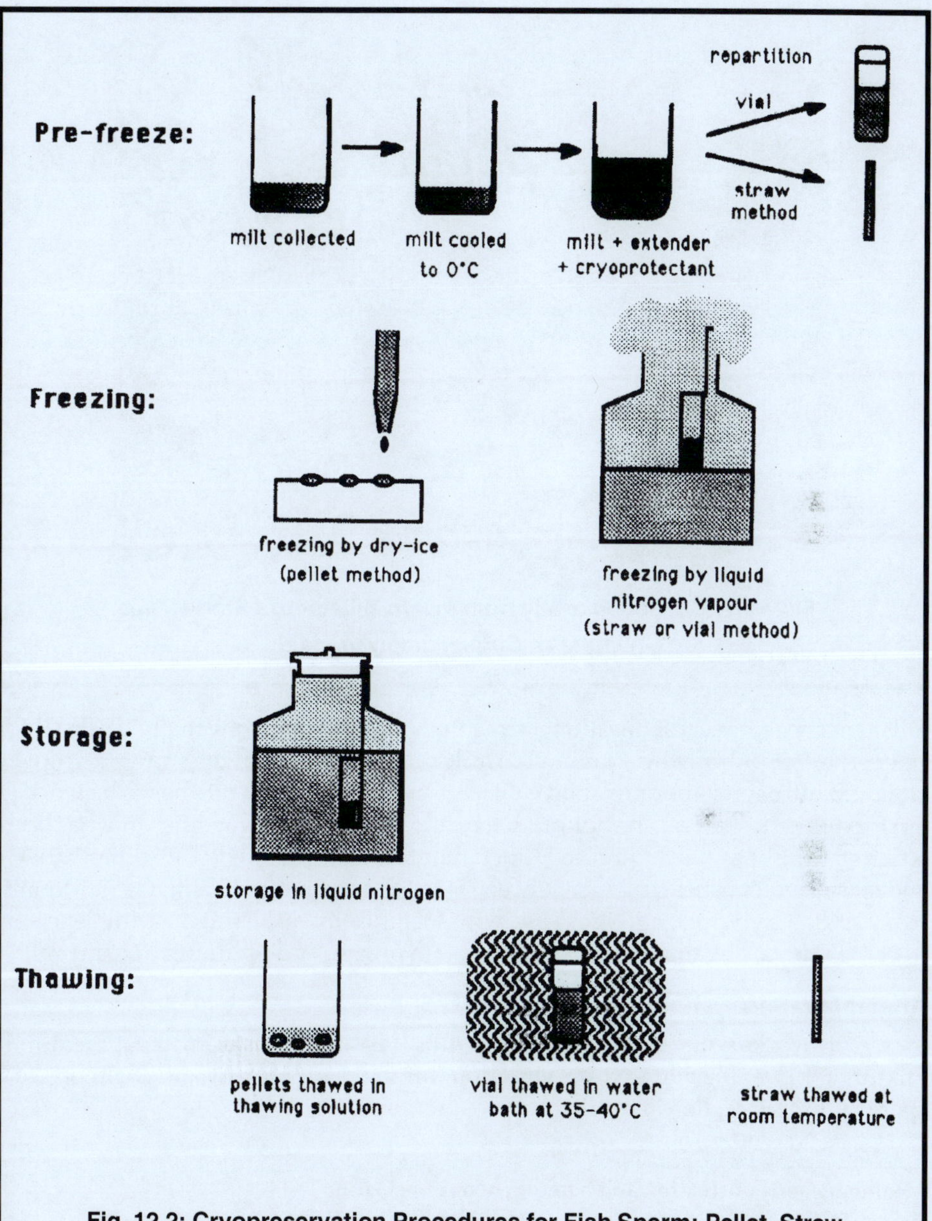

Fig. 12.2: Cryopreservation Procedures for Fish Sperm: Pellet, Straw and Vial Methods.

(Reproduced with permission from Leung, L.K.P. & B.G.M. Jamieson, 1991)

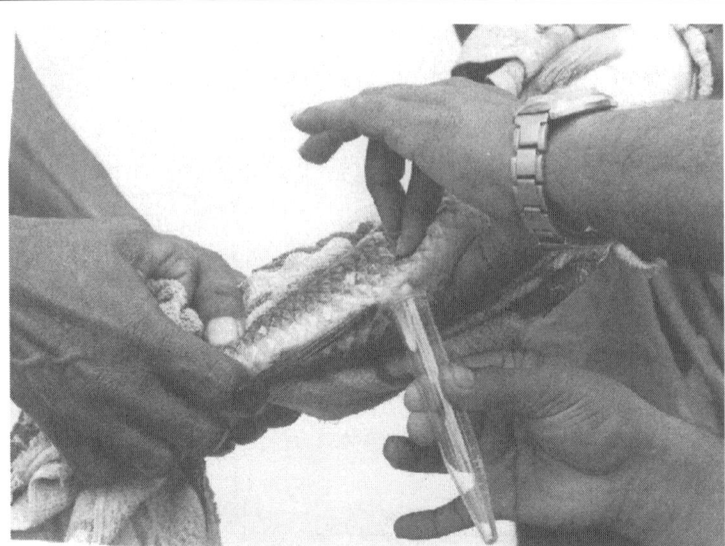

Fig. 12.3: Collection of Milt from a Carp Milter into a Sterile Tube.
(Courtesy, CIFA, Kausalyaganga)

milt as chance for recrystallisation during thawing is less in straw method. (Erdahl *et al*, 1984). After putting the equilibrated milt-diluent mixture in the straw, it is cooled on liquid nitrogen vapour or solid CO_2 ice. Optimum cooling rate has to be worked out for different species. The optimum freezing rate for fresh water fish milt has been reported to be between 20^0 and 160^0C per minute (Scott and Baynes, 1980; Stoss, 1983) and for salt water fishes it may be between 5-150^0C per minutes (Blaxter, 1953; Mounib *et al.*, 1968). Cryo- freezers (Fig. 12.4) with facilities for adjusting varying levels of freezing rates are available at present to obtain precise and optimum freezing rate

Ampoules/vials/ visitubes method

Some workers have used ampoules/vials/visitubes for storing the diluent-milt mixture for freezing and cryopreservation (Mounib, 1978 for Atlantic Salmon/cod sperm, Gupta *et al.*, for carp milt).

*** Homogeneous nucleation and heterogeneous nucleation**

In order to initiate crystallization, a crystal nucleus has to be formed. If the nucleus is produced by random aggregation of molecules it is called homogeneous nucleation. If it is produced by catalyzed aggregation of molecule, it is called heterogeneous nucleation.

 # Vitrification

If water molecules are not allowed to crystallize during cooling they shall solidify as a noncrystalline structure. This process is called vitrification. In other words, vitrification is a process of instant molecualr arrest of a high viscosity medium without permitting ice crystal nucleation. The temperature at which vitrification begins is termed as glass transformation temperature.

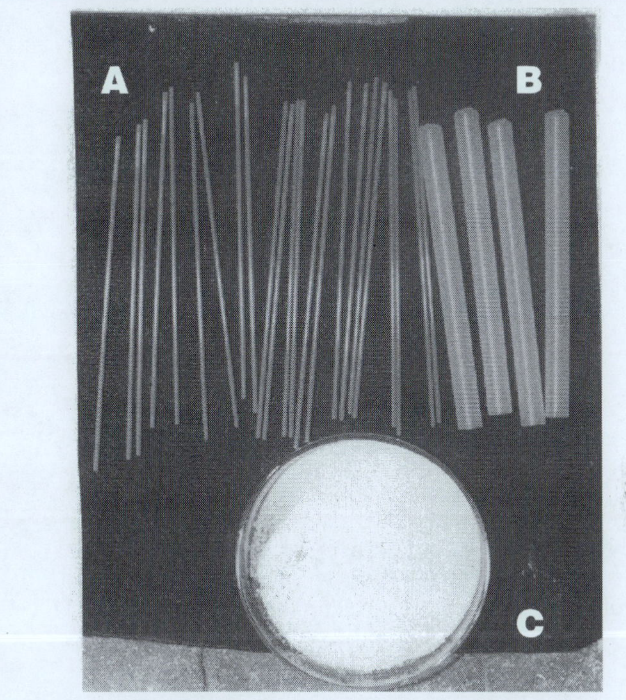

Fig. 12.4: (A) Straw, (B) Vials (visitubes), (C) Sealing Powder Used for Storing Sperms.
(Courtesy, CIFA, Kausalyaganga)

Storage Phase

Frozen milt (pellet/vial/straw) is stored at –196°C using liquid nitrogen kept in cryocans (Fig. 12.5). At this temperature spermatizoa can remain indefenitely without deterioration of cell quality.

Thawing phase

For sperms frozen by pellet method, a thawing solution is required. This solution helps not only in absrobing heat but also in providing a suitable medium for post thaw survival and induction of motility. 1% $NaHCO_3$ or 119m mol $NaHCO_3$ and 120 m mol NaCl have been used as thawing solution for salmonid sperm. One pellet may be placed in 1-2 ml of thawing solution warmed to 25-30°C. The pellets begin to liquefy after 5-10 seconds and this solution can be poured over the egg. The lowest density of sperm cells required for satisfactory fertilization is 3×10^6 sperm cells per egg.

The straws are usually thawed in water bath at 30-40°C. Chao *et al*. (1991) used microwave thawing for straw containing grouper sperm and obtained good results. Fast thawing is preferred as slow thawing may cause crystallisation in sperm cells. The french straws are made to slush, swirling them in water bath at 38 ±2°C for 7-9 seconds. Then 1-2 drops of fresh water is added to the milt. This reactivates the sperm and restores their motility.

Fig. 12.5: Cryofreezer Used for Controlling the Freezing Rate of Milt Mixed with 'Diluent-Extender'.

(Courtesy, CIFA, Kausalyaganga)

Fig. 12.6: Cryocan Used for Cryopreservation of Sperm.

(Courtesy, CIFA, Kausalyaganga)

In case of ampoules/vials 65-70 seconds are required for slush formation. It has been calculated that a vial containing 3 ml thawed, diluted milt can fertilize 1-1. 5 lakh eggs.

Post Thaw Phase and Insemination

Druing this phase, the sperms that have survived the cryopreservation are ready for artificial insemination. It is desirable to estimate the motility of the sperm after thawing. Several factors affect this post thaw fertility of cryopreserved spermatozoa (Table 12.3). Artificial insemination is to be done immediately after thawing as there is reduction in the motility period of spermatozoa after thawing.

Table 12.3: Summary of Factors that Affect Post Thaw Fertility of Cryopreserved Spermatozoa. *(modified from Piironen, J. 1992*)

Phase	*Factors Affecting Post Thaw Fertility*
Pre freezing phase	1. Quantity of donor male (age, Size, health)
	2. Age of the sperm and quality of mllt (milt volume, sperm density, Spermatocrit value, osmolality of milt, organic/ inorganic constituents of milt.)
	3. Type and concentration of cryoprotectant/extenders used
	4. Equilibration time and temperature.
	5. Interval between sperm collection and freezing.
Freezing phase	1. Freezing rate.
	2. Cryoprotectant/extenders used.
	3. Diluent preparation and dilution rate.
	4. Type of freezing container (Pellet/straw/ampoule)
Storage Phase	1. Temperature
	2. Background radiation.
Thawing phase	1. Thawing rate
	2. Thawing solution in case of pellets
Post thawing & insemination	1. Motility induction and duration
	2. Removal of cryoprotectant
	3. Sperm density for insemination
	4. Quality of the donor female for eggs
	5. Incubation.

ADVANTAGES OF CRYOPRESERVATION OF GAMETES

1. In case of seasonal breeders such as major carps, seed can be produced and supplied to farmers at any time of the year as per the market demand, once the cryopreservation technique is standardised.
2. It helps in the selective breeding and in the development of other techniques meant for genetic upgradation of cultivable fishes aimed at generating superior strains that shall exhibit faster growth rate, better disease resistance, better adaptability to extremes of climates and better feed conversion efficiency.

3. It can help in preventing 'inbreeding depression' so often noticed in the present day aquaculture practices.
4. It will be helpful in preserving genetic diversity and in conserving endangered species through the development of gene banks.
5. It can reduce the cost of male brood stock maintenance through the exploitation of a limited stock.
6. This procedure shall be useful in case of fishes where male and female mature at different times.

CRYOPRESERVATION OF INVERTEBRATE LARVAE AND FISH EMBRYOS

In spite of the success achieved in the cryogenic storagc of male gametes of many species of fishes, attempts to cryopreserve inverterbrate and fish embryos/larval stages (Table 12.4) have not succeeded so far in standardising a reliable and reproducible protocol. Presence of impermeable chitinous exoskeletion and numerous yolk granules and varying cryobiological characteristics of different cells and tissues have been cited as the hurdle in developing a protocol for the shrimp larvae. Other studies on the development of a cryopreservation technique for embryos are of Harvey *et al.* (1983), and Zhang *etal* (1996) for Zebra fish and of Steponkus *et al.* (1990) and Mazur *et al.* (1992) for Drosophila. Several scientists have also attempted to overcome the hindrance in cryopreservation due to yolk by removal of internal yolk by microsuction (Nagashima *et al.*, 1995) or by isolation of blastoderm from the yolk for cryopreservation (Harvey, 1983 in Zebra fish). Blastoderm cryopreservation may be used in future for interspecfic cloning.

Table 12.4: Studies on the Cryopreservation of Aquatic Invertebrate Larvae (*Subramoniam, T and R. Arun, 1999*)

Sl. No.	*Species*	*Result*	*Author*
1.	**Rofers.**		
	Brachionus plicatilis	23. 8% of eyed stage embroys survived	Toledo & Kurokura (1990)
	Philodina sp	67% of adults survived from – 196°C	Koehler (1967)
	B. plicatilis	2% survival of adults	King *et al.* (1983)
2.	**Oyster**		
	Crassostrea sp.	48. 8% survival of trochophore larvae	Toledo *et al.* (1989)
3.	**Ragworm**		
	Nereis virens	86% of setiger larvae survived	Olive & Wang (1997)
4.	**Shrimp**		
	Penaeus indicus	63% of nauplius IV at–40°C by slow freze	Subramoniam & Newton (1993)
	P. monodon	85% protozoes I at–3) 70°C by rapid freeze	Arun & Subramoniam (1997)

References

Abraham M., Blanc, N and Yashouv, A. (1966). Oogenesis in five species of grey mullets. (Teleostei, Mugilidae) from natural and land locked habitats. Israel J. of Zool. 15 (3 - 4) 155 - 172.

Abidi, S. A. H. and Thakur, N. K. (1997). Advances in keeping and breeding ornamental fishes Training manual, Central Institute of Fisheries Education, Versova, Mumbai

Alderson, R. & MacNeil, A. J. (1984). Preliminary investigations of cryopreservation of milt of Atlantic salmon (*Salmo salar*) and its application to commercial farming - Aquaculture 43: 351 - 354.

Alikunhi, K. H, K. K. Sukumaran, S. Parameswaran (1962). Induced spawning of Chinese grass carp *Ctenopharyngodon idellus* (C & V) and silver carp, *Hypophthalmichthys molitrix* in pond at Cuttack, India. Tecch. Pap. IPFC, 16, 22.

Alok,D. N. S. Haque and J. S. Killan (1995). Transfer of human growth hormone gene in to Indian major carp (*Labeo rohita*) Ind. J. Anim. Sci. 10: 5-8

Alok Deoraj, T. Krishnan, G. P. Talwar and Lalit C. Garg (1993). Induced spawning of Cat fish, *Heteropneustes fossilis* (Block) using D - lys6 salmon gonadotropin releasing hormone analog. Aquaculture, 115, pp 159 - 167.

Anderson, W. W. (1958). Larval development, growth and spawning of striped mullet (*Mugil cephalus*) along the south Atlantic coast of United States. Fish. Bull. US Fish. Wildl. Serv. 58 (144) 501 - 519.

Angel, C. A. (1992). Report of the seminar on Mudcrab culture and trade. Bay of Bengal programme BOBP/REP/51, 246 pp.

Angell, C. A. (1992). Report of the seminar on Mud crab culture and trade Bay of Bengal programme BOBP/REP/51, 246 PP.

Arun, R. and Subramoniam, T. (1997). Cryo - Lett; 18, 359 - 368.

Atz J. W. (1964). Intersexuality in fishes in' Inter sexuality in vertebrates including man'. (C. N. Armstrong and A. J. Marshall eds) pp 145 - 232, Academic press, NewYork.

Avtalion, R. R. and I. S. Hammerman (1978). Sex determination in Sarotherodon (Tilapia) 1.Introduction to a theory of autosomal influence Bamidgeh. 30 : 110-115.

Axelrod, Herbert, R. and William vorderwinkler (1986). Encyclopedia of tropical fishes, T. F. H. publication, Inc. 24, WestSylvania avenue, Nepture city, N. J. 07753.

Baceetti, B., Burrini, A. G., Callaini, G., Gibertini, G., Mazzini, M. and Zerunian, S. (1984). Fish germinal cells. Comparative spermatology of seven cyprinid species. Gamete research 10, 373 - 396.

Bagarinao. Teodora (1994). The natural life history of milk fish. SEAFDEC Asian Aquaculture XVI (3) Sept. 94 pp 3 - 4.

Balarin, J. D. and R. D. Haller (1982). The intensive culture of Tilapia in tanks, raceways and cages. in James F. Muir Ronald J. Robrts (Ed.) Recent advances in aquaculture. Croom Helm Ltd., Provident house, Bwcrell Row, Beckenham, Kent. BR 3/AT

Balon, E. K. (1975a). Terminology of intervals in Fish development. J. Fish. Res. Board. Can. 32(9) 1663-1670.

Balon, E. K. (1975b). Reproductive guilds of fishes: a proposal and a definition J. Fish. Res. Board Can. 32(6): 821 - 64.

Balon, E. K. (1981). Addition and amendments to the classification of reproductive styles in fishes. Environ. Biol. Fish., 6(3-4): 377 - 89.

Balon, E. K. (1984). Reflections on some decisive events in the early life of fishes. Transactions of the American fisheries society, 113: 178 - 185

Basavaraju, Y., G. Y. Keshavappa, D. Seenappa, K. V. Devraj and K. V. Mohira (1990). Growth of *Labeo rohita* X *Labeo fimbriatus* hybrid in comparison with the parent species. In the second Indian fisheries Forum proceedings, May 27-31,1990.Mangalore, India: 20-22.

Baynes, S. M. and Scott, A. P. and Dawson, A. P. (1981). Rainbow trout, *Salmo gairdneri* Richardson,: spermatozoa- effects of cations and pH on motility. J. Fish. Biol. 19. p. 259 - 267.

Baynes, S. M. & Scott, A. P. (1987). Cryopreservation of rainbow trout spermatozoa: the influence of sperm quality ,egg quality and extender composition on post-thaw fertility. Aquaculture 66: 53-67

Bayrle, H. (1980). Untersuchungen zur Optimierung der Befruchtungsfahigkeit von gefrierkonserviertem Fischsperma. - Dissertation, T. U. Munchen, 201 b.

Bernabe, G (1994). Plankton culture. in "Aquaculture- Biology and ecology of cultured species' No. 54 - 76. Publ. Ellis - Horwood ltd

Bertollo, L. A. C.,Takahashi,C. S. and Filho,O. M. (1983). Multiple sex chromosomes in the genus *Hoplias* (Pisces: Erythrinidae). Cytologia 48: 1-12.

Bhatnagar,G. K. (1967). Observations on the spawning frequency and fecundity of certain Bhakra reservoir fishes. Ind. J. Fish. 11(4);485-502

Bhimachar, B. S. and Tripathi, S. D. (1972). Hypophysation of fishes with particular reference to India. Publ. Directrate of research services, J. N. K. V. V. Jabalpur, M. P.

Blaxter, J. H. S. (1969). Development: Eggs and larvae in Fish Physiology, Vol III (Eds) W. C. Hoar and D. J. Randall, Publ. Academic press, Newyork.

Blaxter, J. H. S. (1953). Sperm storage and cross-fertilization of spring and autumn spawning herring. - Nature (London) 172: 1189 - 1190.

Bolla, S., Holmefjord, I. & Refstie, T. (1987). Cryogenic preservation of Atlantic halibut sperm. - Aquaculture 65: 371 - 374.

Borchard,B. and Schmidt,G. W. (1976). Versuche mit Regenbogenforellensperma, IV Die Tiefkuhlkonservierung. Beobachtungen zum Einsatz beipraktischen und wissenschaftlichen Arbeiten. -Der Fischwirt. 29: 49-51

Breton, B., Jalabert, B. and Fostier, A. (1975). Induction de decharges gonadotropes hypophysaires chez la carpe (*Cyprinus carpio L*) a laide du citrate de cisclomiphene. Gen. Comp. Endocrinol. 25, 400 - 404.

Bromhall, J. D. (1954). A note on the reproduction of grey mullet, *Mugil cephalus.* Hong Kong univ. Fish. J. 1, 19-34

Brooks, M. J., Smitherman, R. O., Chappel, J. A., and Dunham, R. A., (1982). Sex-weight relation in blue, channel and white cat fishes : Implications for brood stock selection. Prog. Fish - Cult., 44:105-107.

Buss, K. and Fox. H. (1961). Modification for the jar culture of trout eggs. Progressive fish culturists. 23 (3) 142 - 144.

Carpenter, M.B. and de Roos, R. (1970). Seasonal morphology and histology of the androgenic gland of the cray fish. *Orconectes nais,* Gen. Comp. Endocr. 15: 43.

Chakrabarty, R. D. and D. S. Murty (1972). Life history of Indian major carps, *Cirrhinus mrigala* (Ham) *Catla catla* (Ham) and *Labeo rohita* (Ham) J. Inland Fish Soc. India. 4: 132 - 161.

Chambeyron, F. & Zohar, Y. (1990). A diluent for sperm cryopreservation of gilthead seabream, *Sparus aurata*. - Aquaculture 90: 345 - 352.

Chandhuri, H. *et al.* (1978). Observation on artificial fertilization of eggs and the embryonic and larval development of milk fish, *Chanos chanos* (Forskal) Aquaculture, 13, 95 - 113.

Chang, J. P. and R. E. Peter (1983). Effect of pimozide and (D-Ala6 des Gly10) Lutenising hormone releasing hormone ethyl amide on serum gonadotropin concentration, germinal vesicle migration and ovulation in female gold fish, *Carassius auratus*. Gen. Comp. Endocr. 55, 89 - 95.

Chaturvedi, S.K. (1976). Spawning Biology of Tor mahseer, *Tor tor* (Ham) J. Bombay Nat. Hist. Soc. 73(1): 336-344.

Chao, N.H. (1982). New approaches for cryopreservation of sperm of grey mullet, *Mugil cephalus*. - In: Richter, C. C. J. & Goos, H. J. Th. (Compilers): Reproductive physiology of fish. Proc. Int. Symp. Reprod. Physiol. Fish, Wageningen, the Netherlands, Aug. 2 - 6, 1982, Pudoc Press, Wageningen, pp. 132 - 133.

Chao, N. H., Chao, W. C., Liu, K. C. & Liao, I. C. (1987). The properties of tilapia sperm and its cryopreservation. - J. Fish Biol. 30:107-118.

Chao, N. H., Chen, H. P. & Liao, I. C. (1975). Study on cryogenic preservation of grey mullet sperm. - Aquaculture 5:389 - 406.

Chao, N. H., Tsai, H. P. & Liao, I. C. (1991). Cryopreservation of the grouper, *Epinephalus malabaricus*, sperm. - In: Scott, A. P., Sumpter, J. P., Kime D. E. & Rolfe, M. S. : Reproductive physiology of fish, Proceedings of the 4th International Symposium on the Reproductive physiology of Fish. Univ. of East Anglia,Norwich,U. K.,7-12 July 1991,p. 271

Chao, T. M. and M. Chow (1990). Effect of methyl testosterone on gonadal development of *E. tauvina* (forsskal) Singapore J. Pri. Ind. 18(1) 1 - 14.

Chao, T. M., Lim, L. C. and L. T. Khoo (1993). Studies on the breeding of brown marbled grouper (*Epinehelus fuscoguttatus* forskal) in Singapore, in T. M. L. Conference proceedings, No. 3.Fin fish hatchery in Asia Eds. Cheng Cheng Lee, Mao-Sen Su, and I. Chiudiao pp. 143-156.

Charles C. Davis (1955). The kinds of plankton organisms in "The marine and freshwater plankton", Michigan state university press. Pp 225 -230

Chaudhuri *et al.* (1978). Observation on artificial fertiliztion of eggs and the embryonic and larval development of milk fish *Chanos chanos* (Corskal) Aquaculture 13, 95-113.

Chaudhuri, H. (1955). Successful spawning of the carp minnow *Esomus danricus* by pituitary gland injection. Study of the life history and bionomics. In: Contributions to the techniques of pond fish culture in India D. Phil thesis Univ. Calcutta.

Chaudhuri, H. and K. H. Alikunhi (1957). Observations on the spawning in Indian carps by hormone injections. Curr. Sci. 26 (12) 381 - 382.

Chaudhuri, H. *et al.* (1976). Notes on the external sex characters of *Chanos chanos* (Forskal) spawners. Fish. Res. J. Philippines 1(2) 76 - 80.

Chaudhuri, H.S.B. Singh, K.K. Sukumaran and P.C. Chakrabarti (1967). Note on natural spawning of grass carp and Silver carp in induced breeding experiments. Sci and Cult. 33 (11) 493 - 494.

Chaudhuri, H., S.B. Singh and K.K. Sukumaran (1966). Experiments on large scale production of fish seed of Chinese grass carp, *Ctenopharyngoden idellus* and silver carp, *Hypophthalmichthys molitrix* by induced breeding in ponds in India. Proc. Ind. Acad. Sci. (B) 63 (2) 80 - 95.

Chellam, A., T. S. Velayudhan, S. Dharmaraj and A. C. C. Victor. (1978). Pearl Oyster farming. CMFRI Bullettin 39: 72 - 77.

Chen, F. Y. *et al.* (1977). Artificial spawning and larval rearing of grouper, *E. tauvina* (Forskal) in Singapore Singapore J. Pri. Ind., 5 (1), 1 - 21.

Chen, F. Y., (1969). Preliminary studies on the sex determining mechanism of *Tilapia mossambica* Peters and *T. hornorum*. Int. Assoc. Theor. Appl. Limnol. Proc. 17:719-724.

Chen, T. and Powers, D. A., (1990). Transgenic fish for aquaculture. Genetic engineering, 13: 331-369.

Chen, T. R., (1969). Karyological heterogamety of deep - sea fishes. Postilla no. 130.

Chen. T. P. (1969). Hybridization and culture of hybridis. FAO Fish. Cult. Bull (33):6.

Chen. T. P. (1976). Aquaculture practices in Taiwan, Fishing news books, Farnham, U. K. 16pp.

Cheong, L. and F.Y. Chen (1980). Preliminary studies on raft method of culturing green mussels, *Perna viridis* (L) in Singapore. Singapore J. Prim. Ind. 8(2): 119-133.

Chimits, P. (1955). Tilapia and its culture : a preliminary bibliography. FAO Fish. Bull 8(1):1-33.

Chourrout, D., Guyomard, R., Houdebine, L. M. (1986). High efficiency gene transfer in rainbow trout (*Salmo gairdneri*) by microinjection into egg cytoplasm. Aquaculture 51: 143-150.

Chu, S. P. (1942). The influence of the mineral composition of the medium on the growth of plankton algae, Methods and Culture medium Journal of ecology, 30: 284 - 325. Reproductive Physiology of Fish. Univ. of East Anglia, Norwich, U. K., 7-12 July 1991, p. 271.

Churrout, D. and J. Itskovich (1983). Three manipulations permitted by artificial insemination in Tilapia : induced diploid gynogenesis, production of all triploid populations and intergeneric hybridization p. 246-255. In L. Fishelson and Z. Yaron (comps) Procedings of the international symposium on Tilapia in aquaculture 8-13, May 1983. Nazareth, Israel, Telaviv University , Israel.

Clemens, H. P. and Sneed, K. E. (1957). Spawning behaviour of the channel cat fish, *Ictalurus punctatus*, U. S. Fish and wild life service, special scientific report - Fisheries No. 219, Washington, D. C.

Clemens, H. P. and T. Inslee (1968). The production of unisexual brood of *T. mossambica* sex reversed with methyl testosterone Trans. Am. Fish. Soc. 97:18-21.

Cloud, J. G., Miller W. H. & Levanduski, M. J. (1990). Cryopreservation of sperm as a means to store salmonid germ plasm and to transfer genes from wild fish to hatchery populations. - Prog. Fish-Cult. 52: 51-53.

Cognie, F, Billard, R. & Choa, N. H. (1989). La cryoconservation de la laitance de la carpe, *Cyprinus carpio*. - J. Appl. Ichthyol. 5: 165 - 176.

Conand, C. (1990). The Fishery resources of Pacific island countries Part 2 Holothurians FAO fisheries technical paper No. 272. 2 Rome FAO 143p.

Cook, H. L. and Murphy. M. A. (1966). Culture of larval penaeid shrimp, Transactions of American Fisheries Society, 98: 751 - 754

Coser, A. M., Godinho, H. & Ribeiro, D. (1984). Cryogenic preservation of spermatozoa from *Prochilodus scrofa* and *Salminus maxillolsus.* - Aquaculture 37: 387-390.

Cridland, C.C. (1962). Laboratory experiments on the growth of *Tilapia* sp. Effect of light and temperature on the growth of *T. zilli* in aquaria. Hydrobiologia 20(2): 155-166.

Cure, K., J. F. Rees, S. Pyatiratitivorakul, P. Menasveta and P. Sorgeloos (1992). Culture performance and stress resistance of penaeid post larvae fed HUFA enriched Artemia, Third Asian Fisheris forum, Singapore. 65.

Das,S. K. and B. B. Jana (1996). Pond fertilization through in organic sources: an over view. Ind. J. Fish, 43 (2): 137 - 155.

Davis, K. B., B. A. Simco, C. A. Goudie, N. C. Parker, W. Cauldwell and R. Snellgrove(1990). Hormonal sex manipulation and evidence of female homogamety in channel catfish. Gen. Comp. Endocrin. 78:218-223.

DeMontalambert, G., Bry, C. & Billard, R. (1978). Control of reproduction in northern pike. - Am. Fish. Soc., Spec. Publ. 11: 217 - 225.

DeMontalembert, G., Jalabert, B., and Bry, C (1978a). Precocious induction of maturation and ovulation in northern pike (*Esox lucius*) Ann. Biol. Anim. Biochim. Biophys. 18, 969 - 975.

Desai, V. R. (1973). Studies on fishery and biology of Tor tor (Ham) from river Narmada. II Maturity, fecundity and larval development. Proc. Ind. Nat. Sci. Acad. 39(2): 228-248

Detlaf, T. A. (1972). Fertilization in fishes and the problem of polyspermy. Publ. National marine fisheries service, Natural oceanic and atmospheric administration, US Dept of commerce and National science foundation Washington, D. C.

Devaraj, K. V. (1989). Aquarium Fishes Hand book for aquarium fish hobbyists. Publ. Sri K. V. trust, Chikballapur, Karnataka P. 8-13

Doi, M., Hoshino, T., Yasuhiko, T. & Ogawasara, Y. (1982). Activity of the sperm of the bluefin tuna *Thynnus thynnus* under fresh and preserved conditions. Bull. Jap. Soc. Sci. Fish. 48(4): 495-498

Donaldson E. M; Hunter, G. A. and Dye, H. M. (1981). Induced ovulation in Coho salmon (*Oncorhynchus kisutch*)11. Preliminary study of the use of the antiestrogen, tamoxifen. Aquaculture 26, 143 - 154

Donaldson, E. M; Hunter , G. A. and Dye, H. M. (1978). Induced ovulation in the cohosalmon (*Oncorehynchus kisutch*) using salmon pituitary preparation, gonadotropin releasing hormone and an antiestrogens West. Reg. Conf. Gen. Comp. Endocrinol. 1978 Abstract page 24.

Donaldson, M. and G. A. Hunter (1983). Induced final maturation,ovulation and spermiation in cultured fish, in Fish physiology vol. IX B (Eds) W. S. Hoar, D. J. Randall and E. M. Donaldson; pp 351 - 403. Academic press, New York.

Dube, G. P. (2000). Bundhs are Fish hatcheries closest to nature- Their genesis, growth and impact. Fishing chimes 19 (10 & 11) pp 49 - 53.

Dunsford,H. S. (1911). The report on fish and fishery of Punjab. Government press,Lahore

Dupree, H. (1995). Channel cat fish (*Ictalurus punctatus)* in Brood stock management and egg and larval quality (Ed. By) Niall R. Bromage and Ronald J Roberts. Blackwell science ltd, London pp 220 - 241

Dupree, H. and Huner, J. V. (1984). Nutrition, Feeds and feeding practices is Dupree, H. K. and Huner, J. V., eds Third report to the fish farmers; the status of warm water fish farming and progress in fish farming research. U. S. Fish and Wild life service, Washington D. C.

Dupree, H. K., Green, O. L., and Sneed, K. E. (1965). Techniques for the hybridization of catfishes. South eastern Fish cultural laboratory, Marion, Alabama, unpublished.

Durbin, H., Durbin, F. J. & Stott, B. (1982). A note on the cryopreservation of grass carp milt. - Fish. Mgmt. 13(3): 115-117.

El. Gamal, A.A., Smitherman, R.O., Behrends, L.L. (1988). Viability of red and normal coloured *Oreochromis aureus* and *O. niloticus* hybrids. In: R.S.V. Pullin *et al.* (eds). Second international symposium on Tialpia in aquaculture ICLARM conference proceedings 15, 623p.

Emmerson, W.D. (1983). Maturation and growth of ablated and unablated *Penaeus monodon.* Fabricius Aquaculture. 32: 235-241.

Epler, P. (1981). Effect of steroid and gonadotropin hormones on the maturation of carp ovaries III Effect of steroid hormone, on the carp oocyte maturation in vitro. Pol. Arch. Hydrobiol. 28, 103-110.

Erdahl, A.W., Erdahl, D.A. & Graham, E.F. (1984). Some factors affecting the preservation of salmonid spermatozoa. - Aquaculture 43: 341-350.

Erdahl, D.A. & Graham, E.F. (1978). Cryopreservation of salmonid sperm. - Cryobiology 15: 362-364.

Erdahl, D.A. & Graham, E.F. (1980). Preservation of gametes of freshwater fish. - IX Int. Congr. Anim. Reprod. Artif. Insem. (Proc.), Madrid, pp. 317 - 326.

Filho, O. M., Bertollo, L. A. C. and Junior, P. M. G. (1980). Evidences for a multiple sex chromosome system with female heterogamety in *Apareiodon affinis* (Pisces, Parodontidae). Caryologia 33 :83-91.

Fishelson, L (1962). Hybrids of 2 species of fishes of the genus, Tilapia (Cichhdae, Teleostei) Fish. Bull., Haifa 4(2):14-19.

Fletcher, G. L., Shears, M. A. ; King, M. J., Davies, P. L. and Hew, C. L. (1988). Evidence for antifreeze protein gene transfer in Atlantic Salmon. Can. J. Fish. Aquat. Sci. 45, pp. 352-357.

Fleteher, G. and Davies, P. L. (1991). Transgenic fish for aquaculture, Genetic Engineering 13, 331-369.

Galman, O. R, Moreau, J. and Avtalion, R. R. (1988). Breeding characteristics and growth performance of Philippine red tilapia n. The second international symposium on Tilapia in Aquaculture, ICLARM Conference proceedings, 15 (eds R. S. V. Pullin, T. Bhukaswan, K. Tongathair and J. L. Maclean)pp 169 - 126, Dept of Fisheries, Bangkok and Thailand and ICLARM, Manila, Philippines.

Gardiner, D. M. (1978). Utilization of extracellular glucose by spermatozoa of two viviparous fishes Comp. Biochem. Physiol. 59 p. 165 - 168.

Gerbilskii, N. L. (1965). The present approach to the problem of neurohormonal conlrol of the fish sexual cycle and techniques of hormonal influence applied in fish culture. Seminar on fish culture in inland waters of USSR for FAO group fellowship study.

Gervis, M. H. and N. A. Sims (1992). Biology and Culture of Pearl Oysters (Bivalvia: Pteriidae) ICLARM Stud. Rev. 21. 44p.

Goetz, F. W. (1983). Hormonal control of oocyte final maturation and ovulation in fishes. In Fish physiology (W. S. Hoar, D. J. Randall and E. M. Donaldson eds) Vol. IX A, pp 117 - 170.Academic press, Newyork

Gopakumar, K., S. Ayyappan, J. K. Jena, S. K. Sahoo, S. K. Sarkar, B. B. Satpathy and P. K. Nayak (1999). National fresh water aquaculture development plan Publ. Central Institute of Freshwater Aquaculture, Kausalyaganga.

Gorai, A.K. & D.N. Ray Chaudhuri (1962). Food and Feeding Habits of *Anisops bouvieri (Kirk).* J. Asiat. Soc., 4(3/4): 135-139.

Gordon, M. 1946). Interchanging genetic mechanism for sex determination in fishes under domestication. J. Hered. 37: 307-320.

Grant, C. J. and Spain, A. V. (1975). Reproduction growth and size allometry of *M. cephalus* L. (Pisces: Mugilidae) from North Queensland inshore waters Austr. J. Zool. 2001. 23, 181 - 201.

Guerrero, R. D. and T. A. Abella (1976). Induced sexreversal of *Tilapia nilotica* with methyl tertosterone Fish. Res. J. Philipp. 1:46-49.

Guerrero, R. D. (1975). Use of androgen for the production of all - male *Tilapia aurea* (Steindachner). Trans. Am. Fish. Soc. 104: 342-348.

Guest, W. C., Avault, J. W., Jr. & Roussel, J.D. (1976). Preservation of channel catfish sperm. Trans. Am. Fish. Soc. 105: 469 - 474.

Gupta, S. D. and Rath, S.C. (1991). A preliminary study on quantitative assessment of milt of *Labeo rohita* (Ham) and its cryopreservation over 365 days. Proc. Nat. Symp. Fresh wat. Aqua. Pp. 43-45

——— (1993). Cryogenic preservation of carp milt and its utilization in seed production. Proc. third Indian fisheries forum. Pp. 77-79

Gupta, S. D; S. C. Rath; S. Dasgupta and S. D. Tripathy (1995). A first report on quadruple spawning of *Catla catla* (Ham) Vet. Archiv. 65(5) 143 - 148.

Gupta, S.D., Rath, S.C. and Dasgupta, S. (1995). Fertilization efficiency of cryopreserved spermatozoa over four years at different time intervals after thawing. Geobios. 22; 208-211

Gwo, J. C., Strawn, K., Longnecker, M. T. & Arnold, C. R. (1991). Cryopreservation of Atlantic croaker spermatozoa. - Aquaculture 94: 355 - 375.

Hara, S., Canto, J. T. & Almendras, J. M. E. (1982). A comparative study of various extenders of milkfish, *Chanos chanos* (Forsskal), sperm preservation. - Aquaculture 28: 339 - 346.

Harvel, M. and Place, A. R. (1998). The nutritional quality of live feeds for larval fish. Proceedings of the live feed session, Aquaculture, Canada, 98. Hendry, C. (Eds) No. 98 - 4, p. 6 - 11.

Harvey, B (1983). Cryobiology, 20, 440 - 447.

Harvey, B. (1984). Cryopreservation of *Sarotherodon mossambicus* spermatozoa. - Aquaculture 32: 313-320.

Harvey, B., Kelly, R. N. & Ashwood-Smith, M. J. (1982). Cryopreservation of zebrafish spermatozoa using methanol. - Can. J. Zool. 60:1867- 1870.

Harvey, B., Kelly, R. N. & Ashwood-Smith, M.J. (1983). Permeability of intact and dechorionated zebra fish embryos to glycerol and dimethylsuphoxide. - Cryobiology 20: 432-439.

Hattori, S. (1970). Reproductive aspects of fish resources p 209 - 222. In Ocean development 4.Exploitation of Fishery resources T. Sasaki (Ed) Ocean Dev. Centy. Press. Tokyo.

Hepher, B. and Pruginin, Y. (1981). Tilapia culture in ponds under controlled conditions. . In The Biology and culture of Tilapias (Ed. By R. S. V. Pullin and R. H. Lowe-McConnel) ICLARM Conf. Proc. 7, 185-203.

Hickling, C. F (1952). Culture of Carps and Tilapia hybrids in Uganda. FAO Fish Rep. 44(4):223-229.

Hillier, A. G. (1984). Artificial conditions influencing the maturation and spawning of sub adult, *P. monodon* (Fabricius) Aquaculture. 36: 179 - 184.

Hirayama, K and I Maruyama (1991). Vitamin B12 content as a limiting factor for mass production of rotifer, *B. plicatilis* in Larvi, 91 - Fish and crustaccar larviculture symposium P. Lavens, P. Sorgeloos, E. Jaspers and F. Ollevier (Eds) European aquaculture society special publication No. 15.Gent. Belgium.

Holtz, W., Schmidt-Baulain, R. & Meiners-Gefken, M. (1991). A simple saccharide extender for Cryopreservation of rainbow trout (*Oncorhynchus mykiss*) sperm. - In: Scott, A. P., Sumpter, J. P., Kime, D. E. & Rolfe, M. S. : Reproductive physiology of fish., Proceedings of the 4th International Symposium on the Reproductive Physiology of Fish. Univ. of East Anglia, Norwich, U. K., 7-12 July 1991, pp. 250-252

Houssay, B. A. (1930). Accion sexual de la hipfisis enlos perces y reptiles. Rev. Soc. argent. Biol. 6: 686 - 88.

Hubbs, C. L. (1943). Terminology of early stages of fishes Copeia, 260.

Hubbs, C. L. and Hubbs L. C. (1932). Apparent parthenogenesis in nature in a form of fish of hybrid origin. Science. 76, 628 - 630.

Hubbs, C. L and Hubbs, L. C. (1946). Breeding experiments with the invariable female, strictly matroclinous fish, *Mollienisia (Poecilia) formosa*. Rec. Genet. Soc. Am. 14, 48.

Hubbs,C. L. (1955). Hybridization between fish species in nature. Syst. Zool.,4 (1) : 1-20.

Hudinaga, M. (1942). Reproduction, development and rearing of *Penaeus japanicus* Bate. Jap. J. Zool. 10 (2): 305 - 393.

Huet, Marcel (1986). Text book of Fish culture: breeding and cultivation of fish. 2nd edition. Fishing news book ltd. England.

Hussain Nazar. A., and Masaki Higuchi (1980). Larval rearing and development of the brown spotted grouper, *Epinephelus tauvina* (forskal) Aquaculture, 19 pp. 339-350.

Hviding, E (1993). The rural context of giant clam mariculture in Solomon islands- an anthropological study ICLARM Tech. Rep. 39, 93p

Hyman, L. H. (1992). Rotifera, in "The invertebrates-Acanthocephala, Aschelminthes and Entoprocta. The pseudocoelomate bilateria" Vol. III pp 55 - 151.

Ibrahim, K. H; and Chaudhuri, H. (1966). Preservation of fish pituitary extract in glycerine for induced breeding of fish. Indian J. Exp Biol. 4, 249 - 250.

Ihering von R (1937). A method for inducing fish to spawn Prog. Fish Cult. 34: 15 - 16.

Ihering von R and S. Wright. (1935). Fisheries investigation in North east Brazil. Trans. Amer. Fish Soc. 65: 267 - 71.

Ihering, Von R. and P. de Azevedo (1934). A curimata do acudes nordestinas (*Prochilodus argenteus*) Arch. Inst. Biol. S. Paulo 5: 143-83.

Jacob, P. K. and Krishna Moorthi, B. (1948). Breeding and feeding habits of mullets Mugil, in Ennove Creek. J. Bombay. Nat. Hist. Soc. 47, 663 - 668.

Jalabert B. and Szollosi, P. (1975). In vitro ovulation of trout oocytes, effect of prostaglandins on smooth muscle like cells of the theca Prostaglandins 9, 765 - 778.

Jalabert, B (1978). Production of fertilizable oocyte from follicles of rainbow torut (*Salmo gairdnerii*) following invitro maturation and ovulation. Ann. Biol. Anim. Biochim. Biophys. 18, 461 - 470.

Jalabert, B. (1976). In vitro oocyte maturation and ovulation in rainbow trout (*Salmo gairdneri*) northern pike. (*Esox lucius*) and gold fish (*Carassius auratus*) J. Fish. Res. Board. Can. 33, 974 - 988.

Jalabert, B., Breton, B. and Fostier, A. (1978b). Precocious induction of oocyte maturation and ovulation in rain bow trout (*Salmo gairdneri*). Problems when using 17a hydroxy -20/3 - dihydro progesterone Ann. Biol. Anim. Biochim., Biophys. 18, 977 - 984.

Jalabert, B., Breton, B., Brzuska, E., Fostier, A. and Wieniawski, J. (1977). A new tool for induced spawning; the use of 17a hydroxy -20/3 dihydro progesterone to spawn carp at new temperature. Aquaculture 10, 353 - 364.

James, P. S. B. R. and K. A. Narasimham (1993). Molluscs' Hand book on aqua farming', Marine products export development Authority pp 71 - 86.

Jamieson Barrie, G. M. (1991). Fish evolution and systematics; evidence from spermatozoa press Syndicate of the University of Cambridge, Cambridge.

Jhingran, V. G. and Gopalakrishnan, V. (1974). A catalogue of cultivated aquatic organisms. FAO Fisheris technical paper (130) 83p

John,G. and P. V. G. K. Reddy, (1989). A note on the *Labeo rohita* (Ham) X *Cyprinus carpio* hybrid. Proc. world sympo. on selection, hybridization and genetic engineering in Aquaculture. Bordeaux 27-30 May 1986 Vol. II Berlin.

John, G., P. V. G. K. Reddy and S. D. Gupta (1984). Artificial gynogenesis in two Indian Major Carps, *Labeo rohita* (Ham) and *Catla catla* (Ham). Aquaculture, 42 : 161-168.

Jose, T. M. and A. G. Sathyanesan (1977). Pituitary Cytology of the Indian carp *Labeo rohita* (Ham) Anat. Anz. 142, 410 - 423.

Joshua, K., A. Sujatha, Elizabeth Carolin, V. Supraba and H. Dinesh kumar (2000). Shrimp hatchery operation and management practices. Fishing Chimes, 19 (10 & 11) pp 123 - 128.

Joshua, K., Subba Rao; A. Sujatha., A. Satyarayana; P. Somanath Rao; B. Trinatha Babu and L. V. Ramana (1993). Shrimp hatchery operation and management. Handbook on aquafarming, Marine products export development Authority, Kochi.

Kagawa, H. and Nagahama, Y. (1981). Invitro effects of prostaglandins on ovulation in gold fish, *Carassius auratus*. Bull. Japn. Soc. Sci. Fish. 47, 1119 - 1121

Kanaujia, D. R. (1995). Emerging technologies in the seed production of *Macrobrachium malcomsonii* in 'Current and emerging trends in aquaculture (Ed). P. C. Thomas, Daya Publishing House, Delhi.

Karmchandani, S. J.,V. R. Desai,M. D. Pisolkar and G. K. Bhatnagar (1967). Biological investigations on the fish and fisheriesof Narmada river(1958-66) Bull. Cent. Inl. Fish. Res. Inst. Barrackpore,10:40pp.

Kathirvel, M., Srinivasagam, S. and Kulasekharapandian, S. (2000). Mud crab hatchery seed production: Recent advances. Fishing chimes Vol. 19, No. 10 & 11 pp 83 - 93.

Kathirvel, M., Srinivasagam, S., Ghosh, P. K. and Balasubramanyam, C. P. (1997). Mud crab culture, CIBA Bulletin No. 10. pp 1 - 4.

Kerby, J. H. (1983). Cryogenic preservation of sperm from striped bass. Trans. Am. Fish. Soc. 112:86-94.

Khan, H. (1938). Ovulation in fish (effect of administrations of anterior lobe of pituitary gland) Curr. Sci. 7(5); 233 - 234.

Khan, H. A. and Jhingran, V. G. (1968). Synopsis of biological data on *Catla catla* (Ham 1822) FAO fisheries synopsis No. 32., FAO, Rome.

Khan, H. A. and Jhingran, V. G. (1979). Synopsis of biological data on mrigal, *Cirrhinus mrigala* , FAO Fisheries Synopsis No. 102, FAO, Rome.

Khan, H. A. and Jhingran, V. G. (1975). Synopsis of Biological data on rohu, *Labeo rohita* (Ham; 1822) p. 14, FAO Fisheries synopsis No. 111,FAO, Rome.

Khoo, H. W., L. H. Ang, H. B. Lim and K. Y. Wong, (1992). Sperm cells as vectors for introducing foreign gene DNA into zebra fish, Aquaculture, 107: 1-19.

Khoo, K. H. (1974). Steroidogenesis and the role of steroids in the endocrine control of oogenesis and vitellogenesis in the gold fish, *C. auratus auratus* Ph. D. Thesis, Uniuersity of British Columbia, Vancouver, B. C. Canada.

King, C. E., Bayne, H. B., Cannon, T. K. and King, A. E. (1983). Hydrobiologia, 104, 85 - 88.

Kirshenblatt, J. D. (1952). The action of steroid hormones on the female Vyun Nauk. Dokl. Akad. Nauk SSSR 83, 629 - 632.

Kirshenblatt, J. D. (1959). The effect of cortisone on ovaries of the loach. Byull. Eksp. Biol. Med. 47, 108 - 112.

Koehler, J. K. (1967). Cryobiology, 3, 392 - 399.

Kohno, H., P. T. Imanto, S. Diani, B. Slamet, and P. Suniyoto (1990). Reproductive performance and early life history of grouper, *E. fuscoguttatus*. Bull. Pen - Perikanan. Special edition 1: 27 - 35.

Koldras, M. and Moczarski, K (1983). Properties of Pike *Esox lucius L.* milt and its cryopreservation. Pol. Arch. Hydrobiol. 30(1): 689-678.

Koldras, M. and Bieniarz, K. (1987). Cryopreservation of crap sperm. Polskie Archiwum. Hydrobiologii, 34: 125-134.

Konardt, A. G. (1968). Methods of breeding the grass carp. *Ctenopharyngodon idella* and silver carp *Hypopthalmichthys molitrix* FAO Fish Rep. (44) 4, 195 - 204.

Kornfield,I. L. (1984). descriptive genetics of Cichlid fishes. In B. . J. Turner (ed)Evolutionary genetics of fishes. Plenum publishing corp., New York, USA.

Korringa, P. (1976). Farming marine organisms low in the food chain: A multidisciplinary approach to edible sea weed, mussel and clam production. Elsevier Scientific Publishing Co., Amsterdam.

Kosswig,C. (1964). Polygenic sex determination. Experientia 20: 190-199

Kossmann,H. (1973). Versuche zur Konservierung des Karpfenspermas (*Cyprinus carpio*) Arch. Fischereiwiss. 24:125-128.

Kowtal, G. V. (1967). Occurrence and distribution of pelagic fish eggs and larvae in the Chilka lake during the year 1964 and 1965. Ind. J. Fish. 14, 198 - 214.

Kulkarni, C. V. (1971). Spawning hebit,eggs and early development of Decean Mahseer, *T. Khudree* (sykes)J. Bom. Nat. Hist. Soc. 67; 510-521

Kulkarni, C. V. and S. N. Ogale (1979). Air transport of mahseer (Pisca:cyprinidae) eggs in moist cotton wool. Aquaculture, 16:367-368.

Kulkarni, C. V. and S. N. Ogale (1986). Hypephysation(induced breeding) of mahseer Tor Khudree (Sykes) Pb. Fish. Bull. VolX(2) 23-26.

Kulkarni, C. V. and S. N. Ogale (1995). Conservation of the mighty mahseer of India, Tata electric company, Bombay house, Bombay: 39 pp.

Kuo, C. M. and Nash, C. E. (1975). Recent progress in the control of ovarian development and induced spawning of grey mullet (*Mugil cephalus L.*) Aquaculture , 5, 19 - 29

Kuo, C. M., C. E. Nash and W. O. Watanabe (1979). Induced breeding experiments with milk fish, *Chanos chanos* (Forskal) in Hawati. Aquaculture. 18, 95 - 105.

Kuo, C. M., Z. H. Shehadeh and K. K. Milisen (1973). A preliminary report on the development, growth and survival of laboratory reared larvae of the grey mullet *Mugil cephalus L.* J. Fish. Biol. 5: 459 - 470.

Kuo, C. M; Shehadeh, Z. H. and Nash, C. E. (1973). Induced spawning of captive grey mullet (*Mugil cephalus L*) females by injection of human chorionic gonadotropin (HCG) Aquaculture, 1, 429 - 432

Kuo, H. (1988). Progress in genetic improvement of red hybrid Tilapia in Taiwan. The second International symposium on Tilapia in Aquaculture ICLARM conference proceedings R. S. V. Pullin; T Bhukaswan, K. Tonguthai, and J. L. Maclean (eds)., 15, p 219 to 221. Dept. of Fisheria, Bangkok, Thailand and ICLARM, Manila, Philippines.

Kurokura, H. and Hirano, R (1980). Cryopreservation of rainbowtrout sperm. Bull. Jpn. Soc. Fish. 46: 1493-1495

Kurokura,H.,Hirano,R.,Tomita,M. and Iwahashi,M. (1984). Cryopreservation of carp sperm. Aquaculture 37: 267-273

Kuronuma, K. (1968). New systems and new fishes for culture in the Far East. FAO Fish. Rep (44) 5, 123 - 142

Lahnsteiner,F.,Weissmann,T and Patzner,R. A. (1991). Gefrierkonsrevierung von. Spermien am Beispeil der Asche (*Thymallus thymallus)* Osterreichs Fischerei 44: 194-200

Lahnsteiner, F., Patzner, R. A. and Weismann, T. (1993). The spermatic duct of salmonid fishes (Salmonidae: Teleostei) Morphology, histochemistry and composition of secretion. J. Fish. Biol. 42 p: 79 - 93.

Lavania, U. C., (1980). Chromosome banding. Science Reporter, September: 580-583.

Langer, R.K., J. Somalingam, and U. K. Maheswari(1992). Availability of mahseer(*Tor tor)*seed in river Narmada at Hoshangabad(MP) and its feeding habit and growth in captivity. (farm pond,Powarkheda). In recent researches in coldwater fisheriesEd. K. L. Seghal Proc. of Nat. workshop on research and development need cold water fisheries ,Today and Tomorrow printers and publishers, New Delhi 145-156.

Lee, J.C. (1979). Reproduction and hybridization of three chichlid fishes, *Tilapia aurea* (steindachner), *T. hornorum Trewavas* and *T. nilotica* (Linnaeus) in aquaria and plastic pools. Ph.D. Dissertation, Auburn University.

Leger, P., D. Grymonpre, E. Van Ballaer and P. Sorgeloose. (1989). Advances in the enrichment of rotifers and Artemia as feed sources in marine larviculture p. 141 - 142. In EAS Spec. Publ. No. 10.Europe. Aquacult. Soc. Bredene, Belgium 344p

Legner, E. F. (1978). Mass culture of *Tilapia zilli* (cichlidae) in pond ecosystems. Entomophaga, 23(1):51-55.

Legendre, M. and Billard, R. (1980). Cryopreservation of rainbow trout sperm by deepfreezing. Reprod. Nutr. Dev. 20:1859-1868

Lessent, P. (1968). Essais d'hybridation dans le genete Tilapia a la station de Rechercties Piscicoles de Bouake Cote d'lvoire. In proceedings of the world symposium on warm water pond fish culture Vol. 4 (Ed. TV. R. Pillai) pp. 148-154.

Leung, L. K. (1987). Cryopreservation of spermatozoa of the barramundi, *Lates calcarifer* (Teleostei: Centropominidae) Aquaculture 64: 243-247.

Leung, L. K. P. and B. G. M. Jamieson (1991). Live preservation of fish gametes in Jamieson, B. G. M. Fish evolution and systematics. Evidence from spermatozoa pp 245 - 269. Cambridge university press, U. K.

Liao, I. C. And T. I. Chen (1984). Gonadal development and induced breeding of captive milkfish in Taiwan in J. V. Juario, R. P. Ferrari and L. V. Benitez (Eds) Advances in milk fish biology and culture. Island publishing house, Inc. Manila, Philippine pp 41 - 52.

Liao, I. C., Y. J. Lu, T. L. Huang and M. C. Lin (1972). Experiment on induced breeding of the grey mullet *Mugil cephalus* Linnaeus in Coastal Aquaculture in Indo Pacific Region pp 213 - 243. Ed. TVR Pillai Publ. Fishing news books ltd.

Liao, I. C; J. V. Juario, S. Kumagi, H. Nakajima, M. Natividad and P. Buri (1979). On the induced spawning and larval rearing of milk fish, *Chanos chanos* (Forskal) Aquaculture, 18, pp 75 - 93.

Lin, H. R. and Peter, R. E (1996). Hormones and spawning in fish. Asian fisheries science. 9.pp 21 - 33.

Lin, L. T. (1984). Studies on the induced breeding of milk fish, *Chanos chanos* Forskal reared in ponds. China Fisheries monthly. 378: 3 - 29. (in Chinese, English abstract)

Lin, L. T. (1985). My experience on artificial propagation of milk fish - studies on natural spawning of pond reared brood stock In. Cheng. Sheug. Lee and I. Chiu Liao (Eds) Reproduction and culture of milk fish, Tung kang Marine laboratory, Tung kang, Taiwan pp 99 - 114.

Linhart, O., Liehman,P. and Rab,P(1988). The first results in cryopreservation of carp sperm. Bul. VUHR Vodnany 2:3-13 (in Czech, with an English summary)

Linhart,O.,Proteau,J. P.,Redondo,C and Billard,R. (1991). Preservation of gametes in Europian cat fish(*Silurus glanis L*). - In Scott, A. P.,Sumpter,J. P.,Kime,D. E. and Rolfe ,M. S. :reproductive physiology of fish., Proceeding of the 4th International symposium on the Reproductive physiology of fish. Univ. of East Anglia, Norwich,U. K. 7-12 July 1991, p. 283

Little, D. C. (1989). An evaluation of strategies for production of Nile tilapia (*Oreochromis niloticus*) fry suitable for hormonal treatment Ph. D. Thesis, Institute of Aquaculture, university of sterling.

Ma. Suzette R. Licop (1988). Hatchery operation and management m Biology and culture of *Penaeus monodon* BRAIS state of the art series no. 2 Aquaculture Department of the South East Asian Fisheries Development center, Philippines.

MacIntosh, D. J. (1985). Tilapia culture: Hatchery methods for *O. mossambicus* and *O. niloticus* with special reference to all male fry production. Institute of aquaculture; University of Sterling, Sterling.

MacIntosh, D. J. and D. C. Little (1995). Nile Tilapia (*Oreochormis niloticus*) in Brood stock management and egg and larval quality (Ed. by) Niall R. Bromage and Ronald J. Roberts (Publ) Black well science Ltd, London. pp - 220 - 241.

Mair, G. C., A. G. Scott, J. A. Beardmore and D. O. F. Skibinski. (1986). A technique for induction of diploid gynogenesis in *Oreochromis niloticus* by suppression of the first mitotic division. Proc. of EIFAC /FAO symposium on selection, hybridization and genetic engineering in aquaculture of fish and shellfish for consumption and stocking. Bordeaux (France) 27-30 May 1986.

Mair, G. C., Abucay, J. S., Skibinski, D. O. F., T. A. Abella and J. A. Breardmore (1997). Genetic manipulation of sex ratio for large-scale production of all male tilapia, *Oreochromis niloticus*.

Majumdar,K. C.,K. Raviden andK. Nasauddin (1997). DNA fingerprinting in Indian major carps and tilapia by BKM 2(8) and M-13 probes. Aquaculture research 28: 129-138

Marichamy, R. and Rajapackian, S. (1984). Culture of larvae of *S. serrata* Mari. Fish. Inf. service. Vol. 158, PP. 13-15.

Marquez, R. M. (1974). Observaciones sobre mortalidat total Y Crescimanto en longitud de la Lisa (*Mugil cephalus*) en lagune de Tamiahua, Mexico. Inst. Nac. Pesca. Sc/3, 1- 16.

Marte, C. L. N. M. Sharwood, L. W. Crim and B. Harvey(1987). Induced spawning of maturing milk-fish (*Chanos chanos* Forskal)with gonadotropin releasing hormone (GnRH)analog administered in various ways. Aquaculture 60 : 303-310.

Mazur, P., Cole, K. W. Hall, J. W., Schreuders, P. D. and Mahowald, A. P. (1992). Science, 258. 1932 - 1935.

McAndrew, B. J., Roubal, F. R., Roberts, R. J; Bullock, A. M and McEwen, I. M. (1988). The genetics and histology of red, blond and associated colour variants, in *O. niloticus* 127 - 37.

McCoy, E. W. and T. Chong peepien (1988). Bivalve mollusc culture reserch in Thailand. ICLARM Technical report No. 19. Pp. 19-30

Meyer, F. P., Sneed, K. E., and Eschmeyer, P. T. eds. (1973). Second report to the fish farmers the status of warm water fish farming and progress in fish farming research. U. S. Fish and wild life service, Washington, D. C.

Mirza, J,A. and Shelton, W. L. (1988). Induction of gynogenesis and sex reversal in silver carp. Aquaculture68: 1-14.

Mishriji. S. Y. and Kubo. T. (1978). The energy metabolism of *T. nilotica*. 11.Actue metabolism at 200 and 260 Bull. Fac. Fish. Hokkaido Univ., 29(4):313-321.

Moczarski (1977). Deep freezing of carp *(Cyprimus carpio)* sperm. Bull. Acad. Pol. Sci. Ser. Sci. Biol., 15: 187-190.

Moretti, A., Pedini Fernandez - Criado, M., Cittolin, G., Guidastri, R. (1999). Manual on hatchery production of sea bass and gilthead sea bream. Vol. I. Rome, FAO. 194p.

Motoh, H. (1979). Larvae of the Decapod crustacea of the Philippines - III Larval development of the giant tiger prawn, *Penaeus mondon* reared in the laboratory. Bull. Japanese Soc. Sci. Fish., 45 (10) 1201 - 1216

Motoh, H. (1981). Studies on the fisheries biology of the giant tiger prawn, *Penaeus monodon* in the Philippines. Technical report No. 7 Tigbauan, SEAFDEC Aquaculture Department, 128p.

Mounib, M. S. (1967). Metabolism of pyruvate, acetate, and glyocylate by fish sperm. Comp. Biochem. Physiol. 20: 987 - 992.

Mounib,M. S(. 1978). Cryogenic preservation of fish and mammalian spermatozoa. J. Reprod. Fertil. 53: 13-18

Mounib,M. S.,Hwang,P. C. and Idler, D. R. (1968). Cryogenic preservation of Atlantic cod (Gadus morhua) sperm. Comp. Biochem. Physiol. 25:703-709

Nagashima, W., Kashiwazak; N., Ashman, R. J., Grupen, C. G., and Nottle, M. B. (1995). Nature, 374 - 416

Nagy,A., M. Bercsergi and V. Csany, (1981). Sex reversal in carp (*Cyprinus carpio*) by oral admininistration of methyltestosterone. Can. J. Fish. Aquat. Sci. 38: 725-728.

Nagy,A., V. Csany, J. Bahos and M. Bercsergi, (1984). Utilization of gynogenesis and sex reversal in commercial carp breeding. Growth of the first gynogenetic hybrids. Aquaculture Hungarica (Szarvas) 4: 7-16.

Nakamura, M (1975). Dosage dependant charges in the effect of oral administration of methyl testosteome on gondal sex differentiation in *Tilapia mossanbica* Bull. Fac. Fish. Hokkaido Univ. 26 : 99- 108.

Nantiyal, P. and M. S. Lal(1985). Fecundity of Garhwal Himalayan mahseer(Tor putitora) J. Bombay Nat. Hist. Soc, 82(2)253-257

Nandeesha, M. C., Ramacharya and T. J. Varghese (1991). Further observations on breeding of carp with ovaprim special publication no. 6 Asian Fisheries society. Indian branch., Mangalore. 11p.

Nash, C.E. and Kuo, C.M. (1976). Preliminary capture, husbandry, and induced breeding results with the milk fish, *Chanos chanos* (Forskal). Proc. Int. Milk fish workshop conf. Tigbauan, Iloilo, Philippines, May 19-22, pp. 121-132.

Nash, C. F., Kuo, C. M. and McConnel, S. C. (1974). Operational procedures for rearing larvae of the grey mullet (*Mugil cephalusl*) Aquaculture, 3, 15 - 24.

Nayak, P. K; A. K. Pandey, B. N. Singh, J. Mishra, R. C. Das and S. Ayyappan (2000). Breeding, larval rearing and seed production of the Asian cat fish, *Heteropneustes fossilis* (Bloch) Central Institute of Freshwater aquaculture, Kausalyaganga, Bhubaneswar. (Publ) Director, CIFA, Kausalyaganga) pp 1-68.

Nayar, K. Nagappan, S. Mahadevan, K. Alagarswami and P. T. Meenakshisundaram (1980). Mussel farming. Progress and prospects. CMFRI Bullettin No. 29.

Nayar, K. Nagappan, S. Mahadevan, K. Alagarswami and P. T. Meenakshisundaram (1980). Mussel farming. Progress and prospects. CMFRI Bulletin No. 29.

Needham James. G and Needham,Paul R. (1941). Guide to the study of Fish water Biology with special reference to aquatic insects and other invertebrate animals and phytoplankton, Comstock publishing company, Inc. Ithaca, Newyork p 88.

Nelson, B. A. (1960). Spawning of Channel catfish by use of hormone Proc. Annual conf. South-east Assoc. Game and Fish Comm., 14:145-148.

Nelson, Joseph & (1984). Fishes of the world. Wiley - Interscience publication, USA.

New, M. B. and S. Singholka (1985). Fresh water prawn farming. A manual for the culture of *Macrobrachium rosenbergii* FAO Fish. Tech. Pap 2(25) Rev. 1.118p

New, M. B. and S. Singholka (1985). Freshwater prawn farming - a manual for the culture of Macrobrachium rosenbergii. FAO Fish. Tech. pap.,(35) Rev. 1:1118p.

Nikolsky, G. V. (1963). The ecology of fishes Pub. Academic press. Inc. (London) Ltd.

Norton, V. M., Nishimura, A., and Davis, K. B. (1976). A technique for sexing channel cat fish. Trans. Am. Fish. Soc. 105:460-462.

Olive, P. J. W. and Wang, W. B. (1997). Cryobiology, 34:284-294

Oren, O. H. (1981). Aquaculture of grey mullets Pub. Cambridge university press, Cambridge.

Oshiro, T. and Hibiya, T. (1975). Presence of ovulation inducing enzymes in the ovarian follicle of loach. Bull. Jpn. Soc. Sci. Fish. 41, 115.

Ott, A.G. (1975). Cryopreservation of Pacific salmon and steelhead trout sperm. Ph. D. Thesis. Oregon state university.,Corvallis,Oreg. 145p

Ott, A.G. and Horton, H. F. (1971a). Fertilization of steelhead trout (*Salmo gairdneri*) eggs with cryopreserved sperm. J. Fish. Res. Board, Canada 28:1915-1918.

Ott, A.G. and Horton, H. F. (1971b). Fertilization of Chinook and coho salmon eggs with cryopreserved sperm J. Fish. Res. Board. Can. 28:745-748

Ozato, K. Inoue, K. and Wakamatsu, Y. (1992). Gene transfer and expression in medaka embryos in C. L. Hew and G. L. Fletcher (eds) Transgenic fish, world scientific publishing, Singapore pp. 27-43.

Palmiter, R. D., Brinster, R. L., Hammer, R. E., Trumbauer, M. E., Rosenfield, MG., Birnberg, N. C. and Evans, R. M. (1982). Dramatic growth of mice that develop from eggs microinjected with metallothionein growth hormone fusion genes. Nature 300, 611-615.

Pandey, S., Stacey, N., and Hoar, W. S. (1973). Mode of action of clomiphene citrate in inclucing ovulation of gold fish. Can. J. Zool. 51, 1315 - 1316.

Pandian, T. J. and K. Varadraj (1987). Techniques to regulate sex ratio and Breeding m 'Tilapia' current Science 56(8) ;. 337-343.

Pandian, T. J. and K. Varadaraj (1988). Tectnique for producing all male and all triploid *O. mossambicus*. 243-249. In R. S. V. Pullin *et al*. (eds.) The second International Symposium on Tilapia in Aquaculture, ICLARM conference proceedings 15, 623 p. Dept. of Fishers Bangkok and ICLARM, Manila.

Pandian, T. J. and L. A. Marian (1994). Problems and prospects of transgenic fish production, Current science 66 (a) 635-649.

Panouse, J. (1943). Eye stalk and ovarian growth in crustaceans. C. R. Acad. sci. 217; 553-555

Parazo, M. M., LMaB Garcia, F. G. Ayson, A. C. Fermin, J. M. E. Almendras, D. M. Reyes, Jr. EM. Avila, JD. Toledo (1998). Sea bass hatchery operation. Aquaculutre extension manual No. 18. South East Asian Fisheries Development center pp 1 - 42.

Park, E.H. and Y.S. Kang (1979). Karyological confirmation of conspicuous ZW sex chromosomes in two species of pacific fishes (Anguilliformes: teleostomii). Cytogent. Cell. gen. 23(1-2): 33-38.

Pathani, S. S. (1978). A note on secondary sexual dimorphism in Kumaon Mahseer, *Tor putitora* (Ham) in Uttar pradesh. Proc. Ind. Acad. Sci. (Anim. sci.) 48: 773-775.

Pathani, S. S. and Das, S. M. (1979). Induced breeding by hypophysation of mahseer, *Tor putitora* in Bhimatal. Sci. cult. 45: 209-211

Payne, A. I. J. Ridgway and J. L. Hamer (1988). The influence of salt concentration and temperature on the growth of *Orecochromis spilurus spillurus, O. mossambicus* and a red tilapia hybridp. 481-487. In R. S. V. Pullin *et al*(ed.) The second international symposium on Tilapia in Aquaculture, ICLARM conference proceedings 15, 623, P. Dept. of Fishers, Bangkok Thiland and ICLARM, Manila.

Pendergrass, P. and Schroeder, P. (1976). The ultra structure of the thecal cell of a teleost, *Oryzias latipes*, during ovulation, invitro. J. Reprod. Fertil. 47, 229 - 233.

Perschbacher, P. W. and R. B. McGeachin (1988). Salinity tolerane of red hybrid Tilapia fry, juvinila and adults. P. 415-419. In R. S. V. Pullin *et al.* (ed.) The second international symposium on Tilapia in Aquaculture, ICLARM conference proceedings 15, 623p. Dept of Fisheres, Bangkok, Thailand and ICLARM, Manila, Philiptus.

Peter Marian (1993). Brachionus, in "Live Feed" Hand book of aquaculture, Part III (Publ.)Marine products Export Development Authority pp. 45 - 54.

Peter Marian (1993). *Daphnia* culture in "Live feed" Hand book of Aquafarming part IV pp. 55 - 61.

Pezold, F. (1984). Evidence for multiple sex chromosomes in the freshwater goby,*Gobionellus shufeldti* (Pisces:Gobiidae). Copei :235-238.

Philippart, J. C. and Ruwet, J. C. (1982). Ecology and distribution of Tilapias, in the Biology and culture of tilapias, ICLARM conference proceedings, 7 (eds R. S. V. Pullin and R. H. Lowe Mc Connel) pp. 15 - 60 International center for living aquatic resources management, Manila, Philippines.

Piironen, J. (1987). factors affecting fertilization rate with cryopreserved sperm of whitefish (*Coregonus muksun* pallas) Aquaculture 66: 347-357.

Piironen, J. (1992). Report on cryopreservation of fish eggs and sperm. Nordiske seminar-og Arbejdsrapporter 1993: 589.

Piironen, J. and H. Hyvarinen (1983). Composition of the milt of some teleost fishes. J. Fish. Biol. 22, 351 - 361.

Pillai, T. V. R. (1988). Aquaculture and practices. 377-392.

Pillai, T. V. R. (1990). Aquaculture: Principles and practices, Fishing new books, Oxford.

Poirier, G. R. and Nicholson, N. (1982). Fine structure of testicular spermatozoa from the channel cat fish, *Ictalurus punctatus.* Journal of ultra structure research 80, 104 - 110.

Primavera, J. H. (1988). Maturation reproduction and brood stock technology in Biology and culture of *Penaeus mondon* BRAIS state of the art series no. 2.publ. Aquaculture department of the South East Asian Fisheries Development Centre, Philippines.

Pruginin, Y. (1967). Culture of carps and Tilapia hybrids in Uganda, FAO Fish Rep., 44(4):223-229.

Pruginin, Y. (1967). Report to the government of Uganda on the experimental fish culture project in Uganda 1965-66, FAO UNDP (TA) Reports : 2446.

Pullin, R. S. V. (1972). The storage of plaice (*Pleuronectes platessa*) sperm at low temperature. - Aquaculture. 1:279-283.

Pullin, R. (1975). Preliminary investigations into methods for controlling the reproduction of captive marine flatfish. Pubbl. Stn. Zool. Napdi. 39., Suppl. 282 - 296.

Pullin, R. S. V. (1982). General discussion on the biology and culture of tilapias in The Biology and culture of tilapias, ICLARM conference proceedings, (eds. R. S. V.

Pullin and R. H. Lowe Mc Connel) pp 15 - 60, ICLARM, Manila, Philippines.

Pullin, R. S. V. (1988). (Ed.) Tilapia genetic resources for aquaculture ICLARM conference proceedings 16, 108p. International centre for living aquatic resources management, Manila, Philippines.

Pullin, R. S. V. (Ed) (1988). Tilapia genetic resources for aquaculture. ICLARM conference proceedings 16, 108p. International center for living aquatic resources management, Manila, Philippines.

Quazim,S. Z. and A. Qayyum (1962). Spwning frequencies and breeding season of some freshwater fishes with special reference to those occurring in the plains of northern India. Ind. J. Fish. 8(1):23-43

Raina, H. S. Shyam Sunder, C. B. Joshi, and Madan Mohan(1999). Himalyan Mahseer. Nat. Res. Cent. Coldwater Fish, Bhimtal(UP) Bull.,1:29p.

Raja Bai Naidu, K. G. (1955). The early development of *Scylla serrata* (Forsk.) de Haan and *Neptunus Sanguinolentus* (Herbst) Ind. J. Fish. PP67-76.

Rajabai Naidu, K. G. (1955). The early development of *Scylla serrata* (Forsk) de Haan and *Neptunus sanguinolentus* (Herbst) Ind. J. Fish. pp 67 - 76.

Rajamani, M., S. Lakshmi Pillai and J. X. Rodrigo. (1999). On the mass production of rotifer with different combinations of fertilizers. Mar. Fish. Infor. Serv; T & E Ser. No. 159 pp 11 - 33

Rana, K. J., Muiruri, R. M., McAdrew, B. J. & Gilmour, A. (1990). The influence of diluents, equilibration time and prefreezing storage time on the viability of cryopreserved *Oreochromis niloticus* (L.) spermatozoa. - Aquaculture Fish. Mgmt. 21: 25-30.

Rana,K. J. and McAndrew,B. J. (1989). The viability of cryopreserved tilapia spermatozoa. Aquaculture. 76: 335-345

Rao, G. R. M., Tripathy, S. D. and A. K. Sahoo (1994). Breeding and seed production of the Asian cat fish, *Clarias batrachus* (Linnaeus) Manual serial No. 3.CIFA, Kausalyaganga.

Rao, K. J. (1995). Emerging technology on the seed production of *Macrobrachium rosenbergii* in 'Current and emerging trends in aquaculture' (ed.) P. C. Thomas, Daya Publishing House, Delhi.

Rao, K. J. and Tripathy, S. D. (1993). A manual on giant fresh water prawn hatchery, CIFA (ICAR) Kausalyaganga.

Rath. S.C., Gupta, S. D. and Subrata Dasgupta (1999). Nonstripping induced spawning and double spawning of grass carp in a hatchery system with foliage free brood diet. Vet. Arch. 69(1) 7 - 15.

Rath, S.C. and S.K. Sarkar (2002). Utilisation of eco-carp hatchery for mass seed production of *cyprinus carpio*. Fishing chimes. 22(8): 36-38.

Ray, L. E. (1978b). Production of Tilapia in Cat fish raceways using geothermal water. In : Smitherman, R. O. Shelton, W. L., and Grover, J. H. (eds.) Culture of exotic fishes. Symp. Proc. Fish cultures section, American Fishers soc. Aulwin, Albama 86-89.

Reddy, P. V. G. K. (1999). Genetic resources of Indian Major carps. FAO Fisheries Technical paper 387.

Reddy, P. V. G. K., George Jhon and R. K. Jana (1987). Induced polyploidy mosaics in Indian carp *Labeo rohita* (Ham). J. Inland Fish. Soc. India 19(1): 9-12.

Reddy, P. V. G. K., H. A. Khan, S. D. Gupta, M. S. Tantia and G. V. Kowtal (1990).On the ploidy of three intergeneric hybrids betweem common carp (*Cyprinus carpio communis* L. and Indian Major carp. Aquaculture Hungarica (Szarvas) Vol. VI :5-11.

Reddy, P. V. G. K., R. K. Jana and S. D. Tripathi (1993). Induction of mitotic gynogenesis in Rohu (*Labeo rohita*) Ham. through endomitosis. The nucleus 36(3): 106-109.

Reddy,P. V. G. K., K. D. Mahapatra, J. N. Saha and R. K. Jana (1998). Effect of induced triplody on the growth of common carp (*Cyprinus carpio* Var. *communis* L.). J. Aqua. Trop. 13 (1): 65-72.

Rees, J. F., N. Chaitanivisuti., P. Menasveta and P. Sorgeloos (1992). Fatty acid composition of *Penaeus mondon* post larvae fed artemia nuaplii enriched with increasing HUFA lvels. Proc. Third Asian Fisheries Forum, Singapore p. 189.

Rishi, K. K. (1976). Karyotypic studies on four species of fishes. the nucleus 19: 95-98.

Robert L. Busch (1985). Channel cat fish culture in ponds: inchannel cat fish culture, Development in Aquaculture and fishery science Vol. 15 Eds. Tucker, C. S. Elsevier Science publishers pp. 13-78.

Rokkones, E., Alestrom, P. Skjervold, H. and Gantvik, K. M. (1989). Microinjection and expression of a mouse metallothionein human growth hormone fusion gene in fertilized eggs. J. Comp. Physiol B. 158, 751-758.

Rosenthal, H., Alderdice, D. F. & Velsen, F. P. J. (1978). Cross-fertilization experiments using Pacific herring eggs and cryopreserved Baltic herring sperm. - can. Fish. Mar. Serv., Tech. Rep. 844.

Sahai, R., R. K. Vijh and P. G. Nair (1989). Application and utility of cytogenetic techniques in the characterisation of animal genetic resources. In: Fish genetics in India, Editors P. Das and A. G. Jhingran. Today and tomorrow printers and publisher, New Delhi: 51-61.

Scheerer, P. D. & Thorgaard, G. H. (1989). Deep - freezing of rainbow trout *Salmo gairdneri* sperm at varying intervals after collection. - Theriogenology 32(3): 439-443.

Schmidt-Baulain, R. & Holtz, W. (1991). Effect of age and stage of spawning season on output fertilizing capacity and freezability of rainbow trout (*Oncorhynchus mykiss*) sperm. - In. : Scott, A. P., Sumpter, J. P., Kime, D. E. & Rolfe, M. S. : Reproductive physiology of fish., Proceedings of the 4th International Symposium in the Reproductive Physiology of Fish. Univ. of East Anglia, Norwich, U. K., 7 - 12 July 1991, p. 287.

Scott, A. P. & Baynes, S. M. (1980). A review of the biology, handling and storage of salmonid spermatozoa. - J. Fish Biol. 17: 707 - 739.

Sebastian, M. J. and Nair, V. A. (1975). The induced spawning of grey mullet, *Mugil macrolepis* (Agas) and large scale rearing of its larvae. Aquaculture. 5, pp 41 - 52.

Seghal,K. L. (1978). Coldwater fish culture in uplands of India. Proc. Summ. Inst. oninland Aquaculture,CIFRI,Barrackpore: 170-179

Selvaraj, G. S. D. (1999). On the occurrence of giant male and female groupers with a note on sex change in groupers. Mar. Fish. Infor. Serv. T & E Ser. No. 159, pp. 13-16.

Sette, O. E. (1945). Biology of Atlantic mackerel (*Scomber scombrus*) of North America Part I. Early life history including growth, drift and mortality of the eggs and larval population U. S. Fish. Wildl. Serv. Fish. Bull. 50: 149 - 237.

Shears, M. A., Fletcher, G. L., Hew, C. L., Gauthier, S. and Davies, P. L. (1991). Transfer, expression and stable inheritance of antifreeze protein genes in Atlantic Salmon (*Salmo Salar*) Mar. Mol. Biol. Biotech. 1: 426-431.

Sherwood, N., Eiden, L., Brown Stein, M., Spiess, J. and Vale, W. (1983). Characterisation of a teleost gonadotropin releasing hormone. Proc. Natl. Acad. Sci. IISA, 80, 2794 - 2498.

Silas, E. G., M. S. Muthu., N. N. Pillai and K. V. George (1978). Larval development: *Penaeus monodon* Fabricius In Larval development of Indian Penacid prawns CMFRI Bullctin no. 28: 2 - 12.

Silas, E. G., K. H. Mohamed., M. S. Muthu. N. N. Pillai., A. Laxminarayana, S. K. Pandian, A. R. Thirunavukkarasu and Syed Ahamed Ali (1985). Hatchery production of penaeid prawn seed, *Penaeus indicus* CMFRI special publication No. 23.

Singh, A. K. and Singh, T. P. (1976). Effect of clomid, sexovid and prostaglandins on induction of ovulation and gonadotropin secretion in a fresh water catfish, He*teropneustes fossilis* (Bloch) Endokrinologie 68, 129 - 136.

Sinha, V. R. P., M. V. Gupta, M. K. Banerjee, D. Kumar (1973). Composite fish culture at kalyani, West Bengal. J. Inland Fish. Soc. India. . 201 - 207.

Slastenzenko, E. P. (1957). A list of natural fish hybrids of the world. Hydrobiology Istanbul 4: 76-97.

Sneed, K. E. & Clemens, H. P. (1956). Survival of fish sperm after freezing and storage at low temperatures. - Prog. Fish-Cult. 18: 99 - 103.

Sneed, K. E., and Clemens, H. P., (1959). The use of human chorioic gonadotropins to spawn warm waters fishes. Prog. Fish-Cult., 21:117-120.

Sneed, R. E. and Clemens, H. P. (1959). The use of human chorionic gonadotropin to spawn warm water fishes. Prog. Fish Cult.,21:117-120

Snow, J. R., Jones, R. O., and Rogers, W. A. (1964). Training manual for warm water fish culture. Bureau of sport Fisheries and Wild life. Warm water inservice training school, Marion, Alabama, Mimeo.

Snyder, D. E. (1976). Terminology for intervals of larval fish development p 41 - 58, In Great Lakes Fish and larvae identification Proceedings of workshop J. Boreman (ed) U. S. Fish Wildl. Serv. Biol. Seru. Prog. FWS/OBS - 76 / 23.

Sorgeloos, P. (1976). The Brine shrimp. *Artemia salina* A bottle neck in mariculture. In Advances in aquaculture (Eds) T. V. R. Pillai and Wm. A. Dill (Publ.) Fishing news books Ltd, England.

Sorgeloos, P. and Pandian, S. K. (1984). Production and use of Artemia in Aquaculture, CMFRI special publication (15) 74p.

Sorgeloos, P. *et al*. (1997). Decapsulation of Artemia cysts: A simple technique for the improvement of the use of brine shrimp in aquacultue. Aquaculture 12 (4) 311 - 115.

Sorgeloos, P., D. A. Bengtson, W. Decleir and E. Jaspers(eds) (1987). The effect of Artemia fed with different diets on the growth and survival of *Penaeus mondon* Fabricius post larvae. Artemia research and its application vol. 3.Ecology, culture and use in Aquaculture, Belgium, 447 - 457.

Sorgeloos, P., E. Bossuyt, P. lavens, P. leger; P. Vanhaecke and D. Versichele (1983). The use of brine shrimp Artemia in crustacean hatcheries and nurseries In. J. P. McVey and J. Robert Moore (eds) CRC Hand book of Mariculture, Vol. 1.Crustacean aquaculture: 71 - 96. CRC press, Florida.

Spotte, S. W. (1970). Fish and invertebrate culture: Water management in closed systems. NewYork, Wiley - Inter - Science, 1- 145.

Stacey, N. E. and Goetz, F. W. (1982). Role of prostoglandins in fish reproduction. Can J. Fish. Aquat. Sci. 39, 92 - 98.

Stacey, N. E. and Pandey, S. (1975). Effects of indomethacin and prostaglandins on ovulation of gold fish. Prostaglandins, 9, 597 - 608.

Stacey, N. E. and Peter, R. E. (1979). Central action of prostaglandins in spawning behaviour of female gold fish, Physiol. Behav. 22, 1191 - 1196.

Stanley, J.G. (1976). Production of hybrid, androgenetic and gynogenetic grass carp and common carp. Trans. Am. Fish. Soc. 105: 10-16.

Stein, H. & Bayrle, H. (1978). Cryopreservation of the sperm of some freshwater telosts - Ann. Biol. Anim. Biochim. Biochim. Biophys. 18:1073-1076.

Stein, H. (1979). Zum Einfluss von Schutzsubstanz und Konferktionierung auf das Berfruchtungsergebnis mit tiefgefrorenem Bachforellensperma (*Salmo trutta fario L.)* - Berl. Miinch. Tierartzl. Wschr. 92:420-421.

Stein, H. (1984). Einfluss verschiedener Gefrier - und Auftaugeschwindigkeiten auf die Befruchtungsfahigkeit von tiefgefrorenem Forellensperma. - Berl. Munch. Tierarzl. Wschr. 97:138-139.

Steponkus, P. L., Myers. S. P., Lynch, D. V., Gardner, G., Bronshteyn, D. V., Leibo, S. P., Rall, W. F., Pitt, R. E., Lin, TT. And McIntyre, R. J. (1990). Nature. 345, 170 - 172.

Stoss, J. & Holtz, W. (1981a). Cryopreservation of rainbow trout (*Salmo gairdneri*) sperm. I. Effect of thawing solution, sperm density and interval between thawing and insemination. - Aquaculture 22: 97-104.

Stoss, J. & Holtz, W. (1981b). Cryopreservation of rainbow trout (*Salmo gairdneri*) sperm. III. Effect of pH and proteins in the diluent and presence of a buffer in the diluent. - Aquaculture 25: 217-222.

Stoss, J. & Holtz, W. (1983a). Cryopreservation of rainbow trout (*Salmo gairdneri*) sperm. III. Effect of proteins in the diluent, sperm from different males and intervals between collection and freezing. - Aquaculture 31: 275-282.

Stoss, J. & Holtz, W. (1983b). Cryopreservation of rainbow trout (*Salmo gairdneir*) dperm. IV. The effect of DMSO concentration and equilibration time on sperm survival, sucrose and KCI as extender components and the osmolality of the thawning solution. - Aquaculture 32: 321-330.

Stoss, J. & Refstie, T. (1983). Short-term and Cryopreservation of milt from Atlantic salmo and sea trout. - Aquaculture 32: 321-330.

Stoss, J. (1983). Fish gamete preservation and psermatozoan physiology. - In: Hoar, W. S., Randall, D. J. & Donaldso, E. M. (Eds.): Fish physiology, Vol. IX. Reproduction, Part B. Academic Press, London, pp. 305-330.

Stoss, J., Buyukhatipoglu, S, & Holtz, W. (1978). Short-term and Cryopreservation of rainbow trout (*Salmo garidneri* Richardson) sperm. - Ann. Biol. Anim., Biochim. Biophys. 18: 1077 - 1082.

Stoss, J., Baker, I. & Donaldson, E. M. (1981). Storage of salmo sperm by Cryopreservation. - 32nd Ann. Norrthwest Fish Cult. Workshop, Vance tyee, Tumeater, Washington, Dec. 1-3, 1981 (mimeogr.)

Strickland, S. and Beers, W. H. (1979). Studies on the enzymatic basis and hormonal control of ovulation. In "Ovarian follicular development and function" (A. R. Midgley and W. A. Sadler, eds) pp 143 - 153, Raven Press, NewYork.

Stuart, G. W., McMurray, J. V. and Westerfield, M.,(1988). Replication, integration and stable germ line transmission of foreign gene sequences injected into early zebra fish embryos. Development 103, 403-412.

Su - Lean Chang, Mao - SenSu and I. Chiudiao (1993). Milk fish fry production in Taiwan T. M. L. conference proceedings 3: 157 - 171.

Subramoniam, T. (1999). Endocrine regulation of egg production in economically important crustaceans. Current science 76 (3)pp350 - 359.

Subramoniam, T. and Newton, S. S. (1993). Cryopreservation of penaeid prawn embryos. Curr. Sci. 65, 176 - 177.

Subramoniam, T., D. Sedlmeier and R. Keller. (1999). Recent advances in the endocrine regulation of reproduction and moulting in crustacea. In "Comparative endocrinology and reproduction" (K. P. Joy, A. Krishna and C. Hakdarm eds). Narosa Publishing House, NewDelhi, India.

Subramoniam. T and R. Arun (1999). Cryopreservation and aquaculture. A case study with penaecid shrimp larvae. Current science. 76(3) 361 - 368.

Sukumasavin, N ,E. Mc Lean and E. M. Donaldson (1992). Orally induced spawning of Thai carp (*Puntius gonionotus,* Bleeker) following co-administration of des Gly12 Arg. GnRH ethylamide and domperidone. Journal of Fish Biology 10, 417 - 479.

Sundararaj, B. I. and Goswami, S. V. (1968). Effect of estrogen, progesterone, and testosterone on the pituitary and ovary of cat fish, *Heteropneustes fossilis* (Bloch) J. Exp. Zool. 169, 211 - 228.

Sundararaj, B. I. and S. V. Goswami (1966). Effects of mammalian hypophysial hormones, placental gonadotropins, gonadal hormones and adrenal corticosteroids on ovulation and spawning in hypophysectomised cat fish, *H. fossilis* (Bloch) J. Exp. Zool. 16: 287 - 296.

Sundershyam and C. B. Joshi (1977). Preliminary observation on the spawning of *Tor putitora*(Ham) in Anji stream,Jammu province during 1969. Ind. J. Fish. 21(1-2) 153-158

Tampi, P. R. S. (1957). Some observations on the reproduction of the milk fish, *Chanos chanos* (Forskal) Proc. Ind. Acad. Sci. (B) 46 (4) 254 - 273.

Tave, D. (1988). Genetics and breeding of Tilapia : a review, p. 285-293. In R. S. V. Pullin *et al*(ed.) The Second International Symposium of Tilapia in Aquaculture, ICLARM conference proceedings 15, 623 p. Dept. of Fishers, Bangkok and ICLARM ,Manila

Thomas, P. and N. Boyd (1989). Dietary administration of LHRH analogue induced spawning in spotted sea trout, *Cynoscion neptaosus*. Aquaculture. 72: 369 - 379.

Thomas, H. S. (1897). The rod in India. WilliamClawes and sons,London:435pp

Thong, L. H. (1969). Contributions 'a l' etude de la biologie des Mugilides (Poissons teleosteens) des côtes du massif Armoricaus, Trav. Fac. Sci. Univ. Rennes (Sir Oceanogr. Biol.) (2) 55 - 182.

Thorgaard, G. H., (1977). Heteromorphic sex chromosome in male rainbow trout. Science 196:900-902.

Toledo, J. D., Kurokura, H. and Kasahara, H (1989). Bull. Jap. Soc. sci. Fish. 55, 1161.

Toledo, J. D. and Kurokura,H (1990). Aquaculture, 91; 385-394

Trewas, E. (1982). Tilapias. Taxonomy and speciation. In the Biology and culture of Tilapias (Ed. by R. S. V. Pullin and R. H. Lowe-McConnel) ICLARM fonf. Proc., 7,3-13.

Trewavas, E. (1982). Generic grouping of Tilapinii used in Aquaculture. Aquaculture 27: 79 - 81.

Trewavas, E (1983). Tilapine fishes of the genera Sarotherodon, Oreochromis and Danakilia. British museum (Natural History) London.

Ucthida, R. N. and King, J. E. (1962) Tank culture of Tilapia. U. S. Fish wildl. Serv. Fish. Bull. 60 : 21-52.

Ueda, H. and Takahashi, H. (1976). Acceleration of ovulation in the loach, *Misgurnus anguillocaudatus* by treatment with clomiphene citrate Bull. Fac. Fish., Hokkaido Univ. 27, 1 - 5.

Uyeno, T. and R. R. Miller (1971). Multiple sex chromosomes in a Mexican Cyprinodontid fish. Nature 231: 452-453.

Vakily, J. M. (1989). The Biology and culture of mussels of the genus Perna. ICLARM studies and reviews 17, 63p. International center for Living Aquatic resources management, Manila Philippines.

Valenti, R. J. (1975). Induced polyploidy in *T. aurea*, (steindachner) by means of temperature shock treatment J. Fish. Biol. 7 : 519-528.

Vanstone, W. E. *et al.* (1977). Breeding and larval rearingof the milk fish *Chanos chanos* (Piscer chanidae)SEAFDEC. Aquculture Department. Tech. Rep. 3.3-17.

Vibert, R. (1975). Repeuplement des eaux a truites. Pisc. Franc., 42: 25 - 48.

Vinogradov, V. K. (1968). Techniques of rearing phytophagous fishes FAO Fish. Rep. 5, 227 - 243.

Wee, K.L. and Tuan, N.A. (1988). Effects of dietary protein level on growth and reproduction in Nile Tilapia (*O. niloticus*) P. 401-410. In: Pullin, R.S.V. *et al.* (eds) Second international symposium on Tilapia in Aquaculture ICLARM conference proceedings, 15, 623p.

Wheeler, P. A. & Thorgaard, G. H. (1991). Cryopreservation of rainbow trout semen in large straws. - Aquaculture 93:95-100.

Winkler, C., Vielkind, J. R. and Schartl, M., (1991). Transient expression of foreign DNA during embryonic and larval development of the medaka fish (*Oryzias latipes*) Mol. Gen. Genet. 226: 129-140.

Withler, F. C. & Lim, L. C. (1982). Preliminary observations of chilled and deepfrozen storage of grouper (*Epinephalus tauvina*) sperm. - Aquaculture 27:389-392.

Withler, F. C. (1982). Cryopreservation of spermatozoa of some freshwater fishes cultured in South and Southeast Asia. - Aquaculture 26: 395-398.

Wohlarth, G. W., G. Hulata, S. Rothbard, J. Hzkowich and A. Halevy, (1983). Comparisons between interspecific Tilapia hybrids for some production trails; 559-569. In L. Fishelson and Z. Yaron (comps) Proceedings of the international symposium on Tilapia m 'Aquaculture' 8-13 May 1983. Nazareth, Isral. Tel Aviv University , Israel

Wohlfarth, G. W. and G. Hulata (1983). Applied genetics of Tilapias. ICLARM studies and reviews 6, 26p. ICLARM Manila, Philippines.

Worthington, A. D., Mac Farlane, N. A. A. and Easton, K. W. (1981). The induced spawning of roach (*Rutilus rutilus*) with carp pituitary extract and oestrogen antagonist Symp. Fish physiol. 3rd 1981 Abstract.

Woynarovich, E. and Horvath. L. (1980). The artificial propagation of warm water fin fishes. A manual for extension. FAO Fish Tech. Paper 201

Yuno, Y and Soo, K. C. (1969). Larval development of *Macrobrachium roserbergii* reared in the laboratory J. Tokyo univ. Fish., 55(2) 179 - 190.

Zarrouk, C. (1996). Contribution a letude duene Cynophycie influence de divers facteurs physiues et chimiues sur la croissance et da photosynthesis de *Spirulina maxima* (Setch. Et, Gardner) Geitler, these, Universite de Paris.

Zhang, P. J., Hayat, M. Joyce, C. Gonzalea-Villosenov, L. I. Lin, C. M., Dunham, A. R., Chen, T. T. and Powers, D. A. (1990). Gene transfer, expression and inheritance of p RSV-rainbow trout-GH cDNA in the common corp, *Cyprinus carpio*(Lin.) Mol. Reprod. Dev., 25: 3-13.

Zhang, S. M. (1990). Inducing polyploidy and Karyological studies in Indian major carps *Catla catla* (Ham.), *Labeo rohita* (Ham.) and *Cirrhinus mrigala* (Ham.) and in eight other freshwater species of fish. A report of young Scientist programme (April 1989-April 1990) at CIFA, Kausalyaganga.

Zang, T. andRawson, D. M. (1996). Cryobiology, 33: 1-13.

Zhao, X, Zhang, P. J. and Wong, T. K. (1993). Application of beakonization : A new approach to produce trangenic fish, Mol. Mar. Biol. Biotchnol 2, 63-69.

Author Index

K

L

M

N

O

P

Q

R

S

T

U

V

W

Y

Z

Species Index

I

K

L

M

N

R

S

T

V

W

X

Z

Subject Index

F

G

H

I

Q

R

S

T

U

V

W

X

Y

Z